Nanomedicine and Applications in Cancer

Nanomedicine represents a paradigm shift in cancer therapeutics, offering targeted drug delivery, reduced systemic toxicity, and enhanced efficacy. This book provides a detailed and systematic overview of nanomedicine formulations and their diverse applications across the cancer treatment landscape.

Beginning with fundamental principles of nanoparticle synthesis, characterization, and application in medicine, the text progresses to explore key nanoparticle platforms. Polymer-based nanoparticles are examined with respect to material properties, production methodologies, and their role in developing tailored cancer therapies. A dedicated section addresses lipid-based nanoparticles, focusing on the complexities of biological delivery, clinical successes, and essential considerations for stability and storage. Self-generating nanoemulsions (SGNEs) are introduced as a strategy to enhance drug bioavailability and overcome resistance, with attention given to regulatory pathways for clinical translation.

This book further elucidates site-specific drug delivery, detailing passive and active targeting mechanisms through various ligands. Diagnostic applications of nanomedicine are also presented, with a focus on fluorescent nanoparticles, their characterization, biomedical uses, clinical trial status, and limitations in comparison to alternative diagnostic modalities.

Critical areas such as cancer immunotherapy, including preventive and therapeutic vaccines and oncolytic viruses, and the burgeoning field of RNA-based therapeutics, highlighting the role of nanotechnology in their stabilization and targeted delivery, are thoroughly discussed. The text culminates with an overview of CAR-T cell therapy, associated challenges, innovative engineering strategies like CRISPR/Cas9, nanoparticle-mediated gene delivery, and relevant clinical advancements. Finally, this book addresses the multifaceted hurdles hindering the clinical translation of cancer-targeting nanomedicines, proposing potential solutions and design considerations to optimize clinical outcomes.

Drugs and the Pharmaceutical Sciences: A Series of Textbooks and Monographs

Series Editor: Anthony J. Hickey, RTI International, Research Triangle Park, USA

The Drugs and Pharmaceutical Sciences series is designed to enable the pharmaceutical scientist to stay abreast of the changing trends, advances and innovations associated with therapeutic drugs and that area of expertise and interest that has come to be known as the pharmaceutical sciences. The body of knowledge that those working in the pharmaceutical environment have to work with, and master, has been, and continues, to expand at a rapid pace as new scientific approaches, technologies, instrumentations, clinical advances, economic factors and social needs arise and influence the discovery, development, manufacture, commercialization and clinical use of new agents and devices.

Emerging Drug Delivery and Biomedical Engineering Technologies: Transforming Therapy
Dimitrios Lamprou

RNA-seq in Drug Discovery and Development
Feng Cheng and Robert Morris

Patient Safety in Developing Countries: Education, Research, Case Studies
Yaser Al-Worafi

Industrial Hygiene in the Pharmaceutical and Consumer Healthcare Industries
Casey Cosner

Cancer Targeting Therapies: Conventional and Advanced Perspectives
Muhammad Yasir Ali and Shazia Bukhari

Molecular Recognition in Pharmacology
Mikhail Darkhovskiy

GMP Audits in Pharmaceutical and Biotechnology Industries
Mustafa EDİK

Purification of Biotechnological Products: A Focus on Industrial Applications
Adalberto Pessoa Jr, Beatriz Vahan Kilikian, and Paul Long

Principles of Research Methodology and Ethics in Pharmaceutical Sciences: An Application Guide for Students and Researchers
Vikas Anand Saharan, Hitesh Kulhari, and Hemant Jadhav

Good Clinical Practices in Pharmaceuticals
Graham P. Bunn

An Introduction to Generative Drug Discovery
Sean Ekins

Viral Oncology: New Approaches to Molecular Cancer Therapeutics
Edited By Umesh Kumar, Deepak Parashar, and Subodh Kumar

A Guide to Particulate Science in Pharmaceutical Product Development
Margaret Louey, Timothy Crowder, and Anthony J. Hickey

Nanomedicine and Applications in Cancer: A Complete Guide to Nanomedicine and Cancer Applications
Imran Saleem and Ahmed AH Abdellatif

For more information about this series, please visit www.crcpress.com/Drugs-and-the-Pharmaceutical-Sciences/book-series/IHCDRUPHASCI

Nanomedicine and Applications in Cancer
A Complete Guide to Nanomedicine and Cancer Applications

Edited by
Imran Saleem and Ahmed AH Abdellatif

CRC Press
Taylor & Francis Group
Boca Raton London New York

CRC Press is an imprint of the
Taylor & Francis Group, an **informa** business

First edition published 2026
by CRC Press
2385 NW Executive Center Drive, Suite 320, Boca Raton FL 33431

and by CRC Press
4 Park Square, Milton Park, Abingdon, Oxon, OX14 4RN

CRC Press is an imprint of Taylor & Francis Group, LLC

© 2026 Imran Saleem, Ahmed AH Abdellatif

ISBN: 9781032853741 (hbk)
ISBN: 9781032853758 (pbk)
ISBN: 9781003517870 (ebk)

DOI: 10.1201/9781003517870

Typeset in Times
by codeMantra

Dedication

To the cause of globally equitable access to medicines and healthcare

Imran Saleem dedicates this book to the loving memory of his mother and sister and expresses his sincere gratitude for the enduring support of his wife, father and brothers.

Ahmed AH Abdellatif dedicates this book to his father, mother, brothers, wife, daughters and sons for their ongoing support.

Contents

Preface

"May everyone be happy. May everyone be healthy. May everyone see. May there be no sorrow or misery."

The convergence of nanotechnology and medicine has ushered in a revolutionary era in the fight against cancer. This interdisciplinary field, known as nanomedicine, offers unprecedented opportunities for precise diagnosis, targeted drug delivery, and innovative therapeutic strategies, holding immense promise for improving patient outcomes and transforming cancer care.

This book, *Nanomedicine and Applications in Cancer: A Comprehensive Guide*, aims to navigate the multifaceted landscape of this exciting domain. It provides a structured and in-depth exploration of the fundamental principles of nanomedicine as applied to oncology, delving into the design, fabrication, and application of nanoscale materials for tackling this complex disease. This comprehensive guide is intended for researchers, clinicians, students, and anyone interested in gaining a deeper understanding of the transformative potential of nanomedicine in the ongoing battle against cancer.

We begin by laying the groundwork with an Introduction to Nanoparticles and Nanomedicine in Cancer Therapy, establishing the core concepts and highlighting the potential of this field. Subsequent chapters explore specific nanoparticle platforms crucial for targeted drug delivery. We examine Polymer-Based Nanoparticle Formulations and Lipid Nanoparticle Formulations, elucidating their unique properties and applications in enhancing therapeutic efficacy while minimizing systemic toxicity. The innovative approach of Self-Generating Nano-Emulsions for Enhanced Anticancer Delivery is then discussed, showcasing novel strategies for improving drug bioavailability and tumor penetration.

A critical aspect of nanomedicine is the ability to selectively target cancer cells, and the chapter on Active Targeting of Nanoparticles as a Smart Way for Treating Cancer provides a comprehensive overview of various targeting ligands and strategies. Beyond therapy, nanomedicine plays a vital role in early and accurate cancer diagnosis, which is explored in detail in Nanomedicine-Based Imaging Techniques in Cancer Diagnosis, highlighting the advancements in early detection and disease monitoring.

This book further delves into cutting-edge therapeutic modalities that are being significantly advanced by nanotechnology. We explore the exciting field of Immunotherapy for Cancer, highlighting how nanoparticles can enhance immune responses against tumors. The rapidly evolving area of RNA Therapeutics for Cancer is also covered, showcasing the potential of nanoscale delivery systems for siRNA, mRNA, and other nucleic acid-based drugs. A dedicated chapter on Chimeric Antigen Receptor (CAR)-T Cell Therapy: State of the Edge and the Recent Strategies Using CRISPR and mRNA examines the forefront of cellular immunotherapy and the role of nanotechnologies in its optimization.

Finally, the crucial aspect of translating laboratory breakthroughs into clinical reality is addressed in the concluding chapter, Clinical Translation of Nanomedicines; Keys to Cancer Targeting. This section focuses on the challenges and opportunities associated with bringing nanomedicine-based cancer therapies to patients.

Through this comprehensive guide, we aspire to provide clarity and insight into the exciting advancements in nanomedicine for cancer. We hope that this book will serve as a valuable resource, illuminating the current state of the field and inspiring future advancements that will ultimately benefit patients worldwide.

Professor Imran Saleem
Professor Ahmed AH Abdellatif

Contributors

Ahmed A.H. Abdellatif
Department of Pharmaceutics
College of Pharmacy
Qassim University
Qassim, Saudi Arabia

Mohamed S. Abdel-Bakky
Department of Pharmacology and
    Toxicology
College of Pharmacy
Qassim University
Qassim, Saudi Arabia

Fatma Ahmed
Department of Zoology
Faculty of Science
Sohag University
Sohag, Egypt

Abdellatif Bouazzaoui
Department of Internal Medicin III
    (Haematology and Internal Oncology)
University Hospital Regensburg
Regensburg; Germany

Ikram A. Burney
Women Health Program
Sultan Qaboos Comprehensive Care &
    Research Center
University Medical City
Muscat, Oman

**Mohamed Abd El Aziz Ahmed
Abd El-Galil**
Fish Diseases and Management Department
Faculty of Veterinary Medicine
Sohag University
Sohag, Egypt

Mahmoud M.A. Elsayed
Department of Pharmaceutics and Clinical
    Pharmacy
Faculty of Pharmacy
Sohag University
Sohag, Egypt

Asal Golchin
Department of Biochemistry
School of Medicine
Urmia University of Medical Sciences
Urmia, Iran

Ishtiaq Ahmed Khan
Jamil-ur-Rahman Center for Genome Research
Dr. Panjwani Center for Molecular Medicine
    and Drug Research
International Center for Chemical and
    Biological Sciences
University of Karachi
Karachi, Pakistan

Saeed A. Khan
Research Institute of Medical and Health
    Sciences
University of Sharjah
Sharjah, United Arab Emirates
and
Department of Pharmacy
Kohat University of Science and Technology
Kohat, Pakistan

Samra Khan
Jamil-ur-Rahman Center for Genome Research
Dr. Panjwani Center for Molecular Medicine
    and Drug Research
International Center for Chemical and
    Biological Sciences
University of Karachi
Karachi, Pakistan

Ella McGovern
School of Pharmacy and Biomolecular Sciences
Liverpool John Moores University
Liverpool, United Kingdom

Masoud Ojarudi
Department of Biochemistry
School of Medicine
Urmia University of Medical Sciences
Urmia, Iran

Shaaban K. Osman
Department of Pharmaceutics and
   Pharmaceutical Technology
Faculty of Pharmacy
Al Azhar University
Assiut, Egypt

Yusef Rasmi
Department of Biochemistry
School of Medicine
Urmia University of Medical Sciences
Urmia, Iran
and
Maternal and Childhood Obesity Research
   Center
Urmia University of Medical Sciences
Urmia, Iran

Mutasem Rawas-Qalaji
Research Institute of Medical and Health
   Sciences
University of Sharjah
Sharjah, United Arab Emirates
and
Department of Pharmaceutics and
   Pharmaceutical Technology
College of Pharmacy
University of Sharjah
Sharjah, United Arab Emirates

Kehinde Ross
School of Pharmacy and Biomolecular Sciences
Liverpool John Moores University
Liverpool, United Kingdom

Imran Saleem
Nanomedicine, Formulation & Delivery
   Research Group
School of Pharmacy and Biomolecular Sciences
Liverpool John Moores University
Liverpool, United Kingdom

Shakir U. Shakir
Department of Pharmacy
Kohat University of Science and Technology
Kohat, Pakistan

Ghareb M. Soliman
Department of Pharmaceutics
Faculty of Pharmacy
Assiut University
Assiut, Egypt

Hesham M. Tawfeek
Department of Industrial Pharmacy
Faculty of Pharmacy
Assiut University
Assiut, Egypt

Mahmoud A. Younis
Department of Industrial Pharmacy
Faculty of Pharmacy
Assiut University
Assiut, Egypt

About the Editors

Professor Imran Saleem is a distinguished Professor of Nanomedicine and leads the Nanomedicine, Formulation and Delivery Research Group within the School of Pharmacy & Biomolecular Sciences at Liverpool John Moores University, UK. He obtained his Pharmacy degree and registration as a practicing pharmacist from the same institution before earning a Ph.D. in Drug Delivery from Aston University, UK, in 2004.

Following his doctoral studies, Professor Saleem pursued a Postdoctoral Research Fellowship at the University of New Mexico, USA, where his research focused on dry powder inhalation for targeted delivery of anticancer agents in lung cancer. His interdisciplinary research integrates formulation science, drug delivery, nanotechnology, polymer science, chemistry, and cell and molecular biology.

Professor Saleem's current research centers on the application and fundamental understanding of novel nanocarrier systems designed to overcome biological barriers for the targeted delivery of diverse therapeutic agents, including small molecules and macromolecules, for both human and veterinary applications.

With over two decades of experience in nanoparticle formulation and drug delivery systems, Professor Saleem is a prominent figure in the field, evidenced by numerous publications in leading peer-reviewed journals, presentations at international conferences, and contributions to book chapters. His leadership continues to drive innovation in nanomedicine and drug delivery, with the potential for significant impact on therapeutic outcomes.

Professor Dr. Ahmed AH Abdellatif is a Professor at the College of Pharmacy, Qassim University, Saudi Arabia, with over two decades of research and teaching experience in pharmaceutical technology and drug delivery systems. He commenced his academic journey in Egypt, earning his Bachelor of Science and Master of Science degrees from Al-Azhar University, where he also served as a lecturer and professor. Registered as a pharmacist with the Egyptian Ministry of Health since 2001, Professor Abdellatif pursued his Ph.D. in Pharmaceutical Technology at the University of Regensburg, Germany, gaining expertise in drug formulation and delivery at the Institute of Chemistry and Pharmacy.

Professor Abdellatif is a leading researcher in nanobiotechnology, specializing in the synthesis and characterization of metal and polymeric nanoparticles and their applications in cell biology and biotechnology. His prolific research output includes over 100 publications, reflecting his comprehensive understanding of modern advancements in drug formulation, nanobiotechnology, and receptor targeting. His work has significantly contributed to the understanding of nanoparticle-based therapeutics for combating cancer and infectious diseases, with the potential to improve patient care globally.

Professor Dr. Ahmed AH Abdellatif's leadership in pharmaceutics, combining a strong educational foundation with practical research applications, continues to make a significant impact both within Saudi Arabia and internationally, driving advancements in pharmaceutical technology.

1 Introduction to Nanoparticles and Nanomedicine in Cancer Therapy

Hesham M. Tawfeek, Mahmoud M.A. Elsayed,
Ahmed A.H. Abdellatif, and Ghareb M. Soliman

1.1 FUNDAMENTALS OF NANOTECHNOLOGY

1.1.1 DEFINITION AND SCOPE OF NANOTECHNOLOGY

The term "nano" originates from Nanos, a Greek word meaning "dwarf." Nanoparticles (NPs) refer to objects having a size that typically falls in the range of 1–1000 nm [1,2]. However, the terms "nanomaterials" or "nanoscale" refer to objects having dimensions that typically range from 1 to 100 nm [3]. Nanotechnology is a science that focuses on the design, development, and application of drugs in nanoscale forms. With the remarkable proliferation in the surface area/volume ratio, the material at this nano-size level results in unique properties, such as enhanced chemical reactivity, electrical conductivity, mechanical strength, and interaction with biological membranes [4]. This type of technology has various applications in several fields, including medicine and pharmaceutical care, electronic design, bioimaging, energy, and materials sciences. This evolutionary technology creates new fields and applications that result in the development of advanced drug delivery systems, nanosensors, and nanoformulations [5,6]. Nanotechnology is an innovative technology transforming many industries, including computing, agriculture, food production, textiles, automotive, healthcare, and medical [7–12]. Its application ranges from developing smaller, faster computer components to innovations in sustainable agriculture, improved food safety, targeted drug delivery, improved diagnostics, and better therapeutic outcomes. In healthcare, nanotechnology has revolutionized diagnostic solutions, enabled targeted and innovative drug delivery, and created novel therapeutic strategies, ultimately leading to significant advancements in diagnosis and disease management. This interdisciplinary technology has broadened the scope of research and applications of NPs, leading to innovative products and solutions. One astonishing example of nanotechnology impact in medicine is the development of several COVID-19 vaccines that utilize NPs to stabilize their active ingredients, such as proteins or nucleic acids. These NPs not only protect the vaccine components but also enhance immunogenicity, thereby significantly improving the vaccines efficacy [13].

1.2 HISTORY OF NANOTECHNOLOGY

The history of nanotechnology can be dated back to ancient civilizations such as Ancient Egypt, Rome, Mesopotamia, India, and the Maya, which unknowingly used nanomaterials in various applications [4]. Although these civilizations lacked the scientific basis to explain nanomaterial properties and applications, their usage of nanoscale materials highlights an early, empirical application of what we know now as nanotechnology. For instance, Egyptians used NPs of gold in glassware to achieve vibrant colors and used lead sulfide NPs in hair dyeing [14]. Romans incorporated gold and silver NPs (AgNPs) with 50–100 nm particle sizes in the famous Lycurgus Cup, giving it

DOI: 10.1201/9781003517870-1

1

its unique light-reflective properties [15]. Similarly, in Mesopotamia, artisans used metal NPs to decorate pottery, creating a glittering effect, and the Mayans achieved rich colors in murals using nanomaterials [16].

The term "nano" was first officially defined to represent one-billionth (10^{-9}) of a unit [4]. It is used in the International System of Units (SI) to describe sizes and quantities in the nanometer range, thus facilitating the standardization of nanoscience. Interest in nanotechnology significantly increased after the National Nanotechnology Initiative (NNI) launch in the United States in 2000. This initiative provided substantial funding and research support and established a framework for advancing the field. Accordingly, there was a dramatic increase in nanotechnology-related research and development, leading to rapid advancements in material science, medicine, electronics, and several other applications [17].

Regarding the history of nanotechnology in drug delivery and medical applications, the pioneering work of Speiser *et al.*, who explored the potential of using NPs as carriers for controlled drug release and targeted delivery in 1978, laid the groundwork for future research and advancement in this field [18]. This was followed by the introduction of bile-salt mixed lecithin micelles and micellar vitamin K in the early 1980s [19,20]. The limited loading capacity of these micelles dictated the use of solvents such as ethanol and surfactants such as Cremophor EL® to increase drug aqueous solubility. For instance, Cremophor EL was used to solubilize anticancer drugs such as paclitaxel (Taxol®, Bristol Myers Squibb) [21]. However, toxicity and side effects associated with Cremophor EL motivated scientists to find alternatives that are safer and more effective as drug delivery systems [22]. Polymeric micelles were first described as drug-delivery systems in the 1980s. They are self-assembly by amphiphilic block copolymers, leading to an exceptional core-shell design. The core is usually hydrophobic and used to encapsulate hydrophobic drugs. The shell is typically composed of hydrophilic polymers such as poly(ethylene glycol) (PEG), which maintain the aqueous solubility of the whole system and increase the circulation time in the blood next to intravenous administration [23]. Their primary advantage lies in their stability against dilution, thanks to their very low critical association concentration, which contrasts with the high critical micelle concentration of surfactant micelles [24,25]. In addition, polymeric micelles demonstrated the ability to improve the bioavailability of water-insoluble drugs [26]. Over time, the field has evolved significantly, leading to various clinically approved polymeric micelle-based formulations designed for improved drug aqueous solubility, targeted delivery, and controlled release [27]. For instance, polymeric micelle formulations of paclitaxel have been approved for commercial use in several countries under the trade names of Genexol® PM and Nanoxel® M, while many other formulations are in various stages of clinical trials [27].

NPs currently under active investigation for drug delivery applications fall under various categories based on the materials used in their preparation, each offering distinct advantages depending on their composition and structure. The primary classifications include:

1. Polymeric NPs, can be defined as biocompatible and biodegradable polymers that provide controlled and sustained drug release. Common types include polymeric micelles, nanospheres, nanocapsules, dendrimers, and carbon nanotubes (CNTs) [28,29].
2. Lipid-based NPs are formed from lipids mimicking natural biological structures such as phosphatidylcholine, making them highly biocompatible. These include liposomes, solid lipid NPs, and nanostructured carriers [30,31].
3. Metallic NPs have unique optical and electronic properties, making them suitable for imaging, drug delivery, and diagnostic applications. These include gold NPs (AuNPs), AgNPs, and silica NPs [32,33]. These NP types offer unique advantages depending on their composition, size, shape, surface properties, and intended application. Additionally, they facilitate the design of highly specialized drug delivery systems for various therapeutic areas, including oncology, gene therapy, and infectious diseases.

1.3 SHAPES AND PROPERTIES OF NPs

NPs intended for drug delivery applications have several structures and sizes to optimize their drug delivery features. The importance of NPs' shape and morphology stems from their influence on determining their pharmacokinetics and interaction with biological systems, influencing their efficacy in delivering their cargo at the right time and place. NPs can be designed in various shapes, including spherical, cylindrical, conical, tubular, rod-like, or spiral, with the spherical shape being the most investigated one (Figure 1.1) [34]. Spherical NPs exhibit uniform distribution and stability, while tubular or rod-shaped NPs can enhance cell membrane penetration and prolong circulation time. Additionally, NPs with a hollow core are attractive for achieving high drug loading capacity, while irregular or asymmetric shapes can improve cellular uptake and targeting specificity. Careful design of these structural attributes allows for precise control over biodistribution, cellular uptake, and drug release profiles, making NPs versatile tools for targeted and controlled drug delivery [34]. Various NP shapes have been explored in drug delivery systems, influencing biological interactions and efficiency. The following section describes some examples of commonly reported NP shapes used in drug delivery, along with their specific applications and advantages:

Spherical NPs: This is the most common shape studied for drug delivery, probably due to its straightforward preparation and characterization methods compared with other shapes [35]. For example, AuNPs and silica NPs are often synthesized as spheres due to their ease of fabrication and predictable pharmacokinetics. Spherical NPs have shown efficient cellular uptake and are widely applied in cancer treatment and imaging studies. Spherical selenium NPs with an average diameter of 100 nm were synthesized and found to cause more pronounced cell death for cancer cells than normal ones, suggesting their potential for treating pancreatic adenocarcinoma [36]. In another study, 5-fluorouracil (5-FU) was loaded into spherical hyaluronic acid (HA)-conjugated silica NPs to increase its efficiency and specificity in the treatment of colon cancer cells [37]. The targeted HA silica NPs had increased 5-FU anticancer efficacy against colo-205 cancer cells, more than four times compared with the non-targeted NPs. The application of spherical AuNPs for nucleic acid vectorization for cancer therapy was recently reviewed [38].

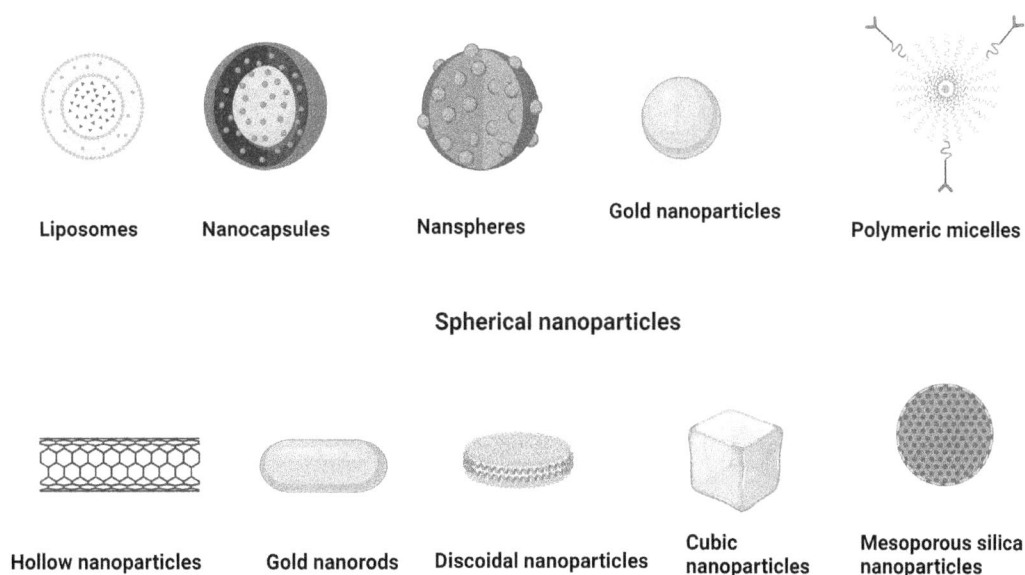

Liposomes   Nanocapsules   Nanspheres   Gold nanoparticles   Polymeric micelles

Spherical nanoparticles

Hollow nanoparticles   Gold nanorods   Discoidal nanoparticles   Cubic nanoparticles   Mesoporous silica nanoparticles

FIGURE 1.1 A schematic illustration of the shapes of nanoparticles (NPs) that are commonly studied for drug delivery applications.

Rod-shaped NPs: Rod-shaped NPs, such as gold nanorods (GNRs), have gained interest as drug delivery systems in the photothermal therapy of cancer due to their unique optical and electronic properties [39]. They have a size in the range of several to hundreds of nanometers. However, GNRs prepared using the cationic surfactant hexadecyltrimethylammonium bromide (CTAB) are cytotoxic, which limits their clinical applications. To increase their biocompatibility, decrease toxicity, and prolong blood circulation times, GNRs were coated with hydrophilic polymers such as PEG [40–42]. It was reported that the PEG coating reduced the GNR clearance rate by the reticuloendothelial system (RES). PEG coatings further enhanced the aqueous solubility and reduced the nonspecific uptake, making these GNRs promising nanocarriers for targeted cancer therapy [41,43]. Other coating materials, such as bovine serum albumin and β-dextran, were also reported to decrease the cytotoxicity of GNRs [42,44].

Cubic NPs: Liquid crystalline NPs (LCNPs), often called cubic NPs, represent a distinct class of nanostructures characterized by their unique internal liquid crystalline organization [45,46]. These NPs are the non-lamellar analog of liposomes [46]. They combine the properties of both liquid crystals and colloidal NPs, resulting in high structural stability and an increased surface area-to-volume ratio. This unique arrangement allows them to efficiently encapsulate various drugs, including hydrophobic, hydrophilic, and amphiphilic [46,47]. LCNPs have shown excellent potential for encapsulating and delivering anticancer drugs due to the increased encapsulation efficiency and sustained drug release [48]. Moreover, their lipid composition makes them biocompatible and biodegradable carriers, enhancing their suitability for drug delivery applications [49]. For instance, the anticancer drug docetaxel was loaded into LCNPs stabilized by a diketopyrrolopyrrole-porphyrin-based photosensitizer conjugated to Pluronic F108 [50]. The NPs had a cubic shape with a size of about 150 nm. The NPs were localized in the lysosomes and mitochondria of the cells and had significantly higher cytotoxicity than the drug alone. In another study, LCNPs were used as a delivery system for doxorubicin. The NPs had pH-triggered drug release and better anticancer efficacy against the glioblastoma T98G cell line than the free drug [51].

Tubular NPs: CNTs are an example of tubular NPs. CNTs are promising nanocarriers for anticancer drug delivery, owing to their distinctive features such as high aspect ratios, extensive surface areas, nanometric particle size, and tunable surface properties. These features make them highly effective for targeted delivery, improving drug loading capacity, and ensuring structural integrity at the nanoscale [52]. However, when inhaled, CNTs have shown several side effects, such as inflammation and immune response, particularly in the lungs, leading to conditions like asbestosis. Fibrosis, particularly in the lungs, has been reported in experimental animals where prolonged exposure has led to a thickening or scarring of connective tissues [53,54]. Due to these risks, further research is ongoing to improve the biocompatibility and safety of CNTs in drug delivery applications. Several studies indicated that surface modifications of CNTs and their length, width, and curvature can significantly enhance their biocompatibility while reducing toxicity. Functionalizing CNTs with biocompatible coatings or modifying their chemical properties improved their aqueous solubility and decreased adverse cell interactions [55]. Furthermore, controlling the dimensions of CNTs can reduce inflammation, cellular apoptosis, and other adverse effects commonly linked to CNT exposure [56]. For example, severe inflammatory responses have been linked to lengthy CNTs, which can be mitigated by reducing length and enhancing surface modifications [57].

Other NP shapes explored for drug delivery include hollow-core NPs such as mesoporous silica NPs (MSNPs). These NPs have hollow cores, which provide a large space for enhanced drug loading capacity. Additionally, the porous structure facilitates controlled and sustained drug release [58]. MSNPs have excellent biocompatibility, chemical stability, and surface modifiability, which make them ideal nanocarriers for therapeutic applications, particularly in cancer chemotherapy [59]. Discoidal NPs, which have a flat disc-like shape, offer unique benefits in adhesion to endothelial cells due to the large surface area available for contact with cells [60]. This shape is advantageous for targeting tumors with leaky vasculature, enhancing the retention and accumulation of therapeutic agents in tumor sites [61].

The importance of the above-discussed NP shape stems from its impact on biodistribution, cellular uptake, and clearance from the body. Thus, shape control is an essential consideration in the design of drug delivery systems to optimize their interaction with biological systems and achieve better therapeutic outcomes. For instance, a previous study showed that spherical and rod-shaped BODIPY NPs were internalized by both HeLa and HepG2 cells. However, the rod-shaped NPs had better cellular imaging capabilities than their spherical counterparts due to their enhanced uptake and stability [62]. Another study comparing the endocytosis of anisotropic $CaCO_3$ NPs (spheres, cuboidal, and elipsoïdal) reported that elongated NPs with the higher aspect ratio had better endocytosis by HeLa cells [63]. Other studies confirmed these findings, where NPs having a high aspect ratio, such as rods and tubes, were shown to have higher cellular internalization than spheres, which was recently reviewed [64]. Rod-shaped MSNPs also showed better endocytosis by the melanoma A375 cell line [65]. It was suggested that this different behavior might be related to the different endocytosis mechanisms for different NP shapes [34]. Thus, clathrin-dependent endocytosis may be the preferred internalization mechanism for spherical NPs, while rod-shaped ones might be internalized by the caveolae-dependent pathway [66]. However, there was no consensus regarding the effect of NP shape on cellular uptake efficiency, whereas other studies showed the opposite effect. For instance, Li *et al.* studied the internalization of PEGylated NPs (spherical, cubic, rod-shaped, and disc-like) having identical surface area, ligand-receptor interaction strength, and grafting density of PEG [67]. They found that the spherical NPs had the fastest internalization rate, followed by the cubic NPs, then rod- and disk-like NPs. These results were attributed to the lower membrane bending energies during the endocytosis of spherical NPs.

Furthermore, Borzęcka *et al.* reported that spherical MSNPs demonstrated higher phototoxicity against bladder cancer cell lines than the corresponding nanorods [68]. The discrepancies observed in these results could be attributed to variations in NP size, differences in the material composition used in NP synthesis, the use of distinct cell types, or the employment of different analytical methods to assess NP uptake [64]. These factors can significantly influence the interaction between NPs and cells, leading to inconsistent findings across different studies. As mentioned earlier, the findings highlight the crucial role of the NP shape in optimizing its performance, cellular uptake, and efficacy as a drug delivery vehicle.

1.4 APPLICATIONS OF NANOTECHNOLOGY IN MEDICINE

The current substantial interest in nanotechnology for medical applications originates from the new possibilities it offers for diagnosing, treating, and preventing diseases at the molecular level. The structures created at the nanometer scale (1–100 nm) have unique physical, chemical, and biological properties. These nanomaterials enable targeted drug delivery, enhance diagnostic precision, support tissue regeneration, and facilitate more accurate medical interventions. They also play a key role in advanced DNA sequencing techniques. Nanotechnology applications cover various medical fields, including oncology, cardiology, regenerative medicine, and the treatment of infectious diseases, among others. Nanomaterials' versatility improves both therapeutic efficiency and patient outcomes by minimizing side effects and offering patient-tailored treatment options [69,70].

Drug delivery is one of the most exciting areas of nanotechnology application in medicine. NPs are engineered to load and direct drugs to disease sites, minimizing side effects and enhancing drug efficacy [2]. For instance, lipid NPs have recently been found to play a crucial role in delivering mRNA vaccines, including the highly effective COVID-19 vaccines [13]. Delivery of chemotherapeutic agents for cancer therapy is an area where NPs have found interesting applications. Anticancer drugs often cause severe side effects by affecting both healthy and cancerous tissues. This results in deleterious effects in several sites in the body, such as blood cells, hair follicles, and gastrointestinal tissues, leading to complications such as fatigue, nausea, infections, and organ damage. These unwanted adverse effects significantly impact patients' quality of life and can even limit the dosage or duration of treatment, increasing the challenge of balancing therapeutic efficacy with safety [71]. In this regard, NPs are being actively investigated to deliver the loaded drugs specifically to the tumor sites by passive and active mechanisms, resulting in enhanced drug efficacy and reduced off-target damage to healthy tissues.

In addition to drug delivery, nanotechnology has found several revolutionary applications in medical diagnostics. The current conventional diagnostic methods have limitations in detecting cancer in the early stages. In this regard, nanotechnology-based diagnostics showed high sensitivity, precision, and the ability to perform multiple measurements simultaneously. This makes nanotechnology a promising tool for detecting early-stage extracellular cancer biomarkers, improving therapeutic outcomes [72]. Furthermore, NPs have played a fundamental role in real-time, non-invasive imaging, enabling the monitoring of tumors and metastatic sites, thus tracking the disease's progression and treatment outcome [73]. These applications, among others, demonstrate the profound impact of nanotechnology on modern medicine, offering patient- and disease-tailored cancer therapy and less invasive diagnosis for many disease conditions. Continued research and innovation in this field are promising to shape the future of cancer therapy and diagnostics.

1.5 OVERVIEW OF CANCER AND CURRENT TREATMENTS

Cancer is still the most critical and virulent disease that attacks humans. It is a major global health problem all over the world and is responsible for a higher incidence of mortality among individuals. It has been estimated that 1,270,800 cancer deaths occurred in 2024 in the European Union, corresponding to Age-Standardized Rates of 123.2/100,000 men (−6.5% versus 2018) and 79.0/100,000 women (−4.3%) [74]. Another study revealed that approximately 1 in 5 men or women develop cancer in a lifetime, whereas around 1 in 9 men and 1 in 12 women die from it [75]. A new study report by the American Cancer Society (ACS) revealed that by 2050, global cancer cases in men are expected to increase by 84%. Cancer deaths for men are expected to increase by 93% [76]. This study emphasized that lung cancer is expected to become the most diagnosed cancer and the leading cause of cancer death in men. This is why new strategies are being sought for disease diagnosis and treatment. Cancer results from the alteration of normal cells to tumor cells that generally progress from a pre-cancerous lesion to a malignant tumor. The interaction between a person's genetic factors and certain carcinogens, like physical, chemical, and biological factors, leads to these cell changes. Ultraviolet and ionizing radiation are among the physical carcinogens. Chemical carcinogens, such as components of tobacco smoke, asbestos, alcohol, aflatoxin, and arsenic, come from drinking contaminated water, and biological carcinogens, such as infections from certain viruses, bacteria, or parasites. WHO emphasized the role of age in the high incidence of cancer. They concluded that the buildup of numerous risk factors and the lower cellular repair mechanism were observed in older adults [77].

It cannot be considered that only one or two possible factors are responsible for this disorder; however, it is a complex combination of factors such as diet and lifestyle, exposure to radiation, and hormonal factors [78]. For instance, people consuming numerous types of foods rich in red and processed meats demonstrated a higher risk of colon cancers, whereas consuming high-fat diets has been linked to breast cancers [79,80]. It has also been noticed that smoking tobacco, consumption of alcohol, eating unhealthy food, physical inactivity, and air pollution are among the risk factors for cancer and other non-communicable diseases [81]. Certain chronic infections contribute to cancer. It was noticed in 2018 that approximately 13% of diagnosed cases of cancers were attributed to carcinogenic infections, like Helicobacter pylori, human papillomavirus (HPV), hepatitis B virus, hepatitis C virus, and Epstein-Barr virus [82]. Different types of cancer can affect one organ or specific cells. Cancer can spread quickly from one tissue to another, which is considered a malignant type of tumor. The term metastasis is a more common term in cancer sciences and refers to the spreading of tumor cells through the bloodstream or lymphatic system to distant areas of the body. Metastatic cancers are often more challenging to treat and more fatal.

The National Cancer Institute reports a list of all types of cancers that are categorized as dangerous. They listed breast cancer as the most common type of hazardous cancer, since more than 300,000 were recorded in the USA in the year 2024. Furthermore, prostate and lung cancer are considered the second most recorded types of cancer, with more than 150 recorded [83]. Moreover, the UK recorded that breast, bowel, lung, and prostate cancers are the most registered types of cancer, more than 53% of all new registered cases in the years (2017–2019) [84–86]. In this study, it was also pointed out that prostate cancer is the most common cancer in UK males, followed by lung cancer. There are different treatment strategies, including the cancer type and its severity. Surgery or radiotherapy are types of cancer treatment that may affect the healing process compared with other types. Systemic treatment of drugs such as chemotherapeutic agents, targeting medications, and immunotherapy can effectively affect the specific site in the body that has cancer [81]. Moreover, phytonutrients are considered effective treatments for cancer due to their antioxidant and anti-inflammatory activity. Flavonoids and anthraquinones are phytonutrients used to protect the human body against numerous malignancies [87,88]. They play a fundamental role in the cell apoptosis cycle arrest. Moreover, they inhibit angiogenesis, enzyme inhibition, and the alteration of nuclear receptors [89,90].

Nowadays, chemotherapeutic agents and radiotherapy have been approved with many number of agents that affect cancer patients with onerous physical and psychological challenges [91,92]. Furthermore, some modalities, which are small molecules, can be used as targeted agents to reach the tumor sites when conjugated with antibody-drug, cell-based therapies, and gene therapy [93].

For designing targeted therapy, drugs are conjugated with suitable carriers such as NPs to strictly target cancer cells without reaching or affecting the surrounding normal cells [94]. They can be small or large molecules. The small-molecule drugs are small enough to internalize a cancer cell via passive or active targeting. Targeting a specific substance inside cancer cells can block it, thus terminating the cancer cell. It was stated that imatinib could treat chronic myelogenous leukemia and other cancers by blocking tumor-activating signals [90,95,96]. Further, prominent molecule drugs such as mAbs cannot fit into a cell because they are too large. They destroy cancer by attacking and terminating proteins or enzymes on the cell's surface and crush tumor growth by disturbing the ligand–receptor interactions. Alemtuzumab is reported to be used in treating chronic leukemias. Trastuzumab is reported to be used in breast cancer treatment [97,98]. Different types of cancer treatment are summarized in Figure 1.2.

It has also studied the availability of therapeutic cancer vaccines. They are classified into patient-specific and non-specific cancer vaccines. The vaccines were derived from the same patient's cancer cells for the patient's particular vaccines. Moreover, the patient's non-specific vaccines can be derived by activating immune systems responsive to the anticancer cells [99]. The mechanism

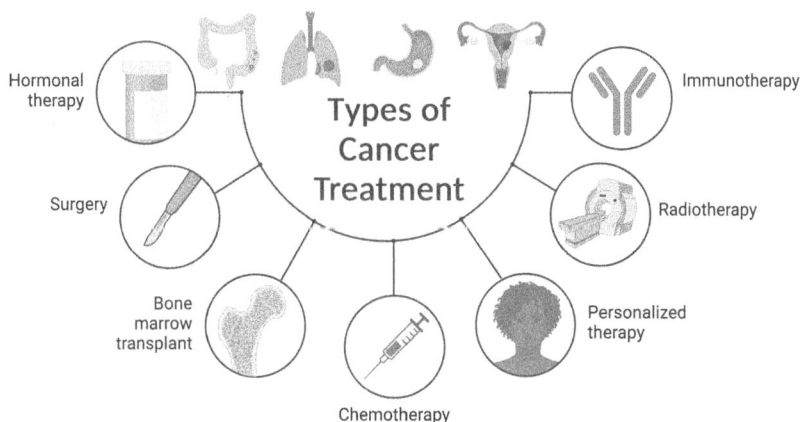

FIGURE 1.2 Different types of cancer treatment tools as examples, including hormonal therapy, surgery, bone marrow transplantation, chemotherapy, personalized therapy, radiotherapy, and immunotherapy. (Created in BioRender (2024). https://BioRender.com/o31k414.)

behind this category is that these vaccines could target specific tumor-associated antigens [100–102]. The following chapters will cover the recent technologies and therapies for cancer in detail.

1.6 NPS AS DRUG DELIVERY SYSTEMS

NPs are unique materials with superior characteristics that make them highly promising for a wide range of medical applications. Their potential has driven significant research and development, particularly in the field of drug delivery. In recent years, nanoparticulate drug delivery systems have shown dramatic advances, enabling researchers to overcome problematic barriers in disease treatment. NPs have different composition types, which exclusively correlate to their properties and functions. There are numerous emerging NPs as drug delivery systems, such as self-assembled carrier-free nanomedicine including pro-drug nano assemblies [103], cell-membrane-based biomimetic delivery systems [104], natural polymeric drug delivery systems [105], lipid NPs [106], inorganic NPs [107], hybrid nanocomposites [108], and theragnostic-based NPs [109]. As a drug carrier system, they are characterized by efficiently targeting certain cell organs with enhanced efficacy and minimal side effects. In addition, they can deliver medications in a controlled manner during the release of their payload, which will maximize therapeutic effect and, finally, significantly improve patient outcome and care, thus enabling more personalized and effective treatment plans [110]. In cancer, the NP carrier systems have opened up new possibilities for delivering drugs to the resistant tumor cells with potentially higher activity and pronounced reduced harmful action to the normal cells [111,112]. Furthermore, NPs provide good stability in the bloodstream when reaching the tumor microenvironment owing to their escape from RES clearances and not being captured by phagocytes. Such a performance is mainly attributed to the change in NP's surface via PEGylation, which confers stealth characteristics to the NPs and allows for more tissue and cell targeting capabilities [113]. NPs can reach the tumor site via an active or passive way, and it has been studied that this targeting mechanism allows for more efficient delivery of therapeutics to the tumor with minimal effect on the normal cells and lower doses of loaded anti-cancer agents [114]. Passive targeting depends on changes in the tumor cells' microenvironment, which precede the formation of leaky vasculature, facilitating the NPs' diffusion and particles up to 40 KDa [114–116]. This behavior is the enhanced permeability and retention effect. Such targeting depends on the type of tumor biology and the carrier characteristics, which determine the drug delivery efficiency. Numerous investigations have been developed to understand the significance of this phenomenon in tumor targeting and to develop appropriate drug delivery systems. On the contrary, active targeting can provide more delivered therapeutics to the tumor cells via decoration of the NPs with certain ligands, which can bind selectively to receptors over-expressed in cancer tissue [90,116]. Different ligands, such as peptides, nucleic acids, and aptamers, have been investigated. Furthermore, a huge number of receptors and antibodies are recognizable. They were efficaciously synthesized and studied *in vitro* and *in vivo*. Designing a ligand-functionalized target drug delivery system requires some essential parameters to be considered in the ligand itself, like final Mwt, targeting affinity, valence, and biocompatibility to ensure drug accumulation in the tumor site. Targeting the angiogenic endothelial cells adjacent to the tumor cells can limit poor delivery of drugs and any anticipated drug resistance and can be adapted to different types of tumors [117,118]. pH-sensitive nanocarriers can also be fabricated to target the mildly acidic tumor microenvironment for different tumor types [119–121]. Nuclear targeting to deliver the drugs deep into cell organelles like nuclei, lysosomes, mitochondria, or endoplasmic reticulum has also been explored. Such a type of targeting can maximize drug response and minimize side effects. It has been investigated that the anticancer effect can be considerably reduced if they do not precisely target the entire cell and the nucleus [122]. To deliver drugs to the nucleus, they should have been able to pass some barriers, beginning from the cell membrane and, finally, cytoplasmic trafficking and nucleus entry, which have become challenging tasks for efficient drug delivery applications. Researchers concluded that the prepared drug delivery system's size is crucial, as a size below 9 nm showed a high nucleus penetration. However, higher NPs sizes from 20 to 200 nm showed little nucleus entry [123]. Interestingly, NPs can be combined with other tools for cancer-fighting, such as chemotherapy, immunotherapy, and radiation therapy, to improve overall efficiency and reduce toxicity [124]. NPs are considered promising techniques for early tumor detection. They are considered versatile and can detect any tumor [125]. Compared to other cancer diagnosis methods and techniques, they showed better selectivity and sensitivity, leading to early tumor detection and better prognostic outcomes [126–128]. In the next chapters, the role of NPs and their carriers will be discussed in depth in tumor fighting and diagnosis.

1.7 TYPES OF NPs USED IN CANCER THERAPY

Different types of NPs have been investigated for cancer therapy and diagnosis. Polymeric NPs, including both natural and synthetic NPs, liposomes, hydrogels, micelles, extracellular vesicles, natural membrane-coated NPs like red blood cell-coated NPs and plateletsomes, viruses, and inorganic NPs including gold, AgNPs, mesoporous silica, and carbon nanomaterials are among the widely used NPs in cancer therapy. Another interesting approach is based on hybrid NPs, which can combine the advantages of two or more materials. This approach can overcome some problems with NPs, such as their accumulation and polymer rigidity. For example, polymeric materials can be modified on the surface of AuNPs to overcome the low rigidity of polymers and improve the biocompatibility of gold-based NPs [129]. Incorporating metal oxide into organic NPs can enhance their biocompatibility and reduce the risk of long-term accumulation [130] (Figure 1.3).

For cancer detection and diagnosis, AuNPs, quantum dots (QDs), and magnetic NPs were used. AuNPs showed a significant surface-enhanced Raman scattering effect, making them valuable in vibrational spectroscopy-based cancer detection methods [131]. QD NPs demonstrate excellent near-infrared absorption capabilities for practical photoacoustic imaging [131,132]. Magnetic NPs, especially those super magnetic iron oxide NPs, could provide a detailed image for tissues and tumor cells via their vital role in magnetic resonance imaging [133]. CNTs and nanorod arrays such as carbon-based NPs demonstrated excellent electrical conductivity for detecting cancer-related proteins [134]. Organic nanofibers contribute to fluorescence-based imaging and the capture of circulating tumor cells [135]. Elnagar N. et al. developed an environmentally friendly and sensitive biosensor for detecting folate receptors. The authors proposed an electrochemical impedimetric biosensor formed by nanofibers of bio-copolymers prepared by the electrospinning technique [136]. Hybrid NPs coated with silicone beads offer a multifunctional platform for cancer detection, carrying imaging agents and allowing for surface modifications for specific targeting [137]. Recently, a new targeted silica NP with fluorescent cyanine 5.5 surrounded by polyethylene glycol chains attached to cyclo-[Arg-Gly-Asp-Tyr] (cRGDY) peptides (cRGDY-PEG-Cy5.5-C) for lymph node mapping of head and neck cancers, colorectal cancer, breast cancer, and melanoma has been developed [138]. The NPs are administered around the tumor site before or during surgery to recognize cancer nodes. Nanochips contribute to this emerging field by identifying specific cancer subtypes and guiding personalized treatment strategies. These nanochips can be functionalized with molecules that can bind to cancer-related biomarkers in the body in response to cancer. Nanochips can detect even smaller concentrations of tumor biomarkers due to their small size and a high surface area-to-volume ratio, which eventually enhances cancer diagnosis [138]. They can detect tumors at early stages, which is so beneficial for improved patient outcomes. Nanochips can also detect tumor markers in body fluids such as urine or blood and reduce the need for invasive techniques [139].

1.8 METHODS OF NP SYNTHESIS

NPs are particles having one or more dimensions of the order of a nanometer. The behavior of NPs is often dominated by the large surface area relative to the particles' volume. Materials frequently show remarkable changes in mechanical, thermal, optical, and catalytic properties when reduced to the nanoscale compared to bulk materials. Studies have shown that the properties of NPs are entirely

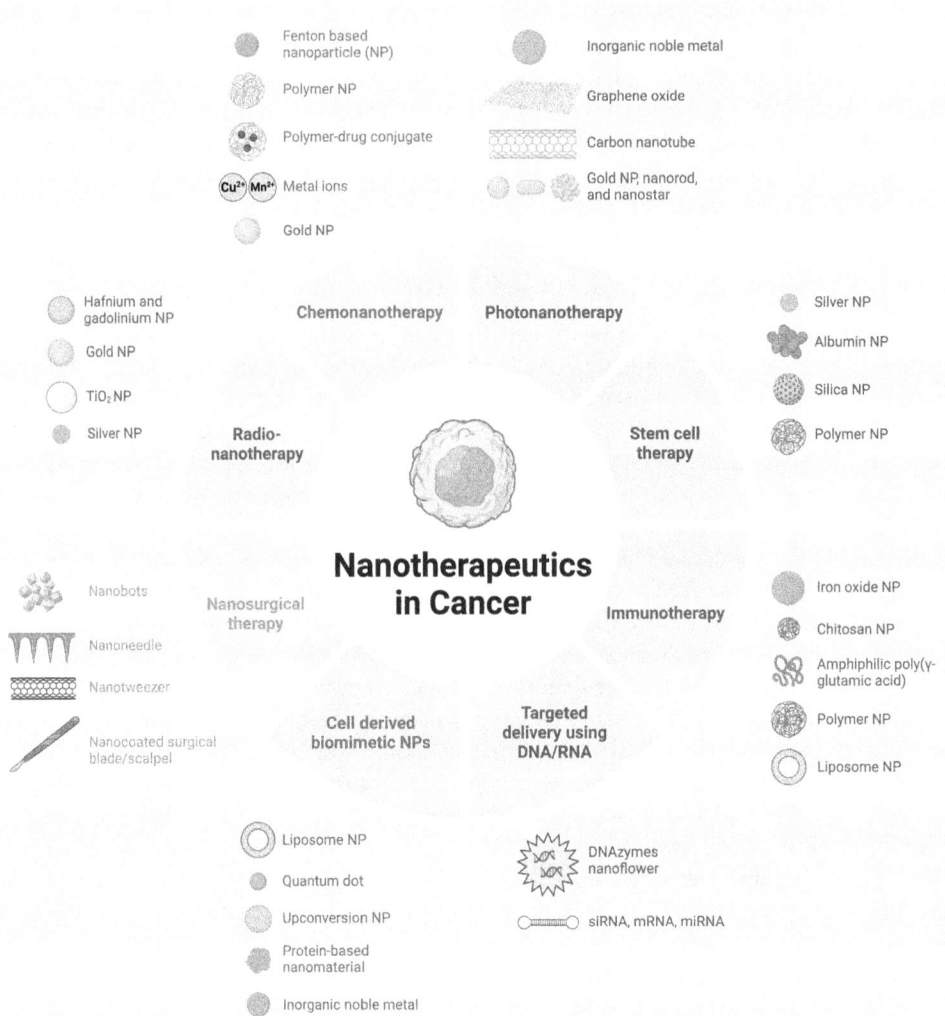

FIGURE 1.3 The different types of nanoparticulate carriers used to fight cancer. These nanoparticles can be metallic, such as gold or silver, polymeric nanoparticles, or lipid nanoparticles. (Created in BioRender. https://BioRender.com/z331183.)

different from those of bulk materials based on the same composition. All these characteristics make NPs with different compositions applicable in various fields, such as electronic devices, materials science, catalysis, energy storage, health management, biotechnology, and health [140–142].

Numerous methods have been developed for creating NPs in different compositions and sizes and in various settings. These NPs can be made with diverse shapes and compositions. At the nanoscale, materials typically have lower melting points and exhibit different phases than those with larger grain sizes. Generally, NPs have markedly distinct properties compared to bulk materials, including differences in melting point, luminescence, and chemical reactivity. Microparticles and NPs find applications in various fields due to their ability to provide high energy input into particle collisions with matter, such as in laser interactions, the small pixel area of recording media, and thin films. The induction of significantly high electric potential by nano- or micron-sized particles makes them useful in fields such as precise fabrication, small structure creation, and modification of microbiological systems. Additionally, magnetite NPs are utilized in several biomedical applications [143–145]. A detailed overview of NP synthesis methods is depicted in Figure 1.4.

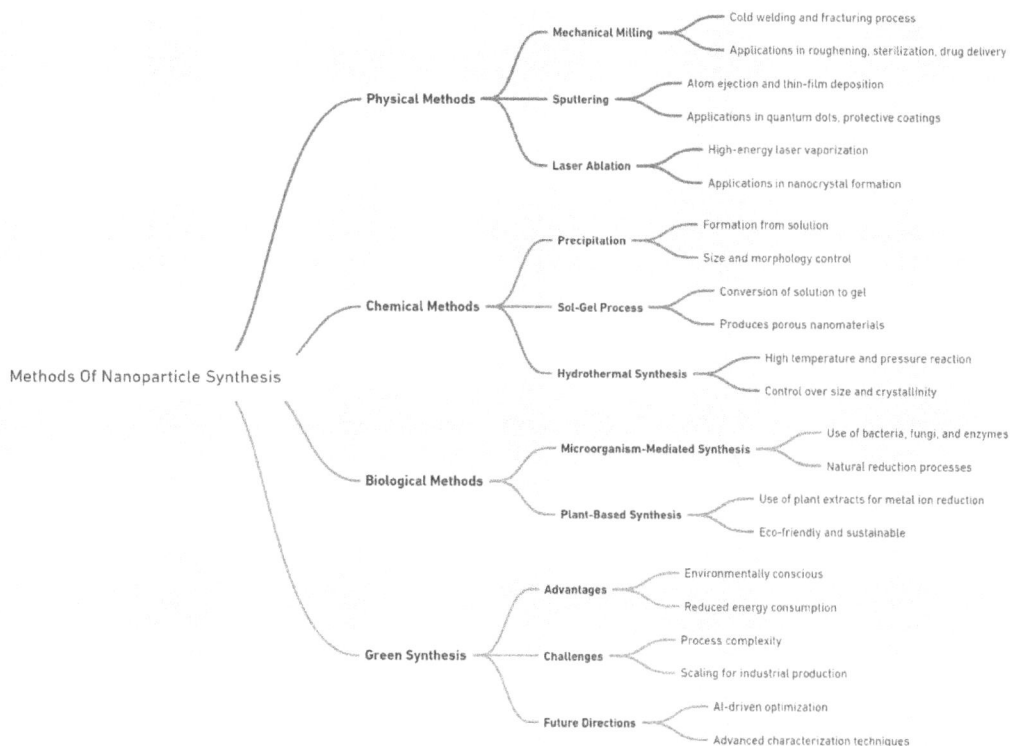

FIGURE 1.4 Detailed explanation of the various approaches to the production of nanoparticles, including physical, chemical, biological, and environmentally friendly synthesis.

1.9 PHYSICAL METHODS OF SYNTHESIS

There are two primary methods for NP synthesis: physical and chemical. We will now discuss physical methods for NP synthesis, including the principles of the methods used and a list of suitable NP materials, as well as each method's relative advantages and drawbacks. Over the past 20 years, interest in processes that result in the reduction of particles to sub-micron sizes has increased due to the unique applications that nanometer-sized materials provide, including high surface area materials for adsorption applications, improved bioavailability of drug compounds, and superior mechanical and electronic properties for toughening materials adversely affected by the presence of particles. Techniques include mechanical milling, which imparts mechanical forces and impurities; particles may become some undefined composition if milling media and milling container materials change; sputtering, a deposition process where atoms are ejected from a solid object into a gas phase using the vapor or gas phase; and laser ablation, in which high-energy lasers vaporize material using femtosecond laser pulses. Milling time, the temperature in the milling chamber, particles, and the existence of oxygen are the parameters in the mechanical milling process that need to be controlled. Adequate milling time is sufficient to produce metal particles with a crystalline structure, ranging in size from 5 nm to 1 µm. To achieve an even smaller size of less than 100 nm, a micro-indenter can be employed. The best utilization of the method is in the mechanical roughening of materials, sterilization, and the controlled delivery of drugs in the form of a dispenser. A comparison of the advantages of synthesis following physical and chemical methods is shown in a table. The chemical method is more effective and efficient for large-scale production if a high-purity NP is preferred. Purified products are better with physical synthesis, though they are more time-consuming and complicated to achieve in large-scale production. Physical methods are superior when there is a high selective effect either by composition or particle size [143,146].

1.9.1 Mechanical Milling

Mechanical milling is well known as an efficient and one of the most used physical methods for synthesizing NPs. In this approach, the mechanical alloying process is described as a process in which the powder particles are repeatedly cold-welded and fractured. Typically, this method comprises two steps: the initial grinding of the bulk materials to produce a sub-micrometer powder and mechanical alloying, which is done using a high-energy ball mill or a shaker mill. During this process, the deformation of the powder particles leads to the formation of fresh surfaces, which makes it very compatible with the formation of metallic, ceramic, or amorphous oxide nanopowders. This process allows for the modification of particle size distribution and the surface properties or characteristics of the obtained powder. The breaking of agglomerated powder particles was repeatedly cold-welded and fractured during mechanical milling. The welding fractures of the agglomerate particles without modification, like the NPs, are the most significant challenge in this study. This limitation will not make the NPs suitable for application. For mechanical alloying, the stationary jar filled with balls and powder is rotated at a certain speed [146–148].

1.9.2 Sputtering

Sputtering is a well-known and relatively sophisticated technique based on the ejection of atoms from a solid target. A stream of energetic particles impacts the target, and under certain conditions, atoms are ejected from the target and deposited on a substrate, leading to the generation of NPs. Besides the type of target material, processing parameters that influence the properties of the synthesized NPs are the pressure and energy of the energetic particles. Sputtering may be performed in various systems and configurations and allows for the deposition of thin films, as well as NPs, with control over the size and composition of the deposited particles. Furthermore, various oxide and nitride-conductive materials and precious or rare metals can be sputtered to produce multinary or platinum-group segment NPs. A short collection of applications of NPs synthesized by sputtering includes QDs in semiconductor technologies and NPs for protective scratch-resistant coatings in the packaging industry [149,150].

1.9.3 Laser Ablation

Laser ablation is a modern method of physical synthesis that uses a high-energy laser beam to vaporize a specific material from a solid target and expand it in the form of plasma. In the rapid cooling process, nucleation and growth of such processed matter occur in the form of NPs. Many energy sources, including a nanosecond, pulsed, or continuous wavelength (or a combination), can be used. The energy can be a UV, visible, or IR laser. The pulse or continuous repetition rate could be a single shot per second or multiple kHz; it will influence the particle size. The repetition rate of the pulse further defines unevenly distributed particle sizes in the case of the kHz order repetition. The corresponding nanomaterials are generally large and small particle size distributions having a medium diameter of about 10 barely distinguishable nm. The fluency of energy applied is a direct factor. It also helps to lock the crystalline or amorphous state [151,152].

1.10 CHEMICAL METHODS OF SYNTHESIS

There are several ways to synthesize NPs, and chemical methods rely on chemical reactions to produce materials of nanoscale dimensions. The precipitation method is based on forming solid NPs from the solution by a chemical reaction, direct evaporation, or a substitution reaction. By controlling reaction conditions, additives, and reactivity of the products in the solvent, the prepared NPs can have a narrower size distribution. The sol-gel process is one of the necessary alternative routes for achieving NPs and nanostructured materials. This process has been utilized in advanced materials because products are derived from sols or solutions. The term "sol-gel" describes the conversion of the solutions into solid networks or gels, which, upon drying, ultimately produce inorganic glass or glass ceramics. During the formation of the inorganic network, the gel solidifies to form a solid system with excellent mechanical properties. Synthesizing materials in the nano-range is achieved by, more specifically, tuning the synthetic chemistry and the parameters of the precursor solution [153–155].

Chemical synthesis offers several advantages: high product purity, the ability to tailor compositions and properties by modifying precursors, solvents, and surfactants, adaptable reaction conditions, scalability, and lower costs compared to physical methods. However, it requires precise control of precursor purity, reaction rates, and particle growth at the molecular level. The solvent plays a crucial role, influencing size distribution and growth, but these factors remain poorly understood. NPs produced via chemical synthesis are highly valued in catalysis and biomedicine due to their surface chemistry, enabling effective study and manipulation of functional groups [146,156].

1.10.1 Precipitation

The principles of metal-based NP synthesis are examined, with a focus on nucleation, particle growth, and the direct reduction of metal salts in solvents. The choice of the synthesis route and conditions significantly impacts NP size and shape, whether for optimizing applications, exploring new phenomena, or studying the effects of size and morphology. Precipitation is a common method, relying on the formation of NPs from dissolved precursors under controlled chemical conditions such as temperature, pH, and reactant concentration. Proper control of these parameters is essential for achieving the desired material properties. The simplicity and scalability of precipitation make it suitable for industrial applications, with slight condition adjustments yielding significant property variations.

1.10.2 Sol-Gel Process

The sol-gel process or precipitation method (in the formation of particles) is based on the transition between the initial state (colloidal solution) and the final state (solid gel or sol) before a solid at the final stage. This method is based on forming a colloidal solution of NPs, followed by a gelation process where the colloid bonding leads to a 3D network of particles. So, the chemistry of the sol-gel process is based on three stages: (1) the sol formation, that is, the particle formation leading to a stable colloidal solution; (2) the gelation, the condensation of particles solidified during a given time to form a 3D network of NPs that leads to the final material; and (3) the drying of the gel to form a porous powder [157–160].

1.10.3 Hydrothermal Synthesis

Various chemical and physicochemical methods have been designed to produce NPs with precise morphologies, defined sizes, and controlled size distributions. Hydrothermal synthesis, conducted in molecular-scale reactors, effectively avoids toxic and insoluble byproducts while achieving high purity, crystallinity, and minimized NP sizes. Key factors such as time, temperature, and surfactants influence the properties and shapes of the resulting materials. While this method enables the creation of pure compounds at moderate temperatures and reduces energy consumption, its scalability for industrial applications remains a limitation.

In hydrothermal methods, reactions occur within a temperature-gradient reactor, promoting the controlled growth of monodispersed NPs through hydrolysis and polycondensation of soluble salts. The process is largely dependent on parameters like temperature, pressure, and the addition of surface-active agents. While water is commonly used as the solvent, alternatives such as ethylene glycol can be employed to dissolve various inorganic compounds and stabilize nanocrystals. Extended reaction times are often required to achieve the desired size and crystallinity of the NPs [161–165].

1.11 BIOLOGICAL METHODS OF SYNTHESIS

While physicochemical approaches to NP synthesis contribute to environmental pollution, the use of toxic materials, and the production of unstable NPs, recent exploration of biological synthesis methods harnesses the power of living organisms to produce a wide variety of NPs. These biological entities include microorganisms and plants. Previously, the term "microorganism-mediated synthesis" was introduced to describe the production of metal/metal oxide NPs with the help of bacteria, fungi, yeast, algae, and viruses. These living organisms produce NPs via primary and secondary metabolites. While the metabolites are responsible for reducing metal ions to the valence zero state metal atoms, the organic material may be involved in capping, stabilization, and NP shape control. At the same time, green synthesis in a broad context occurs when plant extracts are used to reduce metal ions into the corresponding NPs. Initially, the extract can chelate metal ions and leave the metal in a metalloid-free fraction for reduction. The main mechanistic feature of microorganism- and plant extract-mediated synthesis techniques and the stepwise process of NP formation can be represented by a standard protocol in general [166–168].

1.11.1 MICROORGANISM-MEDIATED SYNTHESIS

The microorganism-mediated synthesis of NPs is a fascinating process, wherein extracellular enzymes, proteins, amino acids, or secondary metabolic compounds of algae, bacteria, and fungi are used to form metallic NPs. The extracellular routes have been extensively utilized for synthesizing silver, gold, and other metal oxide and sulfide nanostructures. Microbial enzymes, such as nitrate reductase, have been employed to form NPs. The reduction of the metal in the presence of a microorganism produces metal NPs, often containing extracellular capsular polymers, which aggregate to form large metal or metal oxide composite particles. Gold and AgNPs with different capping polymers are obtained from different microorganisms. The fungal species such as Fusarium oxysporum, Verticillium, and species of Penicillium have been utilized to form Au, Se, Pd, Cu, and S ions [169–171].

The bacterial cells contain enzymes or reductases, which have sequestering potential. Some bacterial cells reduce Ag ions to free radical Ag atoms, forming NPs. Cyanobacteria, Micrococcus luteus, and Mycobacterium species have also been used for synthesizing AgNPs. However, it is difficult to control the size and shape of the NPs as they continue growing and agglomerating with time. There are many distinct advantages of the whole microorganism-mediated synthesis of NPs. The reduction occurs naturally by metabolic processes that usually work at room temperature and attain a higher yield [172,173].

1.12 GREEN SYNTHESIS APPROACHES

Researchers and the general public alike are showing increasing concern for maintaining an adequate supply of NPs and ensuring environmentally conscious production methods. Traditional physical and chemical approaches to NP production have many negative consequences. These non-eco-friendly methods typically rely on non-renewable raw materials, employ toxic and non-biodegradable reagents, and consume excessive energy. In contrast, green synthesis stands out as the most promising platform for NP production. Green synthesis processes, such as bioproduction techniques utilizing microorganisms and plant extracts, offer numerous advantages over traditional methods for producing nanostructured materials. Not only do these green synthesis methods provide superior outcomes, but they also align with the principles of green chemistry, thus embodying a sustainable approach to NP production [174,175].

The production of NPs is increasingly focused on reducing energy consumption and adopting sustainable, environmentally friendly methods. The goal is to create high-value NPs at low costs using non-hazardous materials to minimize waste and environmental impact. There is a critical need for efficient, low-cost, and clean synthesis protocols that yield well-structured NPs, though process complexity remains a challenge. Advancing green synthesis through research can improve productivity, expand applications, and establish regulations for sustainable NP production. As technology evolves, the shift toward sustainable practices aligns with global efforts to combat climate change, fostering economic growth and environmental preservation. By prioritizing green practices, we can protect the planet and ensure a sustainable future. [176–178].

1.13 CHARACTERIZATION OF NPs

An NP is a minute piece of material with at least one dimension less than 100 nm. They are large networks of atoms possessing properties quite different from bulk material due to the properties and behavior of atoms present at or near the material's surface, with a concentration that is large enough to produce a surface effect. Surface atoms can carry unpaired electrons and possess dangling bonds or incomplete coordination with an energy state near the valence or conduction band edges. In bulk materials, many atoms forming NPs are involved in their vicinity, and the surfaces become far apart, which makes it difficult for the physical or electrical signal to pass through quickly. Consequently, electronic behavior can be significantly altered for the atoms on the nanoscale. The behavior difference causes electrical interaction between nanostructures and their environments [141,179,180].

The shift of band positions and the creation of localized energy levels near the band edges in NPs give them remarkable optical, electrical, and magnetic properties. NPs can absorb light of a specific wavelength and can be excited if the energy of the photons is comparable to the band gap energy of the NPs. When excited, NPs primarily transmit, reflect, and scatter light, and only a very small part of the photon excitation is dissipated by the emission of conduction electrons, which results in the enhanced interaction of light with the matter. These unique characteristics and distinguishable inherent functions of NPs make them suitable for use in biomedical and environmental applications, production of efficient solar cells or hydrogen from water, energy storage, transistors, and various sensors [181,182] (Table 1.1).

1.14 PHYSICAL CHARACTERIZATION

The behavior and aggregation of nanoscale particles in any dispersion or dry powder form, in liquid matrices or solid-state films, among other forms, can only be understood once the physical properties of fabrication size, morphology, and surface characteristics are characterized. A broad spectrum of characterization plans is routinely employed for each NP property. Focusing on the related physical characterizing properties of size, morphology, varying crystalline structures, and key surface defects that can serve as chemically active sites, as well as the relationship of these physical characteristics to the physical properties desired of the various nanomaterials, is the primary goal of this chapter. Indeed, the increased surface area available and the associated surface, size, and interfacial characteristics of NPs are all that one would hope to contribute to the exciting physical properties that could lead to breakthrough commercial applications in electronics, in transparent conductive coatings, in catalysis for fuel cells worldwide, and in display technologies. However, only through comprehensive physical characterization can one ascertain the realization of these physics- or chemistry-mediated physical properties [183–185].

1.14.1 SIZE AND MORPHOLOGY

The development and commercial proliferation of NPs (1–100 nm) have demanded new and novel methods to allow for size characterization. To this end, much development has occurred over the

TABLE 1.1

Summary Table for Characterization of NPs

Aspect	Key Characteristics	Techniques	Applications/Implications
1. Physical Characterization	Includes size, morphology, crystalline structure, and surface properties.	1. DLS 2. TEM, 3. SEM, 4. AFM, 5. X-ray diffraction (XRD)	Electronics (e.g., transistors), catalysis (e.g., fuel cells), transparent conductive coatings
2. Chemical Characterization	Focuses on elemental composition, oxidation states, functional groups, and chemical reactivity.	1. XPS 2. FTIR 3. EDS	Stability and reactivity, catalysis, drug delivery systems
3. Optical Characterization	Evaluates light absorption, scattering, and emission properties.	UV-Vis spectroscopy, photoluminescence spectroscopy (PL) Fluorescence microscopy	Plasmonic sensing, imaging, and catalysis
4. Magnetic and Electronic Characterization	Studies magnetic behavior and responses to external magnetic fields. Explores electronic properties, tunneling effects, and surface states.	1. VSM 2. Alternating gradient magnetometry 3. STM	MRI, data storage, magnetic separation Semiconductor devices, battery technology
5. Biological and Toxicological Evaluation	Examine interactions with biological systems, protein adsorption, zeta potential, and cytotoxicity.	1. Cytotoxicity assays (e.g., colorimetric and fluorometric) 2. Protein corona analysis	Drug delivery, nanotoxicity assessment, and cellular interactions

past two decades. For spherical particles, the size determination process usually involves determining the radius. In the case of concentrated dispersions or where the particles are not spherical, additional knowledge is required for the determination of interest. The standard techniques can be grouped into light-based and non-light-based methods. Light-based techniques include static and dynamic light scattering (DLS), visibility, dynamic imaging, and spectral methods [186–188].

The DLS technique measures the diffusivity of particles undergoing Brownian motion due to collisions with other particles. The average hydrodynamic diameter is determined using the Stokes-Einstein equation, which connects the diffusivity to the hydrodynamic radius of the particle through the measurement of the diffusion coefficient of the particle in suspension. DLS is the technique of choice for providing size measurements of particles in a monodisperse or highly polydisperse sample set using a fast and simple instrumental set [189,190]. Transmission electron microscopy (TEM) and high-resolution TEM (HRTEM) are the most potent techniques for nanoscale imaging and microstructural characterization of NPs. TEM can provide morphological, crystallographic, and chemical information at the atomic level. It is widely used for the studies of size, shape, aspect ratio, agglomeration, crystalline surface, alloy composition, core/shell structure, and interfaces between different regions of NPs and is ideal for providing direct morphological and structural evidence of the NPs [191,192]. Scanning electron microscopy (SEM) is useful for studying the morphology of larger NPs. It provides information about the size and shape of NPs in various ranges. These micrographs are also useful for understanding the nature and how these particles interact with each other, as well as the powder particles' size, shape, porosity, and flow properties. Initially, the samples are mounted properly and coated with a thin layer of gold. The overall preparation is then called metallization [178,193]. The atomic force microscopy (AFM) belongs to the scanning probe microscopy family. It is the second of the SPM systems invented after the scanning tunneling microscope. In the early 1980s, it was used first to characterize colloidal particles using AFM. This technique has become one of the most powerful tools for the characterization of particles at the nanoscale. Its great advantage is its ability to capture 3D images of the samples. As with any other microscopy technique, preparation of the sample is required [194–196]. High-resolution electron microscopy (HRTEM) characterizes an NP on an atomic scale. The interpretation of images may provide atomic locations, and it could also be observed that the agglomeration or partial melting of the particles occurs. The HRTEM image is a projection of the NP, with lighter atoms appearing more intensely. The image contrast is caused mainly by the variation of the incident electron wave, which is scattered by the electron density inhomogeneity [179,197].

1.14.2 Techniques for Crystallinity Analysis

X-rays are traditionally used for material research. X-rays can penetrate most materials and have wavelengths on a scale perfect for the boundary conditions set by typical distances between atoms in condensed matter. The X-ray absorption fine structure, also known as extended X-ray absorption fine structure, is an energy-dependent measurement of photoelectron scattering from the sample. The excitation energy is scanned across an absorption edge of the specific element under investigation [179,194]. X-ray diffraction is the most reliable technique for the characterization of the crystalline phases of a sample, and it can be combined with other methods for the size determination of the crystalline structure. The powder X-ray diffraction pattern of nanocrystalline materials is generally like that of the bulk powder of the corresponding materials, except that the diffraction peaks of the crystalline phase of the samples become broadened as a direct consequence of the finite size of the constituting grains or domains. The obtained diffraction pattern contains information about the presence and quantification of the crystalline phase, the lattice constants and crystal structure, and the mean crystallite size, that is, the average size of the ordered domains [198].

1.15 CHEMICAL CHARACTERIZATION OF NPs

NPs often display different optical, electronic, and thermal properties from those in their bulk states, mainly because size dictates a material's energy band structure. There are several types of NPs, of which metallic NPs, ceramic NPs, and polymeric NPs are some of the most common ones.

To turn nanoscience into a predictable field of research and exploit the potential intentionally, more detailed insights derived from chemical characterization should become crucial. If the smallest particle units are used to influence a desired property, it is necessary to characterize them in detail. NPs significantly influence the respective behaviors of nanostructured materials, mainly their interactions with the surrounding environments and agents. Furthermore, a strong influence of surface groups like –OH, –COOH, or –NH$_2$/–SH is found, responsible for stability, reactivity, and bioavailability. For future approaches in tailoring NPs to the respective applications, the need to know these surface groups and their ratios is increasingly important. As the fields of catalysis and material science illustrate so well, the variation in growth and surface defects demands high proficiency in detailed techniques, according to which nanoscale particles and surfaces can be characterized to ensure as much reproducibility as possible. This is the reason for the developments in sophisticated experimental standard operating procedures to establish structurally well-characterized catalysts and materials with reproducible catalytic behavior [141,199].

Surface chemistry techniques form the basis to understand the effects due to the interaction of NPs with their surroundings, thus producing accurate and reliable information about the surface properties or the chemical state of the surface. Indeed, the surface area could play a role in the behavior of NPs in different areas like separation science, including, to mention but a few, chromatography, solid phase extraction, or drug–protein interaction. Surface techniques can be analyzed on two different levels: the physical and the chemical level. These techniques are widely employed for the characterization of different materials and have been used in a wide range of applications [200–202].

X-ray photoelectron spectroscopy (XPS) is a powerful technique for analyzing the elemental compositions of materials, particularly the surface of materials at various depths. It also provides information on the chemical state of the elements present in the material, primarily identifying the oxidation states with respect to the surface of the material. XPS is carried out by producing photoemission of core-level electrons by irradiating the sample with an X-ray source. The energy and angle of the emitted photoelectrons are analyzed to provide a high-resolution chemical analysis that is achieved at the low part to a few parts per million (ppm) level of detection. The interaction depths with an X-ray source for the XPS technique are on the order of 1–10 nm. Thus, XPS can provide information on the outermost top few nanometers of the sample [200].

Fourier transform infrared spectroscopy (FTIR) provides detailed information on vibrational modes and the underlying molecules of solid and liquid materials. IR is electromagnetic radiation located in the 4000–400 cm^{-1} region. Frequencies in the area of 4000–700 cm^{-1} are typical for most vibrations of the molecular functional groups of solids and liquids, and this region is of particular interest for the qualitative evaluation of IR absorption spectra [200,201]. FTIR can provide valuable insights into NPs and their surrounding capping. It can give important information about the chemical nature of the surface functionalization of NPs and polymeric organic coatings. Furthermore, it can provide indications of chemical reactions between an organic coating and the metal surface. However, the interpretation of the FTIR spectra can sometimes be quite unclear and very difficult, mainly due to the complexities in the spectra [196].

Energy-dispersive X-ray Spectroscopy (EDS) is an elemental microanalysis technique that is often used in conjunction with SEM. When an electron beam interacts with a solid sample, inelastically scattered electrons can eject inner-shell electrons; high-energy electrons then fall into the vacancy, resulting in X-ray emission. For surface analysis, EDS is valuable for identifying the elemental composition of NPs. Consequently, a spectrum taken of NPs provides the overall surface analysis. EDS is also of value in detecting trapped components either on the surface or encapsulated in the particle, including contaminants that appear as minor peaks when compared to the major core spectrum. For such trapped contaminants, EDS can distinguish between the encapsulation and adsorption of the contaminant. EDS is also used to study NP morphologies and allow for quantitative and qualitative analysis of the distribution of elements in microscopy images [197,203].

1.16 OPTICAL CHARACTERIZATION OF NPs

Optical characterization represents one of the most critical real-time methods of NP characterization. For a fundamental understanding of the electromagnetic and light scattering properties of individual NPs, one has to start with the Mie theory from classical electrodynamics. Many other modern and classical methods could provide essential information in the optical characterization of NPs. In many applicative cases, the plasmonic and vibrational properties of the NPs can be easily recognized by the light scattering spectrum [204].

The ability of an NP to absorb or scatter light is one of its most intriguing properties, making optical characterization one of the most widely used techniques for such investigations. Among the various optical characterization methods, UV-Vis spectroscopy is an essential technique for analyzing light absorption, mainly in the UV and visible regions of NPs. This section explains the fine details of UV-Vis spectroscopy as applied to NPs. Furthermore, it gives insights into the fundamentals of UV-Vis spectroscopy, including how it works, the kind of information it provides, and some practical aspects of the technique that one should be aware of. The discussion also presents examples of how UV-Vis spectroscopy can be applied in various contexts to determine the properties of metallic NPs. Finally, we discuss potential applications of the technique for investigating biological particles that contain coenzymes and proteins [205].

Since they operate in the visible-infrared region of the electromagnetic spectrum, one of the appealing features of metallic NPs, for example, gold, is their bright colors. This specificity is derived from their local surface plasmon resonance. UV-Vis spectroscopy is a unique tool for researching and developing metal-based NPs for technological applications. Many applications of metallic NPs benefit from their localized surface plasmon resonance. Their characteristic properties, like positions and intensity, can be obtained using UV-Vis spectroscopy. The characteristic absorbance at approximately 520 nm and the linear relationship between increasing extinction and NP concentration have been noted. The peak wavelength position and extinction have been determined to depend on various parameters such as NP shape, size, medium, and interparticle distance. This property has recently brought applications such as colorimetric and plasmonic sensing, imaging, and plasmon catalysis [206].

One of the most powerful characterization techniques used to understand the photophysical properties of semiconductor NPs is photoluminescence spectroscopy. This section will explain a simplified version of PL and the related concepts. When a system absorbs a photon with an energy level larger than that of the ground state of that system, the particle gains enough energy to enter the excited state. The type of excited state that can be generated starting from the ground state energy level and with respect to the excitation energy is determined by three different processes: (1) intra-band or inter-band transitions; (2) direct or indirect transitions; and lastly, (3) excitonic excitation processes [206,207].

Fluorescence microscopy is an optical technique widely applied to visualize small particles such as NPs. It enables the visualization of particles in situ and in vitro with submicron resolution. An essential use of fluorescence microscopy is to help targeted NPs be visualized in a biological context. The particles are equipped with fluorescent dyes or conjugates, which can either be incorporated into the particles during the synthesis or used as a probe adsorbed or attached to the particle surface. Fluorescent probes can be precise and can sometimes visualize just the presence of NP surfaces, which are the particle cortex or just the insides of cells decorated with NPs. Thus, the method has proven to be an essential tool that can be applied to achieve new biological understandings of NP behavior. The favorable ability of fluorescence microscopy to achieve qualitative or quantitative assessments of NP dynamics, such as particle endocytosis and exocytosis, distribution, kinematics of particle translocation, and interactions of the particles with biological and liquid systems, is another area where fluorescence microscopy is unrivaled. Another benefit of fluorescence microscopy is the low concentration of particles that can be visualized. The NPs can be visualized at or even below a concentration of 5 nM or 0.5 ng/mL. Fluorescence microscopy is also the single method for particles where the dispersive concentration of particles is not limited by the M parameter. The M parameter is an absorption cross-section times the quantum yield [179,197].

1.17 MAGNETIC AND ELECTRONIC CHARACTERIZATION OF NPs

Magnetic characterization is an important technique for studying the behavior of magnetic NPs and helping to understand their responses to an external magnetic field used in different biological applications such as data storage, bio-separations, MRI, sensors, and so forth. There are several techniques for the magnetic characterization of NPs. The faster the measurement and the simpler the method, the better the results obtained in a shorter time. Some of the techniques are discussed below. Vibrating sample magnetometry (VSM) provides values of either the magnetic moment using a conversion factor or the magnetization directly. Hysteresis loops are obtained by changing the value of the applied magnetic field and measuring the magnetization at each step [208,209].

An alternating gradient magnetometer (AGM) consists of two superimposed fields. A large and homogeneous magnetic field is used to magnetize particles while varying the face of the trapped field in the sample. After turning off the strong field, a pure trapping field is left, and in this case, the measured magnetization is due to the interaction of the trapped uniform field and the magnetic moment of the particles. Therefore, AGM measures the applied field required to overcome the sample's magnetization to erase the record in the sample, which is the first magnetic field oriented in the opposite direction of the original field. The attractive sampling volume of the NPs in AGM is proportionate to the trapping field squared. Thus, the sensitivity of AGM is inversely proportional to the square of the strength of the field, which is a disadvantage compared to the SQUID in AGM [210–212].

VSM is one of the most frequently used techniques for magnetic characterization of NPs. This technique is based on Faraday's law of electromagnetic induction. In VSM, the properties of a substance are characterized by the voltage induced in an electrical conductor that responds to a change in the magnetical field. VSM measures the sample's magnetic properties in terms of a magnetization curve or M(H) curve by changing the external magnetic field to different field strengths and then recording the corresponding induced voltage in the electrical conductor.

One of the routinely used magnetometric methods for the investigation of magnetic properties of NPs is alternating gradient magnetometry. A unique characteristic of AGM is its exceptionally high sensitivity, based on the magnetic force acting on the sample immersed in a fluid if an external alternating magnetic field exhibits a gradient. As AGM tends to attract particles toward the bottom of the capillary, preferably small samples of 1 mm OD are magnetically investigated. Consequently, AGM is mainly suitable for measurements on NPs [213,214].

Characterizing the electronic properties of an NP, whether as a free system or as part of a larger matrix, can also contribute a great deal to understanding the general techniques of material analysis used. This can also include sensors or batteries, or the electronic behavior of NPs as part of semiconductor devices. Therefore, this section is focused on electronic characterization techniques, based on which the following chapters will be elaborated [215].

Scanning tunneling microscopy (STM) is based on the process of quantum mechanical tunneling of electrons between the tip of an atomically sharp conducting needle and the surface of a metallic or semiconducting medium. It can also offer structural information within 0.5–0.1 nm dimensions. Spin-polarized STM can also provide local information on a surface's electronic structure and spin distribution or at interfaces. The spatial resolution is determined by the mechanical properties of the tip, not the electron wave function. STM gives atomically resolved images, including the surface topography and information on the structure of surface electronic states obtained by tunneling into the semiconductor conduction band. The differences between mechanical, chemical, and electronic etching are described. It provides details of the underlying principles of the surface analysis technique and its strengths and weaknesses. Applications are given for clean semiconductors, but treated and patinated alloys are also addressed [216].

XPS, also known as electron spectroscopy for chemical analysis, can provide information on a sample's surface core or valence levels. XPS analyzes the energy of emitted electrons following X-ray bombardment by ejecting one-carrier, Auger electrons, or even the photoelectrons themselves. Many element detections can be obtained from the binding energies of the core levels. Several well-known techniques can, with excellent sensitivity, present elements from P2 to U7 sensitivity comparable with PIXE. The fundamental principles of XPS are explained along with practical aspects of measurements and data interpretation, as well as an explanation of the physical processes behind the emergence of the photoemission spectra. The advantages and the limiting factors of XPS are also considered. There is an emphasis on the requirements of a surface analysis technique for NPs and their characterization, focusing on XPS. An overview of the fundamental principles of XPS technology and operational matters is provided, appraising the method in terms of its strengths and possible limitations. This complements an appreciation of the technique that has been part of a sequence of instruments for characterizing thin organic films [217].

1.18 BIOLOGICAL AND TOXICOLOGICAL EVALUATION OF NPs

The use of NPs has significantly grown in areas such as cosmetics, medical treatments, food processing, and drug delivery. With this growth, concerns about potential human and environmental exposure have also increased, particularly through absorption in the digestive system and subsequent distribution in the body. Evaluating the toxicological effects of NPs is crucial to ensure compliance with safety standards and address risks associated with their unique properties, such as size, shape, and surface reactivity. Additionally, the environmental and biological accumulation of NPs underscores the urgent need for proper evaluation and regulation [218,219].

A review of the literature reveals significant gaps in understanding the synthesis and evaluation of NPs, as well as a lack of global standards for assessing their biological and toxicological effects. This is particularly true for the long-term impacts of biologically active NPs, resistant nanodevices, and those producing free radicals or oxides. Addressing these gaps requires collaboration among toxicologists and experts in risk management. Research and discussions must prioritize the development of evaluation guidelines and explore key areas needing further study [220].

The therapeutic application and handling of NPs often involve protein adsorption, which influences their biological identity in tissues. This adsorption can cause aggregation, immune uptake, allergenic reactions, or rapid clearance by the immune system. A key parameter for characterizing NPs is the zeta potential, which indicates their surface charge and greatly impacts behavior in biological environments. Positively charged NPs are typically less stable than negatively charged ones due to interactions with salts in biological systems. While positively charged NPs are favored for drug delivery, negatively charged NPs offer safety advantages. Zeta potential also aids in optimizing cellular uptake and assessing the success of surface modifications, providing valuable insights for biological and toxicological evaluations [221,222].

Upon exposure to biological fluids, NPs rapidly adhere biomolecules onto their surface, thus forming a so-called protein corona. This phenomenon is of great significance for the interactions between NPs and the biological environment. In several cases, it has even been shown that the substitution of all serum proteins by human serum albumin does significantly alter the toxicity profiles of the corresponding metal NPs. More detailed studies have shown that the protein corona is composed of both a hard corona and a soft corona. Besides the formation of a characteristic surface layer of proteins, NPs can also adsorb lipoprotein complexes with hydrophobic phospholipids that form discoidal or mature particles and can greatly affect the cellular uptake of the particles. Because all these sequential reactions with particles are irreversible, the interactions with proteins depend on the relative binding forces between naked particles, soft corona, and proteins [223,224].

Evaluating the biological impacts of NPs requires carefully chosen and validated techniques to ensure biocompatibility and predict toxicological effects. Beyond traditional cytotoxicity assays, newer methods assess NPs under conditions resembling target tissues and are tailored for drug delivery applications. While many tests focus on cultured cells, some have progressed to small animal models, narrowing the gap between in vitro and in vivo evaluations [225].

Cytotoxicity evaluation using cell culture and various assays is regarded as the first step in the biological and toxicological evaluation of NPs. Cytotoxicity assays provide the key fundamental information about the effects of NPs on cellular processes such as cell viability, cell number proliferation, and the reducing capacity of cells or cellular health. Cytotoxicity assays give insights into possible in vitro effects that have the potential to cause in vivo toxicity and can be subacute, subchronic, and chronic. There are a variety of assays that are commercially available as kits, which include colorimetric, fluorometric, and enzyme-based methods. The most commonly used one is the assay with a colorimetric endpoint readout. They are selected based on the type of cells being studied and the mechanism of action of NPs [226].

1.19 CONCLUSION

In conclusion, pharmaceutical nanotechnology is a frontier in medication delivery, especially cancer treatment. This chapter describes nanotechnology's history and its importance in solving conventional medication formulation problems. Drug solubility, stability, and bioavailability can be improved by nanoparticles, making therapeutic treatments more effective.

Developing safe and effective drug delivery systems is complicated by exploring nanoparticle varieties, synthesis methodologies, and characterisation methods. Through toxicity assessments, laboratory research to clinical application is carried out with the utmost care for patient safety and regulatory norms.

Nanomedicine can transform cancer treatment in the future by providing tailored and targeted treatments. Using nanoparticles' unique properties, medical professionals and researchers can create more effective and safer treatments. Pharmaceutical nanotechnology is transforming cancer treatment and improving patient outcomes globally. Nanotechnology in pharmaceuticals will change oncology's future as research pushes limits.

REFERENCES

1. Petros RA, DeSimone JM. Strategies in the design of nanoparticles for therapeutic applications. *Nature Reviews Drug Discovery*. 2010;9(8):615–627. doi: 10.1038/nrd2591.
2. Soliman GM. Nanoparticles as safe and effective delivery systems of antifungal agents: Achievements and challenges. *International Journal of Pharmaceutics*. 2017;523(1):15–32. doi: 10.1016/j.ijpharm.2017.03.019.
3. Khan I, Saeed K, Khan I. Nanoparticles: Properties, applications and toxicities. *Arabian Journal of Chemistry*. 2019;12(7):908–931. doi: 10.1016/j.arabjc.2017.05.011.
4. Joudeh N, Linke D. Nanoparticle classification, physicochemical properties, characterization, and applications: A comprehensive review for biologists. *Journal of Nanobiotechnology*. 2022;20(1):262. doi: 10.1186/s12951-022-01477-8.
5. Malik S, Muhammad K, Waheed Y. Nanotechnology: A revolution in modern industry. *Molecules*. 2023;28(2). doi: 10.3390/molecules28020661.
6. Fraceto LF, Grillo R, de Medeiros GA, et al. Nanotechnology in agriculture: Which innovation potential does it have? [Perspective]. *Frontiers in Environmental Science*. 2016;4. doi: 10.3389/fenvs.2016.00020.
7. Taha TB, Barzinjy AA, Hussain FHS, et al. Nanotechnology and computer science: Trends and advances. *Memories – Materials, Devices, Circuits and Systems*. 2022;2:100011. doi: 10.1016/j.memori.2022.100011.
8. Singh T, Shukla S, Kumar P, et al. Application of nanotechnology in food science: Perception and overview [Mini Review]. *Frontiers in Microbiology*. 2017;8. doi: 10.3389/fmicb.2017.01501.
9. Utsev T, Tiza TM, Mogbo O, et al. Application of nanomaterials in civil engineering. *Materials Today: Proceedings*. 2022;62:5140–5146. doi: 10.1016/j.matpr.2022.02.480.
10. Shah MA, Pirzada BM, Price G, et al. Applications of nanotechnology in smart textile industry: A critical review. *Journal of Advanced Research*. 2022;38:55–75. doi: 10.1016/j.jare.2022.01.008.
11. Shafique M, Luo X. Nanotechnology in transportation vehicles: An overview of its applications, environmental, health and safety concerns. *Materials*. 2019;12(15). doi: 10.3390/ma12152493.
12. Malik S, Muhammad K, Waheed Y. Emerging applications of nanotechnology in healthcare and medicine. *Molecules*. 2023;28(18). doi: 10.3390/molecules28186624.
13. Guerrini G, Magrì D, Gioria S, et al. Characterization of nanoparticles-based vaccines for COVID-19. *Nature Nanotechnology*. 2022;17(6):570–576. doi: 10.1038/s41565-022-01129-w.
14. Walter P, Welcomme E, Hallégot P, et al. Early Uue of PbS nanotechnology for an ancient hair dyeing formula. *Nano Letters*. 2006;6(10):2215–2219. doi: 10.1021/nl061493u.
15. Barber DJ, Freestone IC. An investigation of the origin of the colour of the Lycurgus Cup by analytical transmission electron microscopy. *Archaeometry*. 1990;32(1):33–45.
16. Heiligtag FJ, Niederberger M. The fascinating world of nanoparticle research. *Materials Today*. 2013;16(7):262–271. doi: 10.1016/j.mattod.2013.07.004.
17. US G. https://www.nano.gov/national-nanotechnology-initiative [October 2024].
18. Marty JJ, Oppenheim RC, Speiser P. Nanoparticles-a new colloidal drug delivery system. *Pharmaceutica Acta Helvetiae*. 1978;53(1):17–23.
19. Saletu B, Anderer P, Kinsperger K, et al. Comparative bioavailability studies with a new mixed-micelles solution of diazepam utilizing radioreceptor assay, psychometry and EEG brain mapping. *International Clinical Psychopharmacology*. 1988;3(4):287–323. doi: 10.1097/00004850-198810000-00002.
20. Winn MJ, White PM, Scott AK, et al. The bioavailability of a mixed micellar preparation of vitamin K1, and its procoagulant effect in anticoagulated rabbits. *Journal of Pharmacy and Pharmacology*. 2011;41(4):257–260. doi: 10.1111/j.2042-7158.1989.tb06446.x.
21. Gelderblom H, Verweij J, Nooter K, et al. Cremophor EL: The drawbacks and advantages of vehicle selection for drug formulation. *European Journal of Cancer*. 2001;37(13):1590–1598. doi: 10.1016/S0959-8049(01)00171-X.
22. Brixner D, Holtorf A-P, Bhatia A, et al. Burden of cremophor-related paclitaxel therapy on patients with ovarian cancer: A patient focused study from the United Kingdom, Germany, and Spain. *Journal of Clinical Oncology*. 2022;40(16_suppl):e17576–e17576. doi: 10.1200/JCO.2022.40.16_suppl.e17576.
23. Ahmad Z, Shah A, Siddiq M, et al. Polymeric micelles as drug delivery vehicles. *RSC Advances*. 2014;4(33):17028–17038. doi: 10.1039/C3RA47370H.

24. Aliabadi HM, Lavasanifar A. Polymeric micelles for drug delivery. *Expert Opinion on Drug Delivery.* 2006;3(1):139–162.

25. Aliabadi HM, Mahmud A, Sharifabadi AD, et al. Micelles of methoxy poly(ethylene oxide)-b-poly([epsilon]-caprolactone) as vehicles for the solubilization and controlled delivery of cyclosporine A. *Journal of Controlled Release.* 2005;104(2):301–311.

26. Ghosh B, Biswas S. Polymeric micelles in cancer therapy: State of the art. *Journal of Controlled Release.* 2021;332:127–147. doi: 10.1016/j.jconrel.2021.02.016.

27. Hwang D, Ramsey JD, Kabanov AV. Polymeric micelles for the delivery of poorly soluble drugs: From nanoformulation to clinical approval. *Advanced Drug Delivery Reviews.* 2020;156:80–118. doi: 10.1016/j.addr.2020.09.009.

28. Zielińska A, Carreiró F, Oliveira AM, et al. Polymeric nanoparticles: Production, characterization, toxicology and eotoxicology. *Molecules.* 2020;25(16). doi: 10.3390/molecules25163731.

29. Zare H, Ahmadi S, Ghasemi A, et al. Carbon nanotubes: Smart drug/gene delivery carriers. *International Journal of Nanomedicine.* 2021;16:1681–1706. doi: 10.2147/ijn.S299448.

30. Shah S, Dhawan V, Holm R, et al. Liposomes: Advancements and innovation in the manufacturing process. *Advanced Drug Delivery Reviews.* 2020;154–155:102–122. doi: 10.1016/j.addr.2020.07.002.

31. Viegas C, Patrício AB, Prata JM, et al. Solid lipid nanoparticles vs. nanostructured lipid carriers: A comparative review. *Pharmaceutics.* 2023;15(6). doi: 10.3390/pharmaceutics15061593.

32. Neha D, Momin M, Khan T, et al. Metallic nanoparticles as drug delivery system for the treatment of cancer. *Expert Opinion on Drug Delivery.* 2021;18(9):1261–1290. doi: 10.1080/17425247.2021.1912008.

33. Damani M, Jadhav M, Joshi R, et al. Advances in gold nanoparticles: Synthesis, functionalization strategies, and theranostic applications in cancer. *Critical Review in Therapeutic Drug Carrier Systems.* 2024;41(6):1–56. doi: 10.1615/CritRevTherDrugCarrierSyst.2024046712.

34. Ponchel G, Cauchois O. Shape-controlled nanoparticles for drug delivery and targeting applications. In: Vauthier C, Ponchel G, editors. *Polymer Nanoparticles for Nanomedicines: A Guide for Their Design, Preparation and Development.* Cham: Springer International Publishing; 2016. pp. 159–184.

35. Truong NP, Whittaker MR, Mak CW, et al. The importance of nanoparticle shape in cancer drug delivery. *Expert Opinion on Drug Delivery.* 2015;12(1):129–142. doi: 10.1517/17425247.2014.950564.

36. Adam-Dima EI, Balas M, Anastasescu M, et al. Synthesis of homogeneous spherical selenium nanoparticles through a chemical method for cancer therapy applications. *Toxicology in Vitro.* 2024;95:105765. doi: 10.1016/j.tiv.2023.105765.

37. Liu K, Wang ZQ, Wang SJ, et al. Hyaluronic acid-tagged silica nanoparticles in colon cancer therapy: Therapeutic efficacy evaluation. *International Journal of Nanomedicine.* 2015;10:6445–6454. doi: 10.2147/ijn.S89476.

38. Ferreira D, Fontinha D, Martins C, et al. Gold nanoparticles for vectorization of nucleic acids for cancer therapeutics. *Molecules.* 2020;25(15). doi: 10.3390/molecules25153489.

39. Liao S, Yue W, Cai S, et al. Improvement of gold nanorods in photothermal therapy: Recent progress and perspective [Review]. *Frontiers in Pharmacology.* 2021;12. doi: 10.3389/fphar.2021.664123.

40. Kim Y, Kim H, Kang HW. Enhancement of gold nanorods-assisted photothermal treatment on cancer with laser power in stepwise modulation. *Lasers in Surgery and Medicine.* 2022;54(6):841–850. doi: 10.1002/lsm.23549.

41. Niidome T, Yamagata M, Okamoto Y, et al. PEG-modified gold nanorods with a stealth character for in vivo applications. *Journal of Controlled Release.* 2006;114(3):343–347. doi: 10.1016/j.jconrel.2006.06.017.

42. Zong Q, Dong N, Yang X, et al. Development of gold nanorods for cancer treatment. *Journal of Inorganic Biochemistry.* 2021;220:111458. doi: 10.1016/j.jinorgbio.2021.111458.

43. Kesharwani P, Ma R, Sang L, et al. Gold nanoparticles and gold nanorods in the landscape of cancer therapy. *Molecular Cancer.* 2023;22(1):98. doi: 10.1186/s12943-023-01798-8.

44. Zhao S, Luo Y, Chang Z, et al. BSA-coated gold nanorods for NIR-II photothermal therapy. *Nanoscale Research Letters.* 2021;16(1):170. doi: 10.1186/s11671-021-03627-7.

45. Zewail M, Gaafar PME, Ali MM, et al. Lipidic cubic-phase leflunomide nanoparticles (cubosomes) as a potential tool for breast cancer management. *Drug Delivery.* 2022;29(1):1663–1674. doi: 10.1080/10717544.2022.2079770.

46. Leu JSL, Teoh JJX, Ling ALQ, et al. Recent advances in the development of liquid crystalline nanoparticles as drug delivery systems. *Pharmaceutics.* 2023;15(5). doi: 10.3390/pharmaceutics15051421.

47. Madheswaran T, Kandasamy M, Bose RJC, et al. Current potential and challenges in the advances of liquid crystalline nanoparticles as drug delivery systems. *Drug Discovery Today.* 2019;24(7):1405–1412. doi: 10.1016/j.drudis.2019.05.004.

48. Kamya V, Rupa M, Anjna R, et al. Liquid crystalline lipid nanoparticles: Emerging trends and applications in skin cancer. *Pharmaceutical Nanotechnology.* 2024;12:1–8. doi: 10.2174/0122117385312450240816055942.

49. Gowda BHJ, Ahmed MG, Alshehri SA, et al. The cubosome-based nanoplatforms in cancer therapy: Seeking new paradigms for cancer theranostics. *Environmental Research.* 2023;237:116894. doi: 10.1016/j.envres.2023.116894.

50. Jenni S, Picci G, Fornasier M, et al. Multifunctional cubic liquid crystalline nanoparticles for chemo- and photodynamic synergistic cancer therapy. *Photochemical & Photobiological Sciences.* 2020;19(5):674–680. doi: 10.1039/C9PP00449A.

51. Nazaruk E, Majkowska-Pilip A, Bilewicz R. Lipidic cubic-phase nanoparticles—cubosomes for efficient drug delivery to cancer cells. *ChemPlusChem.* 2017;82(4):570–575. doi: 10.1002/cplu.201600534.

52. Son KH, Hong JH, Lee JW. Carbon nanotubes as cancer therapeutic carriers and mediators. *International Journal of Nanomedicine.* 2016;11:5163–5185. doi: 10.2147/ijn.S112660.

53. Sheikhpour M, Naghinejad M, Kasaeian A, et al. The applications of carbon nanotubes in the diagnosis and treatment of lung cancer: A critical review. *International Journal of Nanomedicine.* 2020;15:7063–7078. doi: 10.2147/ijn.S263238.

54. Bonner JC. Carbon nanotubes as delivery systems for respiratory disease: do the dangers outweigh the potential benefits? *Expert Review of Respiratory Medicine.* 2011;5(6):779–787. doi: 10.1586/ers.11.72.

55. Zhang C, Wu L, de Perrot M, et al. Carbon nanotubes: A summary of beneficial and dangerous aspects of an increasingly popular group of nanomaterials [Review]. *Frontiers in Oncology.* 2021;11:693814.

56. Debnath SK, Srivastava R. Drug delivery with carbon-based nanomaterials as versatile nanocarriers: Progress and prospects [Review]. *Frontiers in Nanotechnology.* 2021;3. doi: 10.3389/fnano.2021.644564.

57. Murphy FA, Poland CA, Duffin R, et al. Length-dependent retention of carbon nanotubes in the pleural space of mice initiates sustained inflammation and progressive fibrosis on the parietal pleura. *The American Journal of Pathology.* 2011;178(6):2587–2600. doi: 10.1016/j.ajpath.2011.02.040.

58. Teng Z, Li W, Tang Y, et al. Mesoporous organosilica hollow nanoparticles: Synthesis and applications. *Advanced Materials.* 2019;31(38):1707612. doi: 10.1002/adma.201707612.

59. Li Z, Zhang Y, Feng N. Mesoporous silica nanoparticles: Synthesis, classification, drug loading, pharmacokinetics, biocompatibility, and application in drug delivery. *Expert Opinion on Drug Delivery.* 2019;16(3):219–237. doi: 10.1080/17425247.2019.1575806.

60. Liu Y, Tan J, Thomas A, et al. The shape of things to come: importance of design in nanotechnology for drug delivery. *Therapeutic Delivery.* 2012;3(2):181–194. doi: 10.4155/tde.11.156.

61. Key J, Palange AL, Gentile F, et al. Soft discoidal polymeric nanoconstructs resist macrophage uptake and enhance vascular targeting in tumors. *ACS Nano.* 2015;9(12):11628–11641. doi: 10.1021/acsnano.5b04866.

62. Ma C, Zhang J, Zhang T, et al. Comparing the rod-like and spherical BODIPY nanoparticles in cellular imaging [Original Research]. *Frontiers in Chemistry.* 2019;7. doi: 10.3389/fchem.2019.00765.

63. Parakhonskiy B, Zyuzin MV, Yashchenok A, et al. The influence of the size and aspect ratio of anisotropic, porous CaCO$_3$ particles on their uptake by cells. *Journal of Nanobiotechnology.* 2015;13(1):53.

64. Wang W, Gaus K, Tilley RD, et al. The impact of nanoparticle shape on cellular internalisation and transport: What do the different analysis methods tell us? *Materials Horizons.* 2019;6(8):1538–1547. doi: 10.1039/C9MH00664H.

65. Huang X, Teng X, Chen D, et al. The effect of the shape of mesoporous silica nanoparticles on cellular uptake and cell function. *Biomaterials.* 2010;31(3):438–448. doi: 10.1016/j.biomaterials.2009.09.060.

66. Hao N, Li L, Tang F. Shape matters when engineering mesoporous silica-based nanomedicines. *Biomaterials Science.* 2016;4(4):575–591.

67. Li Y, Kröger M, Liu WK. Shape effect in cellular uptake of PEGylated nanoparticles: Comparison between sphere, rod, cube and disk. *Nanoscale.* 2015;7(40):16631–16646.

68. Borzęcka W, Pereira PM, Fernandes R, et al. Spherical and rod shaped mesoporous silica nanoparticles for cancer-targeted and photosensitizer delivery in photodynamic therapy. *Journal of Materials Chemistry B.* 2022;10(17):3248–3259. doi: 10.1039/D1TB02299G.

69. Emerich DF, Thanos CG. Nanotechnology and medicine. *Expert Opinion on Biological Therapy.* 2003;3(4):655–663. doi: 10.1517/14712598.3.4.655.

70. Doroudian M, O'Neill A, Mac Loughlin R, et al. Nanotechnology in pulmonary medicine. *Current Opinion in Pharmacology.* 2021;56:85–92. doi: 10.1016/j.coph.2020.11.002.

71. Peer D, Karp JM, Hong S, et al. Nanocarriers as an emerging platform for cancer therapy. *Nature Nanotechnology.* 2007;2(12):751–760. doi: 10.1038/nnano.2007.387.

72. Toy R, Bauer L, Hoimes C, et al. Targeted nanotechnology for cancer imaging. *Advanced Drug Delivery Reviews*. 2014;76:79–97. doi: 10.1016/j.addr.2014.08.002.

73. Zhang Y, Li M, Gao X, et al. Nanotechnology in cancer diagnosis: Progress, challenges and opportunities. *Journal of Hematology & Oncology*. 2019;12(1):137. doi: 10.1186/s13045-019-0833-3.

74. Santucci C, Mignozzi S, Malvezzi M, et al. European cancer mortality predictions for the year 2024 with focus on colorectal cancer. *Annals of Oncology*. 2024;35(3):308–316. doi: 10.1016/j.annonc.2023.12.003.

75. Bray F, Laversanne M, Sung H, et al. Global cancer statistics 2022: GLOBOCAN estimates of incidence and mortality worldwide for 36 cancers in 185 countries. *CA: A Cancer Journal of Clinicians*. 2024;74(3): 229–263. doi: 10.3322/caac.21834.

76. Bizuayehu HM, Dadi AF, Ahmed KY, et al. Burden of 30 cancers among men: Global statistics in 2022 and projections for 2050 using population-based estimates. *Cancer*. 2024;130(21):3708–3723. doi: 10.1002/cncr.35458.

77. https://www.who.int/news-room/fact-sheets/detail/cancer.

78. Anand P, Kunnumakkara AB, Sundaram C, et al. Cancer is a preventable disease that requires major lifestyle changes. *Pharmaceutical Research*. 2008;25(9):2097–2116. doi: 10.1007/s11095-008-9661-9.

79. Santarelli RL, Pierre F, Corpet DE. Processed meat and colorectal cancer: A review of epidemiologic and experimental evidence. *Nutrition and Cancer*. 2008;60(2):131–144. doi: 10.1080/01635580701684872.

80. Rather IA, Koh WY, Paek WK, et al. The sources of chemical contaminants in food and their health implications. *Frontiers in Pharmacology*. 2017;8:830. doi: 10.3389/fphar.2017.00830.

81. Kamal N, Ilowefah MA, Hilles AR, et al. Genesis and mechanism of some cancer types and an overview on the role of diet and nutrition in cancer prevention. *Molecules*. 2022;27(6). doi: 10.3390/molecules27061794.

82. de Martel C, Georges D, Bray F, et al. Global burden of cancer attributable to infections in 2018: A worldwide incidence analysis. *The Lancet Global Health*. 2020;8(2):e180–e190. doi: 10.1016/S2214-109X(19)30488-7.

83. https://www.cancer.gov/types/common-cancers.

84. https://publichealthscotland.scot/publications/show-all-releases?id=20468.

85. https://phw.nhs.wales/services-and-teams/welsh-cancer-intelligence-and-surveillance-unit-wcisu/cancer-reporting-tool-official-statistics/

86. https://www.qub.ac.uk/research-centres/nicr/.

87. Liskova A, Samec M, Koklesova L, et al. Flavonoids as an effective sensitizer for anti-cancer therapy: Insights into multi-faceted mechanisms and applicability towards individualized patient profiles. *EPMA Journal*. 2021;12(2):155–176. doi: 10.1007/s13167-021-00242-5.

88. Gupta C, Prakash D. Phytonutrients as therapeutic agents. *Journal of Integrative and Complementary Medicine*. 2014;11(3):151–169. doi: 10.1515/jcim-2013-0021.

89. Bayat Mokhtari R, Homayouni TS, Baluch N, et al. Combination therapy in combating cancer. *Oncotarget*. 2017;8(23):38022–38043. doi: 10.18632/oncotarget.16723.

90. Abdellatif AAH, Hennig R, Pollinger K, et al. Fluorescent nanoparticles coated with a somatostatin analogue target blood monocyte for efficient leukaemia treatment. *Pharmaceutical Research*. 2020;37(11):217. doi: 10.1007/s11095-020-02938-1.

91. Abdellatif AAH, Ibrahim MA, Amin MA, et al. Cetuximab conjugated with octreotide and entrapped calcium alginate-beads for targeting somatostatin receptors. *Science Reports*. 2020;10(1):4736. doi: 10.1038/s41598-020-61605-y.

92. Abdellatif AAH, Abou-Taleb HA, Abd El Ghany AA, et al. Targeting of somatostatin receptors expressed in blood cells using quantum dots coated with vapreotide. *Saudi Pharmaceutical Journal*. 2018;26(8):1162–1169. doi: 10.1016/j.jsps.2018.07.004.

93. Abdellatif AAH, Tolba NS, Alsharidah M, et al. PEG-4000 formed polymeric nanoparticles loaded with cetuximab downregulate p21 & stathmin-1 gene expression in cancer cell lines. *Life Sciences*. 2022;295:120403. doi: 10.1016/j.lfs.2022.120403.

94. Abdellatif AAH, Scagnetti G, Younis MA, et al. Non-coding RNA-directed therapeutics in lung cancer: Delivery technologies and clinical applications. *Colloids Surfaces B Biointerfaces*. 2023;229:113466. doi: 10.1016/j.colsurfb.2023.113466.

95. Seidel JA, Otsuka A, Kabashima K. Anti-PD-1 and Anti-CTLA-4 therapies in cancer: Mechanisms of action, efficacy, and limitations. *Frontiers in Oncology*. 2018;8:86. doi: 10.3389/fonc.2018.00086.

96. Falzone L, Salomone S, Libra M. Evolution of cancer pharmacological treatments at the turn of the third millennium. *Frontiers in Pharmacology*. 2018;9:1300. doi: 10.3389/fphar.2018.01300.

97. Baudino TA. Targeted cancer therapy: The next generation of cancer treatment. *Current Drug Discovery Technologies*. 2015;12(1):3–20. doi: 10.2174/1570163812666150602144310.

98. Lee YT, Tan YJ, Oon CE. Molecular targeted therapy: Treating cancer with specificity. *European Journal of Pharmacology*. 2018;834:188–196. doi: 10.1016/j.ejphar.2018.07.034.

99. Bilusic M, Madan RA. Therapeutic cancer vaccines: The latest advancement in targeted therapy. *American Journal of Therapeutics*. 2012;19(6):e172–e1781. doi: 10.1097/MJT.0b013e3182068cdb.

100. Saxena M, van der Burg SH, Melief CJM, et al. Therapeutic cancer vaccines. *Nature Reviews Cancer*. 2021;21(6):360–378. doi: 10.1038/s41568-021-00346-0.

101. Bouazzaoui A, Abdellatif AAH. Vaccine delivery systems and administration routes: Advanced biotechnological techniques to improve the immunization efficacy. *Vaccine X*. 2024;19:100500. doi: 10.1016/j.jvacx.2024.100500.

102. Bouazzaoui A, Abdellatif AAH, Al-Allaf FA, et al. Strategies for vaccination: conventional vaccine approaches versus new-generation strategies in combination with adjuvants. *Pharmaceutics*. 2021;13(2). doi: 10.3390/pharmaceutics13020140.

103. Li G, Sun B, Li Y, et al. Small-molecule prodrug nanoassemblies: An emerging nanoplatform for anticancer drug delivery. *Small*. 2021;17(52):e2101460. doi: 10.1002/smll.202101460.

104. Xia Z, Mu W, Yuan S, et al. Cell membrane biomimetic nano-delivery systems for cancer therapy. *Pharmaceutics*. 2023;15(12). doi: 10.3390/pharmaceutics15122770.

105. Abdellatif AAH, Alturki HNH, Tawfeek HM. Different cellulosic polymers for synthesizing silver nanoparticles with antioxidant and antibacterial activities. *Science Reports*. 2021;11(1):84. doi: 10.1038/s41598-020-79834-6.

106. Tawfeek HM, Mekkawy AI, Abdelatif AAH, et al. Intranasal delivery of sulpiride nanostructured lipid carrier to central nervous system; in vitro characterization and in vivo study. *Pharmaceutical Development and Technology*. 2024;29(8):841–854. doi: 10.1080/10837450.2024.2404034.

107. Tawfeek HM, Younis MA, Aldosari BN, et al. Impact of the functional coating of silver nanoparticles on their in vivo performance and biosafety. *Drug Development and Industrial Pharmacy*. 2023;49(5):349–356. doi: 10.1080/03639045.2023.2214207.

108. Aldosari BN, Abd El-Aal M, Abo Zeid EF, et al. Synthesis and characterization of magnetic Ag-Fe$_{(3)}$O$_{(4)}$@ polymer hybrid nanocomposite systems with promising antibacterial application. *Drug Development and Industrial Pharmacy*. 2023;49(12):723–733. doi: 10.1080/03639045.2023.2277812.

109. Hosseini SM, Mohammadnejad J, Najafi-Taher R, et al. Multifunctional carbon-based nanoparticles: Theranostic applications in cancer therapy and diagnosis. *ACS Applied Biomaterials*. 2023;6(4):1323–1338. doi: 10.1021/acsabm.2c01000.

110. PourGashtasbi G. Nanotoxicology and challenges of translation. *Nanomedicine*. 2015;10(20):3121–3129. doi: 10.2217/nnm.15.131.

111. Begines B, Ortiz T, Perez-Aranda M, et al. Polymeric nanoparticles for drug delivery: Recent developments and future prospects. *Nanomaterials*. 2020;10(7). doi: 10.3390/nano10071403.

112. Talelli M, Hennink WE. Thermosensitive polymeric micelles for targeted drug delivery. *Nanomedicine*. 2011;6(7):1245–1255. doi: 10.2217/nnm.11.91.

113. El-Readi MZ, Abdulkarim MA, Abdellatif AAH, et al. Doxorubicin-sanguinarine nanoparticles: Formulation and evaluation of breast cancer cell apoptosis and cell cycle. *Drug Development and Industrial Pharmacy*. 2024:1–15. doi: 10.1080/03639045.2024.2302557.

114. Attia MF, Anton N, Wallyn J, et al. An overview of active and passive targeting strategies to improve the nanocarriers efficiency to tumour sites. *Journal of Pharmacy and Pharmacology*. 2019;71(8):1185–1198. doi: 10.1111/jphp.13098.

115. Abdellatif AAH. A plausible way for excretion of metal nanoparticles via active targeting. *Drug Development and Industrial Pharmacy*. 2020;46(5):744–750. doi: 10.1080/03639045.2020.1752710.

116. Abdellatif AAH, Aldalaen SM, Faisal W, et al. Somatostatin receptors as a new active targeting sites for nanoparticles. *Saudi Pharmaceutical Journal*. 2018;26(7):1051–1059. doi: 10.1016/j.jsps.2018.05.014.

117. Seon BK, Haba A, Matsuno F, et al. Endoglin-targeted cancer therapy. *Current Drug Delivery*. 2011;8(1):135–143. doi: 10.2174/156720111793663570.

118. Sakurai Y, Akita H, Harashima H. Targeting tumor endothelial cells with nanoparticles. *International Journal of Molecular Science*. 2019;20(23). doi: 10.3390/ijms20235819.

119. Karimi M, Eslami M, Sahandi-Zangabad P, et al. pH-Sensitive stimulus-responsive nanocarriers for targeted delivery of therapeutic agents. *Wiley Interdisciplinary Reviews in Nanomedicine and Nanobiotechnology*. 2016;8(5):696–716. doi: 10.1002/wnan.1389.

120. Imtiyaz Z, He J, Leng Q, et al. pH-Sensitive targeting of tumors with chemotherapy-laden nanoparticles: Progress and challenges. *Pharmaceutics*. 2022;14(11). doi: 10.3390/pharmaceutics14112427.

121. Chu S, Shi X, Tian Y, et al. pH-Responsive polymer nanomaterials for tumor therapy. *Frontiers in Oncology*. 2022;12:855019. doi: 10.3389/fonc.2022.855019.

122. Mizutani H, Tada-Oikawa S, Hiraku Y, et al. Mechanism of apoptosis induced by doxorubicin through the generation of hydrogen peroxide. *Life Sciences*. 2005;76(13):1439–1453. doi: 10.1016/j.lfs.2004.05.040.

123. Dam DH, Lee JH, Sisco PN, et al. Direct observation of nanoparticle-cancer cell nucleus interactions. *ACS Nano*. 2012;6(4):3318–26. doi: 10.1021/nn300296p.

124. Elbeltagi S, Abdel Shakor AB, H MA, et al. Synergistic effects of quercetin-loaded $CoFe_{(2)}O_{(4)}$@ Liposomes regulate DNA damage and apoptosis in MCF-7 cancer cells: Based on biophysical magnetic hyperthermia. *Drug Development and Industrial Pharmacy*. 2024;50(6):561–575. doi: 10.1080/03639045.2024.2363231.

125. Singh R. Nanotechnology based therapeutic application in cancer diagnosis and therapy. *3 Biotech*. 2019;9(11):415. doi: 10.1007/s13205-019-1940-0.

126. Marchetti C, De Felice F, Romito A, et al. Chemotherapy resistance in epithelial ovarian cancer: Mechanisms and emerging treatments. *Seminars in Cancer Biology*. 2021;77:144–166. doi: 10.1016/j.semcancer.2021.08.011.

127. Jia S, Zhang R, Li Z, et al. Clinical and biological significance of circulating tumor cells, circulating tumor DNA, and exosomes as biomarkers in colorectal cancer. *Oncotarget*. 2017;8(33):55632–55645. doi: 10.18632/oncotarget.17184.

128. Formoso P, Muzzalupo R, Tavano L, et al. Nanotechnology for the environment and medicine. *Mini-Reviews in Medicinal Chemistry*. 2016;16(8):668–675.

129. Wang B, Hu S, Teng Y, et al. Current advance of nanotechnology in diagnosis and treatment for malignant tumors. *Signal Transduction and Targeted Therapy*. 2024;9(1):200. doi: 10.1038/s41392-024-01889-y.

130. Huang R. Peptide-mediated drug delivery systems for targeted glioma therapy. *Nanomedicine: Nanotechnology, Biology and Medicine*. 2016;12(2). doi: 10.1016/j.nano.2015.12.247.

131. Lu Y, Lin L, Ye J. Human metabolite detection by surface-enhanced Raman spectroscopy. *Materials Today Bio*. 2022;13:100205. doi: 10.1016/j.mtbio.2022.100205.

132. Younis MA, Tawfeek HM, Abdellatif AAH, et al. Clinical translation of nanomedicines: Challenges, opportunities, and keys. *Advanced Drug Delivery Reviews*. 2022;181:114083. doi: 10.1016/j.addr.2021.114083.

133. Park SJ, Han SR, Kang YH, et al. In vivo preclinical tumor-specific imaging of superparamagnetic iron oxide nanoparticles using magnetic particle imaging for cancer diagnosis. *International Journal of Nanomedicine*. 2022;17:3711–3722. doi: 10.2147/IJN.S372494.

134. Bura C, Mocan T, Grapa C, et al. Carbon nanotubes-based assays for cancer detection and screening. *Pharmaceutics*. 2022;14(4). doi: 10.3390/pharmaceutics14040781.

135. Hernández SNH, Chauhan G. Nanofibers for cancer sensing and diagnostics. *Materials Today: Proceedings*. 2022;48:66–70. doi: 10.1016/j.matpr.2020.10.164.

136. Elnagar N, Elgiddawy N, El Rouby WMA, et al. Impedimetric detection of cancer markers based on nanofiber copolymers. *Biosensors*. 2024;14(2). doi: 10.3390/bios14020077.

137. Chen HY, Deng J, Wang Y, et al. Hybrid cell membrane-coated nanoparticles: A multifunctional biomimetic platform for cancer diagnosis and therapy. *Acta Biomaterilia*. 2020;112:1–13. doi: 10.1016/j.actbio.2020.05.028.

138. Rashidi N, Davidson M, Apostolopoulos V, et al. Nanoparticles in cancer diagnosis and treatment: Progress, challenges, and opportunities. *Journal of Drug Delivery Science and Technology*. 2024;95. doi: 10.1016/j.jddst.2024.105599.

139. Huber F, Lang HP, Zhang J, et al. Nanosensors for cancer detection. *Swiss Medical Weekly*. 2015;145:w14092. doi: 10.4414/smw.2015.14092.

140. Findik F. Nanomaterials and their applications. *Periodicals of Engineering and Natural Science*. 2021;9(3):62–75.

141. Pearce AK, Wilks TR, Arno MC, et al. Synthesis and applications of anisotropic nanoparticles with precisely defined dimensions. *Nature Reviews Chemistry*. 2021;5(1):21–45

142. Selmani A, Kovačević D, Bohinc K. Nanoparticles: From synthesis to applications and beyond. *Advances in Colloid and Interface Science*. 2022;303:102640.

143. Yaqoob AA, Umar K, Ibrahim MNM. Silver nanoparticles: Various methods of synthesis, size affecting factors and their potential applications–A review. *Applied Nanoscience*. 2020;10(5):1369–1378.

144. Baig N, Kammakakam I, Falath W. Nanomaterials: A review of synthesis methods, properties, recent progress, and challenges. *Materials Advances*. 2021;2(6):1821–1871.

145. Salem SS, Hammad EN, Mohamed AA, et al. A comprehensive review of nanomaterials: Types, synthesis, characterization, and applications. *Biointerface Research in Applied Chemistry*. 2022;13(1):41.

146. Ijaz I, Gilani E, Nazir A, et al. Detail review on chemical, physical and green synthesis, classification, characterizations and applications of nanoparticles. *Green Chemistry Letters and Reviews*. 2020;13(3):223–245.

147. Krishnia L, Thakur P, Thakur A. Synthesis of nanoparticles by physical route. *Synthesis and Applications of Nanoparticles*. Springer; 2022. pp. 45–59.

148. El-Eskandarany MS, Al-Hazza A, Al-Hajji LA, et al. Mechanical milling: A superior nanotechnological tool for fabrication of nanocrystalline and nanocomposite materials. *Nanomaterials*. 2021;11(10):2484.

149. Jagdeo KR. Physical methods for synthesis of nanoparticles. In Sharma A, Oza G (eds.), Nanochemistry: *Synthesis, Characterization and Applications* (1st ed.). CRC Press; 2023. pp. 66–76.

150. Behera A, Aich S, Theivasanthi T. Magnetron sputtering for development of nanostructured materials. In Thomas S, Kalarikkal N, Abraham AR (eds.), *Design, Fabrication, and Characterization of Multifunctional* Nanomaterials. Elsevier; 2022. pp. 177–199.

151. Balachandran A, Sreenilayam SP, Madanan K, et al. Nanoparticle production via laser ablation synthesis in solution method and printed electronic application-A brief review. *Results in Engineering*. 2022;16:100646.

152. Chen L, Hong M. Functional nonlinear optical nanoparticles synthesized by laser ablation. *Opto-Electronic Science*. 2022;1(5):210007-1–210007-25.

153. Mitchell S, Qin R, Zheng N, et al. Nanoscale engineering of catalytic materials for sustainable technologies. *Nature Nanotechnology*. 2021;16(2):129–139.

154. Asiya S, Pal K, Kralj S, et al. Sustainable preparation of gold nanoparticles via green chemistry approach for biogenic applications. *Materials Today Chemistry*. 2020;17:100327.

155. Vanderfleet OM, Cranston ED. Production routes to tailor the performance of cellulose nanocrystals. *Nature Reviews Materials*. 2021;6(2):124–144.

156. Abid N, Khan AM, Shujait S, et al. Synthesis of nanomaterials using various top-down and bottom-up approaches, influencing factors, advantages, and disadvantages: A review. *Advances in Colloid and Interface Science*. 2022;300:102597.

157. Pierre AC. *Introduction to Sol-Gel Processing*. Springer Nature; 2020.

158. Bokov D, Turki Jalil A, Chupradit S, et al. Nanomaterial by sol-gel method: Synthesis and application. *Advances in Materials Science and Engineering*. 2021;2021(1):5102014.

159. Alabada R, Kadhim MM, Sabri Abbas Z, et al. Investigation of effective parameters in the production of alumina gel through the sol-gel method. *Case Studies in Chemical and Environmental Engineering*. 2023;8:100405.

160. Waqar MA, Mubarak N, Khan AM, et al. Sol-gel for delivery of molecules, its method of preparation and recent applications in medicine. *Polymer-Plastics Technology and Materials*. 2024;63(12):1564–1581.

161. Mengqi Z, Shi A, Ajmal M, et al. Comprehensive review on agricultural waste utilization and high-temperature fermentation and composting. *Biomass Conversion and Biorefinery*. 2021;13(7):5445–5468.

162. Sahu A, Poler JC. Removal and degradation of dyes from textile industry wastewater: Benchmarking recent advancements, toxicity assessment and cost analysis of treatment processes. *Journal of Environmental Chemical Engineering*. 2024;12(5):113754.

163. Cheng F, Zhang Y, Zhang G, et al. Eliminating environmental impact of coal mining wastes and coal processing by-products by high temperature oxy-fuel CFB combustion for clean power Generation: A review. *Fuel*. 2024;373:132341.

164. Nie P, Zhao Y, Xu H. Synthesis, applications, toxicity and toxicity mechanisms of silver nanoparticles: A review. *Ecotoxicology and Environmental Safety*. 2023;253:114636.

165. Azanaw A, Birlie B, Teshome B, et al. Textile effluent treatment methods and eco-friendly resolution of textile wastewater. *Case Studies in Chemical and Environmental Engineering* 2022;6:100230.

166. Singh SK, Sharma A, Sharma L, et al. Green Synthesis of algal nanoparticles: Harnessing nature's biofactories for sustainable nanomaterials. In Bhardwaj AK, Srivastav AL, Rai S (eds.), *Biogenic Wastes-Enabled Nanomaterial Synthesis: Applications in Environmental Sustainability*. Springer; 2024. pp. 257–284.

167. Kumari S, Raturi S, Kulshrestha S, et al. A comprehensive review on various techniques used for synthesizing nanoparticles. *Journal of Materials Research and Technology*. 2023;27:1739–1763.

168. Holghoomi R, Kharab Z, Rahdar A, et al. Harnessing the power of green synthesis of nanomaterials for anticancer applications: A review. *Coordination Chemistry Reviews*. 2024;513:215903.

169. Yang Y, Waterhouse GI, Chen Y, et al. Microbial-enabled green biosynthesis of nanomaterials: Current status and future prospects. *Biotechnology Advances*. 2022;55:107914.

170. Srividya D, Seema JP, Navya H. Microbial metallonanoparticles—An alternative to traditional nanoparticle synthesis. In Singh P, Kumar V, Bakshi M, Hussain CM, Sillanpää M (eds.), *Environmental Applications of Microbial Nanotechnology*. Elsevier; 2023:149–166.

171. Patil MP, Kim G-D. Microorganism-mediated functionalization of nanoparticles for different applications. In Kumar V, Guleria P, Dasgupta N, Ranjan S (eds.), *Functionalized Nanomaterials I*. CRC Press; 2020. pp. 279–298.

172. Grabowski Ł, Wiśniewska K, Żabińska M, et al. Cyanobacteria and their metabolites-can they be helpful in the fight against pathogenic microbes? *Blue Biotechnology.* 2024;1(1):4.
173. Ferreira ML, Behrmann ICL, Daniel MA, et al. Green synthesis and antibacterial-antibiofilm properties of biogenic silver nanoparticles. *Environmental Nanotechnology, Monitoring & Management.* 2024;22:100991.
174. Dikshit PK, Kumar J, Das AK, et al. Green synthesis of metallic nanoparticles: Applications and limitations. *Catalysts.* 2021;11(8):902.
175. Samuel MS, Ravikumar M, John JA, et al. A review on green synthesis of nanoparticles and their diverse biomedical and environmental applications. *Catalysts.* 2022;12(5):459.
176. Madani M, Hosny S, Alshangiti DM, et al. Green synthesis of nanoparticles for varied applications: Green renewable resources and energy-efficient synthetic routes. *Nanotechnology Reviews.* 2022;11(1):731–759.
177. Jain K, Takuli A, Gupta TK, et al. Rethinking nanoparticle synthesis: A sustainable approach vs. traditional methods. *Chemistry–An Asian Journal.* 2024;19(21):e202400701.
178. Kumar D, Seth CS. Green-synthesis, characterization, and applications of nanoparticles (NPs): A mini review. *International Journal of Plant and Environment.* 2021;7(01):91–95.
179. Khan Y, Sadia H, Ali Shah SZ, et al. Classification, synthetic, and characterization approaches to nanoparticles, and their applications in various fields of nanotechnology: A review. *Catalysts.* 2022;12(11):1386.
180. Kambhampati P. Nanoparticles, nanocrystals, and quantum dots: What are the implications of size in colloidal nanoscale materials? *The Journal of Physical Chemistry Letters.* 2021;12(20):4769–4779.
181. Kang MS, Lee SY, Kim KS, et al. State of the art biocompatible gold nanoparticles for cancer theragnosis. *Pharmaceutics.* 2020;12(8):701.
182. Ansari AA, Parchur AK, Chen G. Surface modified lanthanide upconversion nanoparticles for drug delivery, cellular uptake mechanism, and current challenges in NIR-driven therapies. *Coordination Chemistry Reviews.* 2022;457:214423.
183. Low LE, Siva SP, Ho YK, et al. Recent advances of characterization techniques for the formation, physical properties and stability of Pickering emulsion. *Advances in Colloid and Interface Science.* 2020;277:102117.
184. Rathinavel S, Priyadharshini K, Panda D. A review on carbon nanotube: An overview of synthesis, properties, functionalization, characterization, and the application. *Materials Science and Engineering: B.* 2021;268:115095.
185. Serafin J, Dziejarski B. Activated carbons—Preparation, characterization and their application in CO_2 capture: A review. *Environmental Science and Pollution Research.* 2024;31(28):40008–40062.
186. Ali IAM, Ahmed AB, Al-Ahmed HI. Green synthesis and characterization of silver nanoparticles for reducing the damage to sperm parameters in diabetic compared to metformin. *Scientific Reports.* 2023;13(1):2256.
187. Ahmed Naseer MI, Ali S, Nazir A, et al. Green synthesis of silver nanoparticles using Allium cepa extract and their antimicrobial activity evaluation. *Chem International.* 2022;8(3):89–94.
188. Saddik MS, Elsayed MM, Abdelkader MSA, et al. Novel green biosynthesis of 5-fluorouracil chromium nanoparticles using harpullia pendula extract for treatment of colorectal cancer. *Pharmaceutics.* 2021;13(2):226.
189. Caputo F, Vogel R, Savage J, et al. Measuring particle size distribution and mass concentration of nanoplastics and microplastics: Addressing some analytical challenges in the sub-micron size range. *Journal of Colloid and Interface Science.* 2021;588:401–417.
190. Bélteky P, Rónavári A, Zakupszky D, et al. Are smaller nanoparticles always better? Understanding the biological effect of size-dependent silver nanoparticle aggregation under biorelevant conditions. *International Journal of Nanomedicine.* 2021;16:3021–3040.
191. Pu Y, Niu Y, Wang Y, et al. Statistical morphological identification of low-dimensional nanomaterials by using TEM. *Particuology.* 2022;61:11–17.
192. Wang S, Chen Z, Chen R, et al. Superior strength-ductility synergy in additively manufactured CoCrFeNi high-entropy alloys with multi-scale hierarchical microstructure. *Journal of Alloys and Compounds.* 2024;1006:176268.
193. Bagheri AR, Aramesh N, Hasnain MS, et al. Greener fabrication of metal nanoparticles using plant materials: A review. *Chemical Physics Impact.* 2023;7:100255.
194. Harish V, Ansari M, Tewari D, et al. Cutting-edge advances in tailoring size, shape, and functionality of nanoparticles and nanostructures: A review. *Journal of the Taiwan Institute of Chemical Engineers.* 2023;149:105010.

195. Tripathi N, Goshisht MK. Recent advances and mechanistic insights into antibacterial activity, antibiofilm activity, and cytotoxicity of silver nanoparticles. *ACS Applied Bio Materials*. 2022;5(4):1391–1463.

196. Khan M, Khan S, Omar M, et al. Nickel and cobalt magnetic nanoparticles (MNPs): Synthesis, characterization, and applications. *Journal of Chemical Reviews*. 2024;6(1):94–114.

197. Hossain N, Mobarak MH, Mimona MA, et al. Advances and significances of nanoparticles in semiconductor applications–A review. *Results in Engineering*. 2023;19:101347.

198. Wan K, He J, Shi X. Construction of high accuracy machine learning interatomic potential for surface/interface of nanomaterials—A review. *Advanced Materials*. 2024;36(22):2305758.

199. Sable H, Kumar V, Singh V, et al. Strategically engineering advanced nanomaterials for heavy-metal remediation from wastewater. *Coordination Chemistry Reviews*. 2024;518:216079.

200. Lee S, Sim K, Moon SY, et al. Controlled assembly of plasmonic nanoparticles: From static to dynamic nanostructures. *Advanced Materials*. 2021;33(46):2007668.

201. Francis DV, Abdalla AK, Mahakham W, et al. Interaction of plants and metal nanoparticles: Exploring its molecular mechanisms for sustainable agriculture and crop improvement. *Environment International*. 2024;190:108859.

202. Dezfuli AAZ, Abu-Elghait M, Salem SS. Recent insights into nanotechnology in colorectal cancer. *Applied Biochemistry and Biotechnology*. 2024;196(7):4457–4471.

203. Solangi NH, Karri RR, Mubarak NM, et al. Mechanism of polymer composite-based nanomaterial for biomedical applications. *Advanced Industrial and Engineering Polymer Research*. 2024;7(1):1–19.

204. Hammed V, Bankole AA, Akinrotimi O, et al. Silver nanoparticles (AGNPs): A review on properties and behavior of silver at the nanoscale level. *International Journal of Science and Research Archive*. 2024;12(2):1267–1272.

205. Jouyban A, Rahimpour E. Optical sensors based on silver nanoparticles for determination of pharmaceuticals: An overview of advances in the last decade. *Talanta*. 2020;217:121071.

206. Kaur B, Kumar S, Kaushik BK. Recent advancements in optical biosensors for cancer detection. *Biosensors and Bioelectronics*. 2022;197:113805.

207. Hosseingholian A, Gohari S, Feirahi F, et al. Recent advances in green synthesized nanoparticles: From production to application. *Materials Today Sustainability*. 2023;24:100500.

208. Shukla S, Khan R, Daverey A. Synthesis and characterization of magnetic nanoparticles, and their applications in wastewater treatment: A review. *Environmental Technology & Innovation*. 2021;24:101924.

209. Shabatina TI, Vernaya OI, Shabatin VP, et al. Magnetic nanoparticles for biomedical purposes: Modern trends and prospects. *Magnetochemistry*. 2020;6(3):30.

210. Roberts AP, Heslop D, Zhao X, et al. Unlocking information about fine magnetic particle assemblages from first-order reversal curve diagrams: Recent advances. *Earth-Science Reviews*. 2022;227:103950.

211. Egli R. Magnetic characterization of geologic materials with first-order reversal curves. In *Magnetic Measurement Techniques for Materials Characterization*. 2021. pp. 455–604.

212. Torres D. In Franco V, Dodrill B (eds.), *Magnetic Switching Behavior of Permalloy and Samarium Cobalt Nanocaps on Drop-Casted Nanosphere Templates*. California State University; 2024.

213. Barmpatza AC, Baklezos AT, Vardiambasis IO, et al. A review of characterization techniques for ferromagnetic nanoparticles and the magnetic sensing perspective. *Applied Sciences*. 2024;14(12):5134.

214. Frandsen BA, Read C, Stevens J, et al. Superparamagnetic dynamics and blocking transition in Fe 3 O 4 nanoparticles probed by vibrating sample magnetometry and muon spin relaxation. *Physical Review Materials*. 2021;5(5):054411.

215. Ye N, Zhao G, Wu W, et al. An improved method for estimating the saturation magnetization and core size distribution of magnetic nanoparticles based on M–H Curve. *IEEE Access*. 2024;12:160450–160463.

216. Araujo JF, Arsalani S, Freire Jr FL, et al. Novel scanning magnetic microscopy method for the characterization of magnetic nanoparticles. *Journal of Magnetism and Magnetic Materials*. 2020;499:166300.

217. Yıldırım M, Acet Ö, Acet BÖ, et al. Innovative approach against cancer: Thymoquinone-loaded PHEMA-based magnetic nanoparticles and their effects on MCF-7 breast cancer. *Biochemical and Biophysical Research Communications*. 2024;734:150464.

218. Egbuna C, Parmar VK, Jeevanandam J, et al. Toxicity of nanoparticles in biomedical application: nanotoxicology. *Journal of Toxicology*. 2021;2021(1):9954443.

219. Ahmad A, Imran M, Sharma N. Precision nanotoxicology in drug development: Current trends and challenges in safety and toxicity implications of customized multifunctional nanocarriers for drug-delivery applications. *Pharmaceutics*. 2022;14(11):2463.

220. Schmeisser S, Miccoli A, von Bergen M, et al. New approach methodologies in human regulatory — toxicology–Not if, but how and when! *Environment International*. 2023;178:108082.

221. Maillard APF, Espeche JC, Maturana P, et al. Zeta potential beyond materials science: Applications to bacterial systems and to the development of novel antimicrobials. *Biochimica et Biophysica Acta (BBA)-Biomembranes.* 2021;1863(6):183597.
222. Midekessa G, Godakumara K, Ord J, et al. Zeta potential of extracellular vesicles: toward understanding the attributes that determine colloidal stability. *ACS Omega.* 2020;5(27):16701–16710.
223. Kopac T. Protein corona, understanding the nanoparticle–protein interactions and future perspectives: A critical review. *International Journal of Biological Macromolecules.* 2021;169:290–301.
224. Bilardo R, Traldi F, Vdovchenko A, et al. Influence of surface chemistry and morphology of nanoparticles on protein corona formation. *Wiley Interdisciplinary Reviews: Nanomedicine and Nanobiotechnology.* 2022;14(4):e1788.
225. Tirumala MG, Anchi P, Raja S, et al. Novel methods and approaches for safety evaluation of nanoparticle formulations: A focus towards in vitro models and adverse outcome pathways. *Frontiers in Pharmacology.* 2021;12:612659.
226. Petersen EJ, Barrios AC, Henry TB, et al. Potential artifacts and control experiments in toxicity tests of nanoplastic and microplastic particles. *Environmental Science & Technology.* 2022;56(22):15192–15206.

2 Polymer-Based Nanoparticle Formulations for Targeted Anticancer Drug Delivery

Shaaban K. Osman, Ahmed A.H. Abdellatif,
Mohamed A. A. Abd El-Galil, and
Fatma Ahmed

2.1 INTRODUCTION

Nanomedicine has been effectively applied for cancer treatment by chemotherapy, immunotherapy, and radiotherapy [1]. The polymer-based drug delivery systems, constructed using carriers of polymeric nanoparticles (PNPs), have shown considerable therapeutic efficacy for chemo-gene therapy of carcinoma [2,3]. In addition, therapeutic PNPs have been used for synergistic cancer treatment by employing the co-delivery of polymers, drugs, and genes [4–6]. Furthermore, PNPs have been employed for cancer imaging [7].

The PNP carriers are biodegradable and have modulated physical characteristics of the shape, size, charge, and pattern that promote and optimize cell internalization of the nanoscale therapeutics directly into tumors and their passage across the blood-brain barrier (BBB) [8]. Nanoparticulate drugs are injectable colloids or nano-gels, suitable for active and passive targeting, and have the unique advantage of realizing stimuli-tailored delivery [9,10]. In addition, nano-gels offer controllable swelling and viscoelasticity, as they are highly solvated and possess both liquid and solid-like characteristics [10]. Furthermore, this system provides several new strategies for drug internalization into tumor sites and offers long circulation time and high drug cumulative dosing for enhanced cancer therapy. Moreover, the PNPs perform multiple tasks for tumor targeting and diagnosis [11]. The enabling pathways of the PNPs for the fast recognition and internalization of their cargo into cancer cells were reviewed in detail by Sun and his coworkers [12]. Their physicochemical characteristics and capability of tumor-specific localization were reviewed on various occasions [13].

Solvent evaporation and emulsion polymerization (conventional emulsion or surfactant-free) are the frequently used methods for lipid-coating PNP fabrication [14]. More recently, hyperbranched PNPs were reported as specific delivery platforms targeting the nuclei; this, in turn, enhanced the objectives of cancer therapy [15]. Noteworthily, nano-gel carriers of biopolymer-based NPs (bio-PNPs) offer high drug-loading and -release properties and active and passive targeting, with fast response to biological stimuli, leading to low toxicity. These features make them suitable as vascular drug carriers [10]. In addition, their stability, storage capacity, biodistribution, target selectivity, and internalization into cells are important issues that affect the efficacy of intravenous delivery of injectable PNPs [16]. Recently, vascular delivery of siRNA cancer therapeutics was facilitated by different strategies, which prolonged the siRNA survival and movement in the circulatory system until reaching the targeted tissues [17].

Here, the methods of PNP preparation, including polymerization, nanoprecipitation, emulsification, solvent evaporation, freeze-drying, and spray-drying techniques, were discussed. Additionally, the characterization methods for PNPs, including the microscopy for the morphological characterization and the DLS analysis for particle size and zeta potential determinations, were highlighted. Toxicity, release, and stability of the modified PNPs were also mentioned. In addition, the different

DOI: 10.1201/9781003517870-2

types of synthetic and natural polymers used for PNP fabrication in different cancer therapy applications are surveyed. Furthermore, the natural resources of the biopolymers used for cancer therapy were surveyed, and the fish-based polymers were considered.

2.2 POLYMERIC NP FABRICATION

Researchers are now interested in biodegradable nanocomposites, and several biodegradable polymer matrices with enhanced characteristics, such as sustained release, controlled release, and delayed release, have been developed [18]. The polymers used in pharmaceutical and biological applications can be divided into several groups, as represented in Figure 2.1.

2.3 THE COMMON TECHNIQUES FOR PNP FABRICATION

Several methods are suitable for PNP preparation, including solvent evaporation, nanoprecipitation, emulsification, spray-drying, freeze-drying, self-assembly, salting out, supercritical fluid, polymerization, and dialysis techniques [19] (Figure 2.2).

For example, polylactic glycolic acid (PLGA)-based NPs and polylactic acid (PLA)-based NPs were prepared by the solvent evaporation technique and loaded with different anticancer agents such as Sorafenib, Paclitaxel, and doxorubicin (DOX) [20–23]. Also, the nanoprecipitation technique was used for Paclitaxel-loaded PLGA-NP preparation [24]. Carriers of chitosan (CS) NPs for drug loading and release were best prepared by the ionotropic gelation method, while the electrostatic interaction method was used for DOX-loaded polyurethane NP preparation [25]. Additionally, the dialysis technique was used for the hyaluronic acid NPs (HANPs) [26].

Possible applications of PNPs for drug delivery systems have been extensively considered in pharmaceutical and medical research [20]. The Food and Drug Administration (FDA) has approved and signified several biodegradable and biocompatible polymers for laboratory-to-clinical application through oral and parenteral administration. The most common methods used for NPs preparation are arranged under either bottom-up or top-down methods. Through the bottom-up methods, the NP preparation starts from the polymer molecules and includes polymerization (microemulsion, precipitation, and interfacial polymerization). The top-down methods employ the polymers to

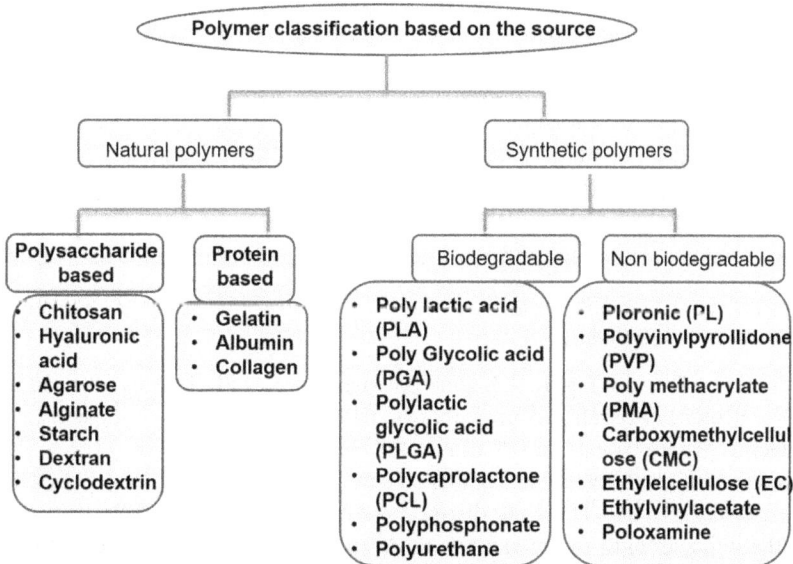

FIGURE 2.1 A schematic representation of polymer classifications.

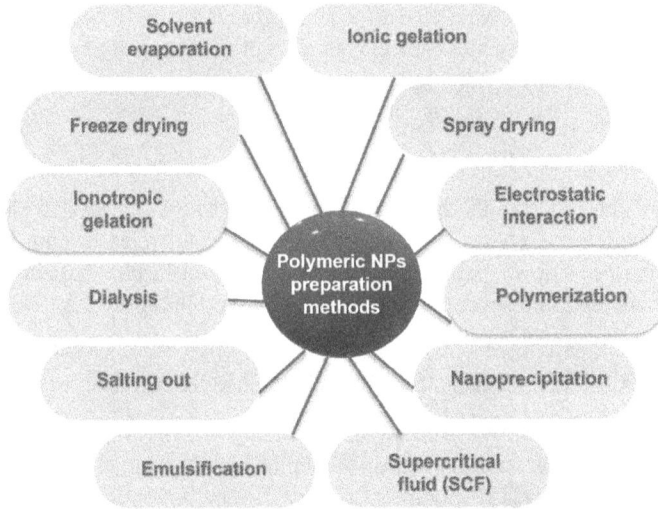

FIGURE 2.2   A schematic representation of polymeric NPs preparation techniques.

modify the NPs, using different techniques including salting out, emulsion diffusion, nanoprecipitation, and emulsion evaporation [21].

2.3.1 Self-Assembly Technique

It is an effective technique for producing PNPs. The self-assembled NPs contribute to improving the strategies for cancer treatment and show promising tendencies as drug carriers for targeted drug delivery [27]. Utilizing diverse forces such as hydrogen bonding, electrostatic interactions, hydrophobic interactions, stereo complexation, and host/guest interactions is frequent in self-assembled PNPs. Notably, precisely adjusting one or multiple forces of these interactions can create various self-assemblies with a broad spectrum of structures and functions. Diverse self-assembly strategies used for the fabrication of PNPs suitable for anticancer drug delivery systems were reviewed by Chen and his coworkers [9].

Recently, copolymeric nanostructures of self-assembled polymers were fabricated and worked efficiently as anticancer drug carriers. Amphiphilic copolymeric nanocarriers were constructed from different building blocks of hydrophilic/hydrophobic polymers. Polymerosomes, liposomes, hydrogen bonding, complexation (electrostatic, stereo-complexation, hydrophobic interaction, and host/guest interactions), and polymeric micelles are frequent preparation methods that use different building blocks to fabricate different copolymeric nanoplatforms for the anticancer drug delivery systems [9,28]. Recently, assemblies of linear amphiphilic copolymers formed a branched architecture called "star polymers" that showed considerably enhanced performance in drug delivery [29].

2.3.2 Salting-out Technique

In this technique, as shown in Figure 2.3, good polymers such as PLGA are dissolved in an organic solvent (such as acetone or tetrahydrofuran) to formulate the oil phase (water-miscible). On the other hand, the aqueous phase is composed of a saturated electrolyte salt solution (60% w/v; insoluble in organic solvents) with a surfactant. The emulsion can be constructed by adding the oil phase to an aqueous phase under stirring. Then, the distilled water will be added to the formed o/w emulsion to decrease the ionic strength in the electrolyte. In parallel, the hydrophilic organic solvents will migrate from the oil to the aqueous phase, constructing the NPs. Finally, centrifugation will purify the mixture from the salt [30].

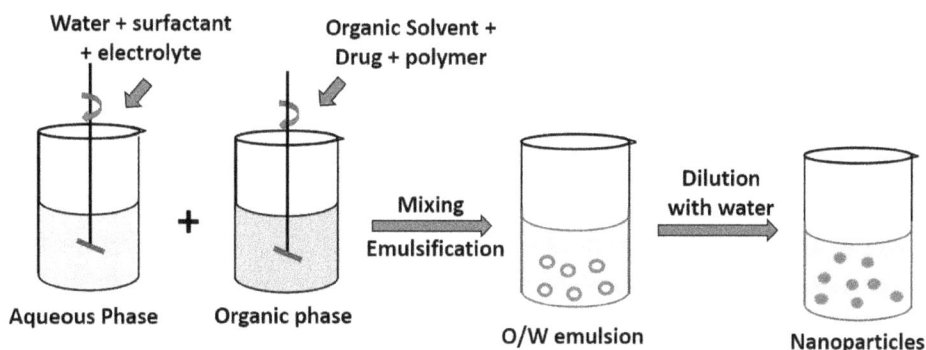

FIGURE 2.3   Preparation of nanoparticles (NPs) by the salting-out method.

2.3.3   EMULSION DIFFUSION TECHNIQUE

This NP formulation technique has two emulsion systems depending on the nature of the encapsulated reagents. The first is a singular emulsion system for oil-soluble (hydrophobic) substance formulation [31]. The second is a double emulsion system for the entrapment of hydrophilic reagents, in which an organic phase containing the polymer must be partially miscible in an aqueous phase [32,33].

The solvent diffusion step is the most important step for fabricating such NPs because the organic phase diffuses out from the oil to the aqueous phase, which hardens the formed particles. There are different kinds of surfactants used for NP fabrication, including di-dodecyl dimethyl ammonium bromide (DMAB) (cationic) [34], polyvinyl alcohol (PVA, non-ionic) [35], and sodium dodecyl sulfate (SDS, anionic) [36]. The kind of surfactant plays an important role in the size of the formulated NPs; for example, both DMAB and PVA were used as surfactants in the preparation of PLGA-NPs [37]. The results indicated that smaller particles were produced in the case of DMAB compared to those produced by using PVA as the surfactant. In addition, the surfactant concentration plays an important role in the final properties of the produced NPs. If an excessive amount of surfactant is used, the drug loading capacity will decrease because of the possible strong interaction between the surfactants and the drug. Also, a low surfactant concentration provides a high polydispersity index (PDI) and particle aggregation [38]. Alternatively, the mono-emulsion can be formulated using probe sonication [39].

2.3.4   NANOPRECIPITATION TECHNIQUE

This technique was introduced for NP formulation by *Fessi's* group [40]. Typically, in this technique, as illustrated in Figure 2.4, a suitable surfactant was dissolved in an aqueous solution to form an aqueous surfactant solution [41].

On the other hand, both the drug and polymer are dissolved in acetone or methanol to produce an organic solution. The polymeric NPs will be formed immediately through rapid solvent diffusion once the organic phase is dropped into the aqueous phase. Finally, the used solvents are removed under vacuum or by stirring overnight. The formed NPs can be collected by centrifugation, washed several times with water, and dried. This is a simple technique usable for hydrophilic and hydrophobic drug entrapment, in which a one-step procedure of adding the organic phase containing the drug and the polymer into the aqueous phase containing the surfactant is required for the NP formulation [42,43].

2.3.5   EMULSION/SOLVENT EVAPORATION TECHNIQUE

The principle of this technique is based on applying a high-shear force for the reduction of the emulsion droplet size [41]. The emulsification process was achieved by adding the organic solvent (such as ethyl acetate) containing the polymer into the aqueous phase, containing the suitable surfactant

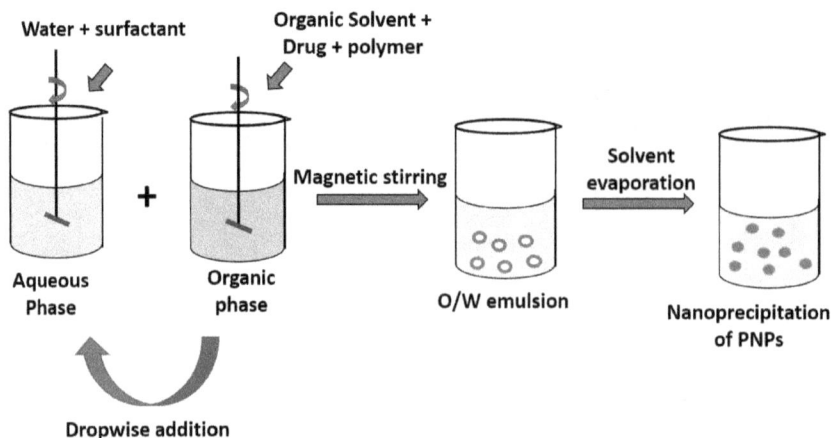

FIGURE 2.4 A schematic illustration of the nanoprecipitation method.

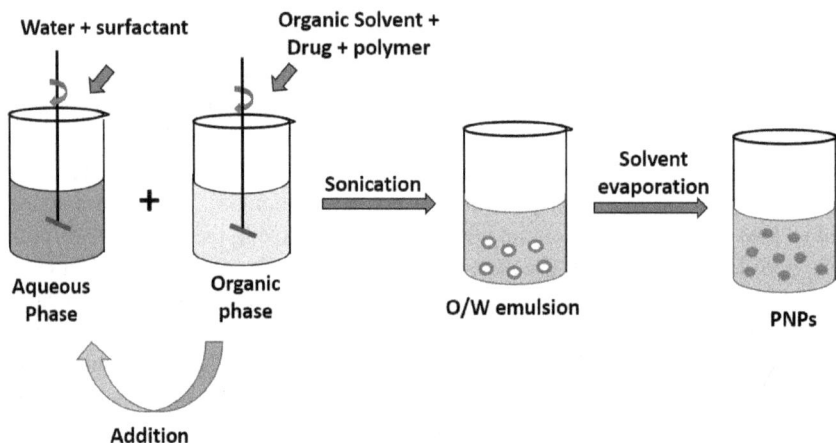

FIGURE 2.5 A schematic representation of the solvent evaporation method.

such as poloxamer 188, polyvinyl acetate, and Tween 80 by the aid of ultrasonication or homogenization at high speed (Figure 2.5). Then, the organic solvent will be allowed to evaporate under low pressure or continuous stirring, producing the modified polymeric NPs. After solvent removal, the solidified NPs are ready to be washed, collected by centrifugation, and kept lyophilized for further use [44–47].

2.3.6 Ionic Gelation Technique

This method is suitable for charged polymers such as CS. The principle of this technique is based on the ionic interaction and crosslinking between the positively charged CS (amino groups) and the negatively charged polyanion, such as tripolyphosphate (TPP) [48]. In the current method, the NP formulation occurs in an aqueous solution without the need for using any organic solvent. Typically, CS NPs were formulated via dropwise addition of the aqueous solution of TPP into the diluted acetic acid solution of CS with continuous stirring at room temperature (Figure 2.6). The physicochemical properties (particle size, loading capacity, and surface morphology) of the produced Cs-NPs were influenced by Cs concentration, TPP concentration, and Cs molecular weight [49]. This method was used for the preparation of Cs-NPs loaded with bovine serum albumin [50].

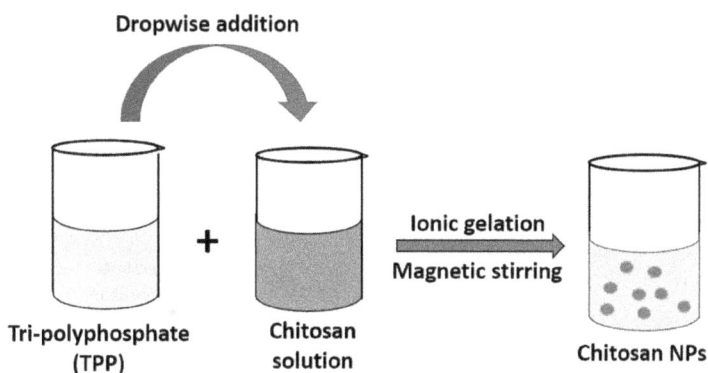

FIGURE 2.6 Preparation of chitosan NPs by ion gelation technology.

2.3.7 REVERSE MICELLAR (MICROEMULSION) TECHNIQUE

The principle of this technique depends on the precipitation of NPs in an aqueous core of reverse micellar droplets, followed by covalent crosslinking with glutaraldehyde. Typically, Cs-NPs were dispersed in an aqueous solution. This aqueous solution will be added dropwise into an organic solution, containing a suitable surfactant such as cetyltrimethylammonium bromide, which was dissolved to form reverse micelles, with continuous stirring. A crosslinking agent will be added to the mixture, under stirring overnight, to ensure the process of cross-linking. Afterward, the solvent was evaporated under a vacuum, and an excess surfactant was added to produce Cs-NPs. The removal of excess surfactant was carried out via precipitation using a suitable salt addition, such as $CaCl_2$. Then, the added salt was removed by centrifugation [51]. The produced NPs were purified by dialysis (to remove unreacted materials) and then lyophilized [52]. Noteworthily, the physicochemical properties of the obtained Cs-NPs vary according to the variation of formulation conditions and other parameters such as the amount of both Cs and crosslinking agent.

2.3.8 COACERVATION/PRECIPITATION TECHNIQUE

Here, the principle of the current technique is based on the change in pH of the polymer solution. For example, the Cs polymer is soluble in a mildly acidic solution of pH less than 6.5, while upon raising the pH to more than 6.5 (alkaline solution), the polymer starts to coacervate or precipitate out. Using a nozzle compressing air, the acidic Cs solution loaded with the drug was injected into an alkaline solution (i.e., ethan-diamine). The prepared Cs-NPs were precipitated by centrifugation, collected by filtration, and purified by washing several times with distilled water. For control of the drug release profile, the crosslinking agent was used to harden the modified NPs. Different parameters affect the physicochemical properties of the modified NPs, including the extent of Cs concentration, the extent of compressed air pressure, and the diameter of the spray nozzle used [30].

2.3.9 SPRAY-DRYING TECHNIQUE

This technique was used for the polymeric NP preparation loaded with either heat-sensitive materials, such as proteins and hormones, or heat-resistant drugs. The principle of this technique is based on the atomization of a solution mixture (containing the drug and excipient) into small droplets, which are blown into a hot air-containing chamber. The dried formulated particles were settled and collected after the solvent evaporation in the hot air (see Figure 2.7).

The formulation parameters, such as the nozzle size, the temperature of the inlet air, the atomization pressure, and the flow rate, will affect the physicochemical properties of the produced NPs [30]. In this method, the NPs are generally aqueous suspensions of the used polymer plus one drying

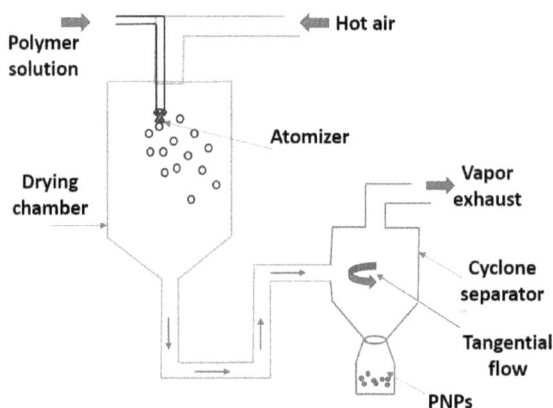

FIGURE 2.7 Preparation of chitosan particulate systems by the spray-drying method.

auxiliary compound such as lactose, mannitol, and polyvinylpyrrolidone (PVP) [53]. To prevent possible aggregation, the modified dried NP powder should be stored below the glass transition temperature of the used polymer. This technique was used for the preparation of different anticancer NPs. For example, Rezazadeh et al. [54] modified mixed polymeric micelles, which are composed of tocopherol succinate-polyethylene glycol 1000 and 5000 Da and loaded with paclitaxel as an anticancer drug. Lactose was used as a drying agent to produce dried NP powders for inhalation. The results indicated that the modified NPs elicited higher cytotoxic efficacy against the A945 (lung cancer cell line) than the unmodified drug.

2.3.10 Lyophilization (Freeze-Drying) Technique

The principle of this technique depends on removing the liquid content of the frozen sample under vacuum, leaving the dried powder content (changing the ice into powder mass without melting). This method is very suitable for NP preparation from temperature-sensitive compounds [55]. In this method, there is a possibility of irreversible powder aggregation due to the high concentration of nanoparticles in the final dried product. Therefore, to overcome such a problem, certain cryoprotectants such as mannitol, trehalose, lactose, and sucrose must be added to improve the resistance of nanoparticles toward the stresses caused by the lyophilization steps (freezing and drying) and, consequently, increase NPs stability during the storage. To prevent the possible aggregation of NPs and to keep the glassy state of the used cryoprotectant, the modified dried NP powder should be stored at a temperature below the glass transition temperature of the used polymer [53]. Recently, the lyophilization technique was used by *Ho and coworkers* [56]. To prepare artesunate (an effective anticancer drug) NPs, PLGA was used as a polymer. Also, they used three kinds of cryoprotectants: dextran, PVP K30, and mannitol. The obtained NPs were spongy-like spherical particles about 200 nm in size.

2.3.11 Supercritical Fluid (SCF) Technique

This method depends on the fluid formation at supercritical conditions (i.e., at temperature and pressure values higher than the critical point). The produced fluid, in its supercritical state, has the properties of gas (diffusivity) and liquid (density) to be comparable to both liquid and gas properties [57]. The most used candidate is supercritical carbon dioxide fluid (SC-CO_2) due to its non-toxicity, easily reachable critical values (Pc = 7.38 MPa and Tc = 31.1°C), and cheapness. In addition, due to it being in a gaseous state at ambient conditions, it can be separated easily without post-process treatments from liquids and solids [58], as illustrated in Figure 2.8.

Techniques, based on supercritical carbon dioxide (SC-CO_2) fluid, are used for the solvent or oil extraction (in the organic phase) from the nano-emulsion. Subsequently, the removal of such

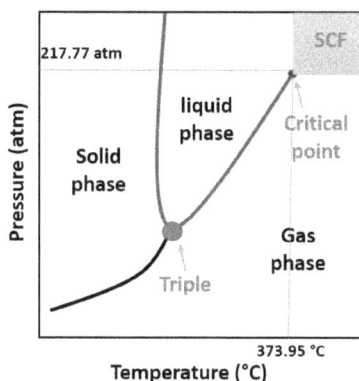

FIGURE 2.8 Phase diagram of water showing different phases as a function of temperature and pressure. The supercritical fluid stage is at 373.95°C, and the pressure is at 217 atm (220.64 bar).

solvent/oil will be accompanied by the precipitation of the encapsulated drug with the coating polymers (in the form of NPs). The modified particles have an ordered size and morphology without any possibility of particle aggregation [59]. This technique was used for the preparation of spherically shaped polycaprolactone (PCL) NPs [60]. Also, the NPs of several anticancer drugs were prepared by SC-CO$_2$ expansion techniques, such as Gambogic acid [61], Cisplatin [62], and Paclitaxel [63].

2.3.12 ELECTROSTATIC INTERACTION TECHNIQUE

In this method, the preparation of NPs is principally based on electrostatic interactions between alginate and piperazine under acidic aqueous conditions. This special negatively charged NP system has been modified to be used as a carrier since the carboxylate groups can be considered as the binding sites for different kinds of molecules, such as proteins and organic molecules. This technique of NP formation was reported by Román et al. for Cisplatin loading and delivery via covalent crosslinking on the surface of alginate/piperazine NPs using carbodiimide chemistry [64]. Also, the electrostatic self-assembly technique was used by *Maciel et al.* [65] for (Cs+pectin+NaCl) NP preparation, and the modified NPs were used for insulin loading and release.

2.3.13 POLYMERIZATION TECHNIQUE

We discussed and described the preparation of polymeric NPs from preformed polymers above. This section describes the preparation of NPs by monomer polymerization through four different methods, as follows [66]:

2.3.13.1 Emulsion Polymerization

Emulsion polymerization is the most common and fastest technique for polymeric NP production. The principle of this technique is based on the building of NPs from their monomers with the aid of suitable initiators. The technique starts with the dispersion of the respective monomers into the continuous phase of an emulsion. In the traditional approach, the emulsion system is composed of water-soluble initiator, water-soluble monomer, surfactant, and water [66].

Noteworthily, the monomer's dispersion into a nonsolvent emulsion forming a continuous organic phase is a considerable part of the technique. Once the monomer is introduced to the continuous phase, it will crosslink with an initiating agent, forming monomeric radicals that initiate the reaction. The radical's propagation continues until a critical chain length is obtained, since the solubility is reduced. The polymerization process continues until there is no precipitation of any new particles. The surfactant was used for particle stabilization [67,68].

The SDS surfactant was used for polyacrylate NP formation via the polymerization of acrylate compounds in an emulsion of styrene and butyl acrylate [69]. Also, microwave irradiation was used by other groups for the construction of styrene emulsion polymerization under microwave radiation, which was achieved by using a surfactant of SDS and an initiator of potassium persulfate. Polystyrene NPs were fabricated by the ultrasonic emulsion polymerization of styrene using surfactants of carboxymethyl cellulose and alkyl poly(ethoxy) acrylate [70].

2.3.13.2 Micro-Emulsion Polymerization

The microemulsion polymerization technique differs from the traditional emulsion polymerization technique, in that the modified particles have a minute size and fewer chains per particle, compared to the parent particles [71]. For this purpose, the water-soluble initiator is introduced to an aqueous phase containing numerous surfactants. However, the initiation of polymer chain formation begins in some microdroplets and continues successively, as it cannot occur in them all simultaneously. Many research articles reported the microemulsion polymerization technique for the construction of different kinds of PNPs [72,73].

2.3.13.3 Mini-Emulsion Polymerization

In this technique, a low molecular weight co-stabilizer and a high shear device (ultrasonicator) are required [66]. The modification of this system requires a mixture of monomer, initiator, co-stabilizer, and surfactant in water [74,75]. The physicochemical properties of the modified PNPs were affected by the kind and concentration of the used co-stabilizer and initiator. This technique was applied for the construction of different PNPs [76,77].

2.3.13.4 Surfactant-Free Emulsion Polymerization

Surfactants were needed to stabilize the NPs obtained by the traditional emulsion polymerization technique. However, the excess of these surfactants must be removed after the end of the process. The removal of such surfactants is time-consuming and economically questionable. Therefore, the formulation of PNPs without using any surfactant will overcome such limitations [78]. Several research articles used surfactant-free emulsion polymerization for the construction of polymeric NPs [79,80]. Biodegradable polymers have gained interest for biomedical applications, including targeted medication delivery, gene delivery, vaccination, and controlled drug delivery, because they decompose naturally, can be absorbed or expelled, and do not require surgery [81]. Many polysaccharides have been investigated for drug delivery and site-specific targeting to the colon, including CS, pectin, chondroitin sulfate, cyclodextrin, and dextrin. PLGA-PEG block copolymers for drug delivery include injectable delivery systems, microparticles, nanoparticles, nanogels, micelles, and hydrogels. Both CS and starch are abundant natural polysaccharides [82]. In addition, biomaterials have been investigated as a polymeric vector for delivering genetic or medicinal cargo. Interestingly, several biodegradable natural polymers, such as dextran, CS, and HA, can transfect cells without changing active targeting, making them tissue- or cell-specific. Unlike HA, dextran sulfate can build up in the liver by attaching itself to receptors on sinusoidal endothelial cells [83–85]. Since CS is known to bond with ciliary or mucous membranes, improving nanoparticle retention over the surface, coating PLGA nanoparticles with CS was done to increase the antiepileptic efficacy by intranasal delivery [86].

2.4 PRECAUTIONS DURING PNP COLLECTION AND PURIFICATION

The collection process of PNPs and conducting laboratory tests may alter their dispersion state. For pure PNP collection, the unwanted impurities must be removed by either centrifugation or filtration before any further experiments [87]. For dense-particle separation, centrifugation is more effective, whereas microfiltration is commonly employed due to its ease of use. The deposition of nanoparticles onto a membrane occurs through either impact or electrostatic forces.

2.5 CHARACTERIZATION OF PNPS

2.5.1 Physicochemical Characterization

The stability and integrity of NPs, as well as the drug loading and release, are influenced by the successful formulation of NPs. Therefore, there are some advanced microscopic techniques for NP morphological characterization, including scanning electron microscopy (SEM), transmission electron microscopy (TEM), and atomic force microscopy (AFM). In addition, the size, size distribution, and zeta potential are characterizable by dynamic light scattering (DLS) spectroscopy, TEM, and AFM. Furthermore, X-ray diffraction spectroscopy (XRD) is usable for the surface chemistry analysis of NPs [41].

2.5.1.1 Microscopy

2.5.1.1.1 Transmission Electron Microscopy

TEM is a microscopical technique in which the beam of an electron is transmitted through the samples and forms an image. Then, the image is magnified and focused on an imaging device. This kind of analysis investigates the size, shape, and purity of the formulated NPs. SEM and TEM analyses are complementary but require different examination principles from sample preparation to investigation. Technically, TEM is more complicated and time-consuming and requires ultrathin samples to suit the electron affinity. It can be performed using a JEOL-2100 electron microscope operating at 200 kV, and the samples are air-dried drops of diluted dispersions of NP samples on carbon-coated copper grids. Once an electron beam is transmitted through an ultra-fine mesh, it contacts the sample, screening its surface characteristics [88].

2.5.1.1.2 Scanning Electron Microscopy

SEM is useful for the examination of the morphological features of the PNPs. The operating principle of SEM involves using high-energy electrons to investigate a sample's surface by focusing on small areas. This technique primarily relies on generating images through the electrons' reflection from the specimen. While SEM offers certain benefits for analyzing morphology and dimensions, it provides only limited insights into size distribution and overall population averages [89]. An electron gun generates electrons, which are then accelerated by an anode and aligned using magnetic lenses. This process creates the electron beam necessary for measurement. The beam is subsequently focused using the objective lens and collected using the condenser electromagnetic lens. Finally, electromagnetic deflector coils guide the beam to scan the sample surface [41].

2.5.1.1.3 Atomic Force Microscopy

The AFM generates an image of interactions of a probe attached to an arm (a needle-like tip; 1–2 μm in length and <100 Å in diameter) with the sample surface. The probe movements provide interactions of its tip with the sample surface, causing a shift in its position. By measuring this displacement, a topographical map of the surface is generated. The AFM can function in two distinct modes, contact and non-contact modes. The contact mode involves gentle physical contact of a sharp tip with the sample surface while recording the arm's positions. The non-contact mode involves just the oscillation of the arm near the sample surface (50–150 Å away) without contact. This method measures variations in the arm's vibration frequency, influenced by the attractive van der Waals forces and surface changes. AFM is a potent technique providing high-resolution topographic images of the sample surfaces [90].

2.5.1.2 X-ray Diffraction

XRD analysis serves as a crucial method for determining the crystal structure of NPs. This non-destructive technique allows researchers to identify the atomic composition at lattice points and the crystal planes and spacing. The interaction between light waves and the crystal's structure produces distinct diffraction patterns, which vary based on the crystal's arrangement and the incident light wavelength. The interference of the rays reflected from different layers within the sample's periodic structure initiates a diffraction phenomenon [91].

2.5.1.3 Fourier Transform Infrared (FT-IR) Spectroscopy

The FT-IR spectroscopy is used for recording the interactions between the active constituents and the additives in the analyzed formula. The functional groups of the analyzed NPs and the individual components can be detected at a certain wavenumber (cm^{-1}). The principle of this analysis depends on the IR radiation of the samples. Some radiation is absorbed, and the rest will be passed or transmitted. These absorption-transmission spectra are characteristic of each molecule, and each is composed of several peaks indicating the vibrational frequency of all bonds between atoms. The position of each peak at a specific wavelength describes the bonding type between atoms. Also, the intensity of the peaks explains the number of materials. The samples were mixed with KBr and compressed to form discs. Spectra were recorded in the range of 4000–600 cm^{-1} with an accumulation of 100 scans and a spectral resolution of 4 cm^{-1} [92].

2.5.1.4 Dynamic Light Spectroscopy

This analysis is useful for particle size and size distribution determination in liquid samples. In addition, it can be used for surface charge detection via zeta potential measurement. Moreover, the molecular weight of certain polymers can be measured by DLS by measuring the hydrodynamic diameter [93]. The advantages of this technique include ease, low cost, speed, large measuring range, accuracy, and precision [94]. Besides, this technique is suitable for size analysis of monodisperse and polydisperse materials. The apparatus used for size measurement is called Zetasizer. This measurement strategy depends on the Brownian motion of the dispersed particles. These particles are subjected to a laser beam, and the intensity of scattered light is detected to translate the size diameter.

2.5.1.5 Electrophoretic Light Scattering (ELS)

ELS is suitable for zeta potential measurements via electrophoresis. Zeta potential values provide information about the stability of the modified colloids. High zeta potential values indicate that all particles in the solution possess large positive or negative charges. This means the presence of a repulsion between the similarly charged particles, and there is no possibility of aggregation (high stability of the system) and vice versa. The acceptable zeta potential value is 30 mV [95]. Noteworthily, zeta potential is mostly affected by different parameters, including the electric field, the polymer nature, and the dispersion medium conditions (e.g., pH, ionic strength, viscosity, and even the type of ions in the particle suspension).

As a typical procedure, particle size (diameter), size distribution (PDI), count rate, and zeta potential of the prepared and dried NPs are measured by DLS using Zetasizer Nano ZS90, Malvern Instruments. The samples for the DLS analysis are a diluted NP suspension or a freeze-dried powder of NPs distributed in distilled water. The intensity of the scattering light is detected at 90° to the incident, and the refractive index is 1.543. The data analysis is performed automatically, and the particle size records are based on the average value of 20 runs. Each sample is measured in triplicate [96]. Accordingly, the drug release kinetics from the nanosphere carriers generally follow a first-order model [97]. While in the case of nanocapsules, the release mechanism generally follows zero-order kinetics [98].

2.5.1.6 Stability of PNP Suspensions

According to the Brownian movement theory, colloidal NPs are characterized by their slow sedimentation and late-phase separation after a few months of storage, even though there is a possibility for particle agglomeration over storage time [99]. The presence of adsorbed surfactants and the adsorption of active molecules on the surface of the nanoparticles are two examples of variables that can affect the stability of the prepared NPs. The possibility of particle aggregation (instability) over time may represent the main challenge and limitation of such NPs applications. Therefore, stability studies (normal or accelerated) of polymeric colloidal suspensions can be carried out to explore and monitor a variety of physicochemical characteristics such as drug content, pH, zeta potential, polymer molar mass distribution, and particle size [100].

Another issue of interest is that the existence of NPs in a liquid form makes them susceptible to microbial contamination, which requires the inclusion of suitable preservatives [101]. So, it is typically advised to use drying techniques like lyophilization (freeze-drying) or spray-drying to avoid these physicochemical and microbiological issues and to produce fine powders of NPs [102,103].

2.5.2 *In Vitro* Release Kinetics

The *in vitro* release of the drug from the incubating polymeric NPs was carried out by using one of the following methods [104].

- Dialysis from dialyzing bags
- Ultracentrifugation,
- Normal filtration, or
- Ultrafiltration-centrifugation,

There are different mechanisms by which the loaded drugs can be released from the modified NPs as follows: [105–107]

- Drug desorption from the surface of NPs and/or the erosion of the polymeric matrix.
- Drug diffusion through the matrix of the nanosphere or the polymeric wall of nanocapsules.
- Combination of both erosion and diffusion mechanisms.

2.5.3 Toxicity of PNPs

There are three different physiological states of any cell, indicating that the cell is alive and active. These states are the cell viability, vitality, and proliferation. Cell viability is indicated by the number of living cells [108]. The living cells can divide and proliferate, and the others cannot because of the toxic or necrotic effects of physical or chemical factors that cause apoptosis (cells are alive without the activity of cellular division). So, the number of nonproliferating cells is not an indicative parameter for the number of dead cells [108].

2.6 CELL VIABILITY DETECTION ASSAYS

Cell viability measurement aims to quantify the overall healthy cells in a sample and evaluate cellular response to various treatments and stimuli [109]. Cell viability can be assessed through numerous methods, including dye exclusion by certain stains (Trypan blue and Eosin) [110], and colorimetric assays, including tetrazolium bromide (MTT), sulforhodamine B (SRB), neutral red uptake (NRU), and resazurin (RES) assays [111,112], and cell proliferation assays (enzymatic and adenosine triphosphate assays) [113,114], as depicted in Figure 2.9.

To adjust the optimal viability assay, different parameters should be taken into consideration, including the kind of cancer cell lines, the culture conditions, and the cost, speed, and availability of the analyzers used. Regarding the calorimetric assays, there are four different assays depending on the dye used in the experiment: MTT, SRB, NRU, and RES assays.

MTT is the most common kind of colorimetric assay that is used for cell viability detection by using a specific MTT reagent, named 3-(4,5-dimethyl-2-thiazolyl)-2,5-diphenyl-2H-tetrazolium bromide [115,116]. The principle of the detection is based on the reduction of the MTT reagent (yellow color) to formazan (purple color) by the metabolic effect of living cells (Figure 2.10). Then, the

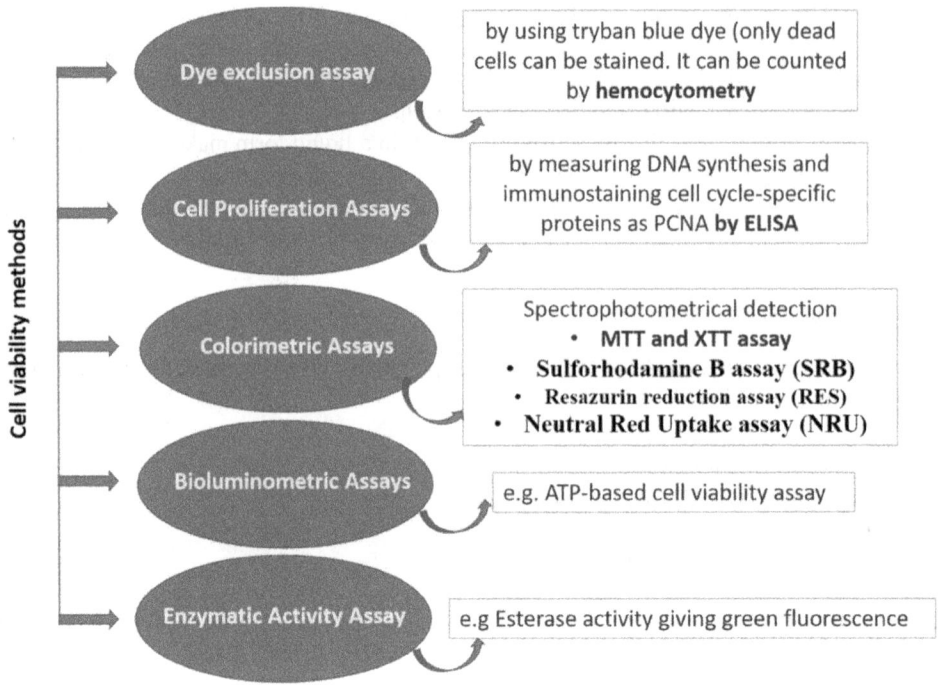

FIGURE 2.9   Schematic representation of different techniques for cell viability measurements.

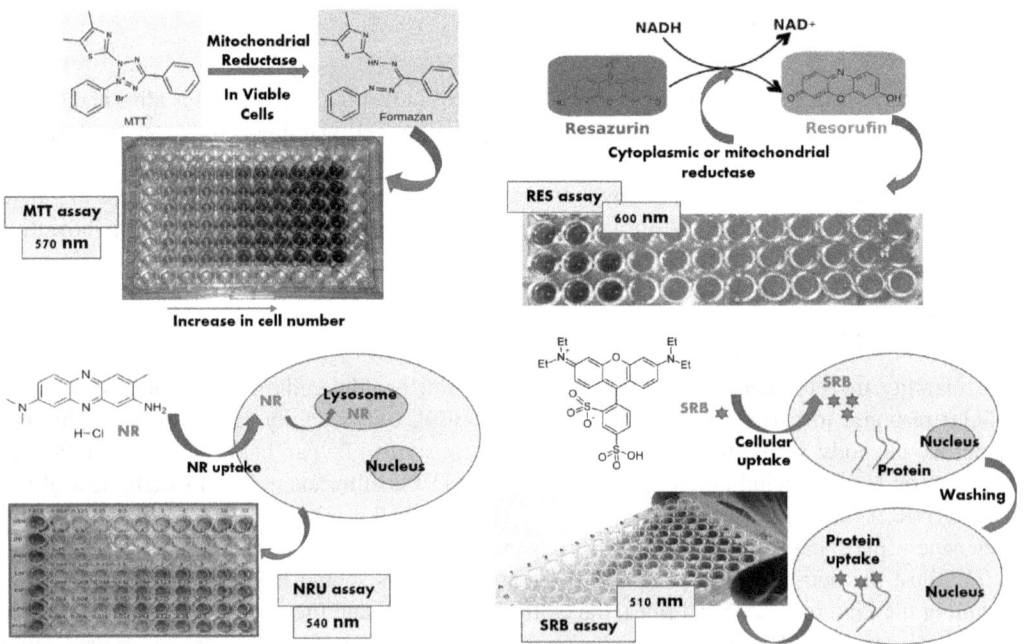

FIGURE 2.10   Representation of the mechanism involved in the colorimetric assays (MTT, SRB, RES and NRU) to determine cell viability.

produced crystals of insoluble Formazan will be dissolved by a suitable solvent such as dimethyl sulfoxide and analyzed spectrophotometrically at 500–600 nm using a microplate reader spectrophotometer [117]. The following equation calculates cell viability in percentage.

$$\% \text{ Cell viability} = \frac{\text{Optical density of sample}}{\text{Optical density of control}} \times 100$$

The extent of color intensity indicates the cell number; therefore, this assay helps in cellular viability and proliferation evaluations.

As a comparison, the MTT assay was considered by many researchers to be the "gold standard" compared to the three commonly used cell viability assays (NRU, RES, and SRB). Table 2.1 and Figure 2.10 represent different cell viability assays, showing the applied dye and the color change by cellular organelles.

For example, the reduction of MTT (yellow) and RES (nonfluorescent blue color) reagents into Formazan (purple) and Resorufin (fluorescent pink color), respectively, occurs in mitochondria or endoplasm of the living cells in the cases of MTT assay and RES assays. In the case of the NR assay, lysosomes are responsible for dye uptake. Regarding the SRB assay, the dye uptake was achieved by the cellular proteins of living and non-living cells. The cell viability can be measured spectrophotometrically at different maximum absorbances depending on the assay. For example, the optimal absorbances for color detection in the MTT, SRB, NRU, and RES assays are 570, 510, 540, and 600 nm, respectively [118–120].

All the MTT, RES, and NR assays depend on the cell metabolic activity, and they help quantify the viability and proliferation of living cells. In contrast, the SRB assay measures total cellular protein content and does not depend on cellular metabolic activity. Consequently, it does not distinguish between living and non-living cells [116]. SRB assay exhibits certain advantages over the MTT assay. For example, due to the independence of cell metabolic activity, the assay steps of SRB are fewer than those of the MTT assay. Also, in the case of MTT assay, the reduction of MTT can be done by other endogenous or exogenous compounds, which can also catalyze this chemical change without any effect on cell viability (assay interference). However, this interference is not possible with the SRB assay. Notably, the SRB assay is the preferred cytotoxicity assay of the National Cancer Institute in the USA [124].

TABLE 2.1
Different Cell Viability Assays

Assay	Stain	Action Inside Cells	Fluorometric Excitation/ Emission Wavelengths	Photometric Absorbance (λmax)	Ref.
			Detection Method		
MTT	Tetrazolium bromide (Yellow)	Formazan (Purple) in mitochondria	—	570 nm using 630 as a reference	[121]
SR-B	SRB stain	Cellular protein (viable and nonviable cells)	EX/Em 488 nm/585 nm	510 nm	[120]
NRU	Neutral red stain	Lysosomal uptake	Ex/Em 530 nm/645 nm	540 nm	[122]
RES	RES (blue nun fluorescent dye)	Resorufin (pink) in the cytoplasm and mitochondria	Ex/Em 550 nm/590 nm	600 nm	[123]

2.7 MAJOR CATEGORIES OF THE PNPs USED FOR CANCER THERAPY

Biodegradable polymers of different types are usable for the preparation of PNPs for drug delivery applications [18]. They are solid colloidal systems usable for constructing polymeric matrices with different therapeutics via entrapping, encapsulation, or adsorption. The PNPs are categorized according to their origin and structure into two types: biological and synthetic.

2.7.1 BIOPOLYMER-BASED NPs

Bio-PNPs of naturally sourced ingredients are biodegradable, bioavailable, and applicable biomedically for drug delivery in a wide range of biological tissues [110,125]. They are abundant, renewable, and favored for pharmaceutical applications and can replace chemotherapy and prevent patients from suffering its drawbacks. In addition, the bio-PNPs are water-soluble and non-toxic, which makes them more favorable than synthetic polymers for biological applications [126]. According to their source, the biopolymers comprise two major types: proteins (albumin, collagen, and gelatin) and polysaccharides (CS, starch, alginate, HA, dextran, and pullulan). For PNP functionalization, the functional groups of their parent polymers can be modified to produce high loading capacity and minimal invasive behavior, which are the key factors for drug-targeting delivery [18,127]. Here, we will display the protein and polysaccharide bio-based PNPs used for anticancer drug delivery applications.

2.7.1.1 Protein-Based PNPs

Protein-based polymers are biodegradable amino acid chains derived from natural resources with minimal bioaccumulation. Albumins, collagens, and gelatins are the most frequent bio-proteins used for drug delivery applications. Two types of albumins, including bovine serum- and human serum-albumins, are highly soluble, nontoxic proteins common in bio-applications for singular and dual anticancer drug delivery [98]. Albumin-based NPs can penetrate the blood-brain barrier, especially by modifying them with cell-penetrating peptides on the surface [127]. Dual drug delivery for targeting intracranial glioma was fabricated using the blood–brain barrier-penetrating albumin NP synthesis [128]. Furthermore, albumin-based NPs were implicated in the diagnostic imaging of tumors by their complexation with metallic NPs [129]. Important information about albumin-based drug carriers and their endogenous and exogenous formulations has been reviewed recently [130].

Collagen is the main protein found in vertebrate tissues; therefore, it is characterized by high biocompatibility and biodegradability for application in drug delivery. However, collagen is unstable and might dissociate during its processing in isolation and purification. By collagen hydrolysis, a different water-soluble protein, "gelatin," is derived. Its improved solubility compromises its efficacy for long-term drug release. Additionally, gelatin is incorporated in thermo-reversible gel preparation, which has broad pharmaceutical and medical applications [131]. Collagen and gelatin are often cross-linked with other polymers or targeting ligands that fortify and stabilize their structure, or modify their properties and control their release profile to match the therapeutic need [127]. Table 2.2 lists the recent applications of some protein-based PNPs in anticancer drug delivery systems.

2.7.2 POLYSACCHARIDE-BASED PNPs

2.7.2.1 Chitosan

The cationic CS is an amino-linear polysaccharide containing 1, 4 glycosidic links holding several d-glucosamine groups attached to N-acetyl-d-glucosamine units (Figure 2.11).

CS could be synthesized by various methods of deacetylation of the natural chitin obtained from fish scales and crustacean shells. It is a biodegradable polymer but elicits poor solubility at high pH levels, restricting its biocompatibility and bio-applications. CS NPs overcome the unfavorable features of the parent CS and offer tiny-sized and highly dispersed particles suitable for a wide range of biological applications useful for the food industry, aquaculture development, and human

TABLE 2.2

Examples of Some Applications of Protein-Based PNPs in Anticancer Drug Delivery and Control Release

Protein Polymer	Nanocomposite	Tumor	Ref.
Albumin	Bovine serum albumin-paclitaxel/fenretinide	Intracranial glioma	[128]
	Bovine serum albumin-anacardic acid/docetaxel/gemcitabine	Breast cancer	[132]
Collagen	Hybrid gel of hydroxyapatite-collagen/paclitaxel	Metastatic cancer	[133]
	Collagen-chitosan NPs/DOX	Cervical cancer cell line (HeLa cells)	[134]
Gelatin	Dendrimer-gelatin NPs/DOX	4T1 Cell line	[135]
	Phenylboronic acid-decorated gelatin NPs/DOX	SH-SY5Y, H22, and HepG2 tumor cell lines	[136]
	Matrix metalloproteinase-2 (MMP-2) - responsive gelatin NPs/DOX	S180-bearing mice	[137]

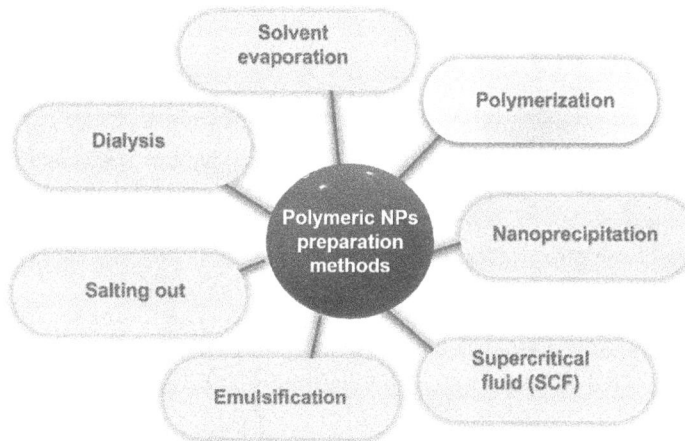

FIGURE 2.11   A schematic representation of the applications of polymeric NPs.

healthcare [110]. CS NP-based composites are well-known as antimicrobial reagents *in vitro* against various aquatic and terrestrial plant and animal pathogens [138–140]. In addition, they are smart, nontoxic immunostimulants and nano-carriers that provide an efficient biodegradable delivery platform for vaccines and drugs *in vivo* into fish tissues for disease prevention and treatment [141–143]. Owing to their mere safety, the PNPs of CS have recently been widely used for human therapy as drug vehicles for the delivery applications of various therapeutics, overcoming different diseases in different ways, oral, topical, nasal, and so forth. They have been effectively administered in tablets, beads, hydrogels, films, microspheres, nanoparticles, nanocomposites, and nanofibers [144]. Promising breast cancer control was reported by CS/5-Fluorouracil hydrogel [145]. In addition, CS NPs are excellent and safe vehicles for the targeted drug delivery systems of breast cancer treatment [146]. Copolymeric NPs of CS NPs and other polymers were constructed to produce more sensitive and efficient delivery systems and/or sustained release effects for anticancer drugs [147,148]. More recently, CS NPs have contributed strongly to human cancer immunotherapy [149,150].

TABLE 2.3

The Most Recent Applications of Cs-NPs in the *In Vitro* and *In Vivo* Therapy and Targeting of Various Types of Human Cancer

PNPs Complex	Tumor	Ref.
Cs-NPs/Battling	Lung cancer	[152]
Berberine/CSNPs		[153]
Cs-NPs/curcumin/docetaxel		[154]
Zein-Cs-NPs/curcumin/berberine	Breast cancer	[155]
RGD peptide-Cs-NPs/raloxifene		[156]
Cs-NPs/curcumin		[157]
Cs-NPs-HA/siRNA	Bladder cancer	[158]
N-deoxycholic acid glycol-chitosan/docetaxel	Gastric cancer	[159]
PLGA-chitosan-zeolites metal organic frameworks/paclitaxel	Prostate cancer	[160]
N-isopropylacrylamide-itaconic acid-chitosan/Erlotinib	Ovarian cancer	[161]
Chitosan-polysorbate 80/Imatinib	Colorectal cancer	[162]
Catechol-chitosan-HA/DOX	Oral cancer	[163]
D-α-tocopherol polyethylene glycol 1000 succinate- chitosan/ Docetaxel	Brain cancer	[163]
Cs-Fe$_2$O$_3$ NPs		[164]
Cs-NPs/Resveratrol-ferulic acid	Epidermoid carcinoma	[165]
Cs-NPs	Cervical cancer cell line (HeLa cells)	[166]
Cs-NPs/siRNA	Non-small cell lung cancer (NSCLC)	[167]

Interestingly, a pH-sensitive "on/off" platform of an optical biosensor was fabricated from mesoporous silica@CS@gold NPs (AuNPs) and tested against cancer [151]. Noteworthily, better tumor inhibition efficacy was observed on the formulations with CS of lower molecular weight [127]. Selective examples of some recent applications of CS NPs in the *in vitro* and *in vivo* therapy and targeting of various types of human cancer are collected in Table 2.3.

Several methods are used for CS NP preparation, including self-assembly, ionic gelation, and polyelectrolyte complexation. The mode of action of CS NPs for the active and passive drug targeting delivery was discussed by *Herdiana and his coworkers* [146], as illustrated in Figure 2.12.

2.7.2.2 Hyaluronan Polymer

It is a linear anionic non-sulfated glycosaminoglycan molecule constructed with a glycosaminoglycan chain possessing repeated subunits from N-acetyl D-glucosamine and D-glucuronic acid (-1,4 and -1,3 glycosidic glucuronic acid) disaccharides [126]. HA is naturally abundant as an extracellular component, especially in the vitreous humor and synovial fluid of all vertebrates and humans, and is synthesized by plasma membranes. Mainly, HA possesses bio-adhesive properties that facilitate cellular uptake, and its negatively charged molecules can prolong the circulation of the encapsulated therapeutics [126]. In addition, it can form controlled drug-release systems since the hyaluronidases are overexpressed in the tumor cells under strong acidic or basic conditions, which helps the degradation of the drug-loading acid and facilitates the release of its cargo. Although this acid is active against cancer cell growth, several considerations and precautions should be taken when using HA, as it also plays a role in the proliferation of tumor cells [127].

The molecular weight of the HA biopolymer controls its function in cancer progression, considering that its size and concentration vary depending on its source. On the one hand, the low molecular weight polymer (oligo-HA; <200 k Da) is more prominent on the periphery of cancer cells, exhibits aqueous solubility, has a high affinity to bind with the abundant transmembrane glycoprotein receptors (i.e., CD44) on the tumor cell surfaces, which are associated with their overexpression, and

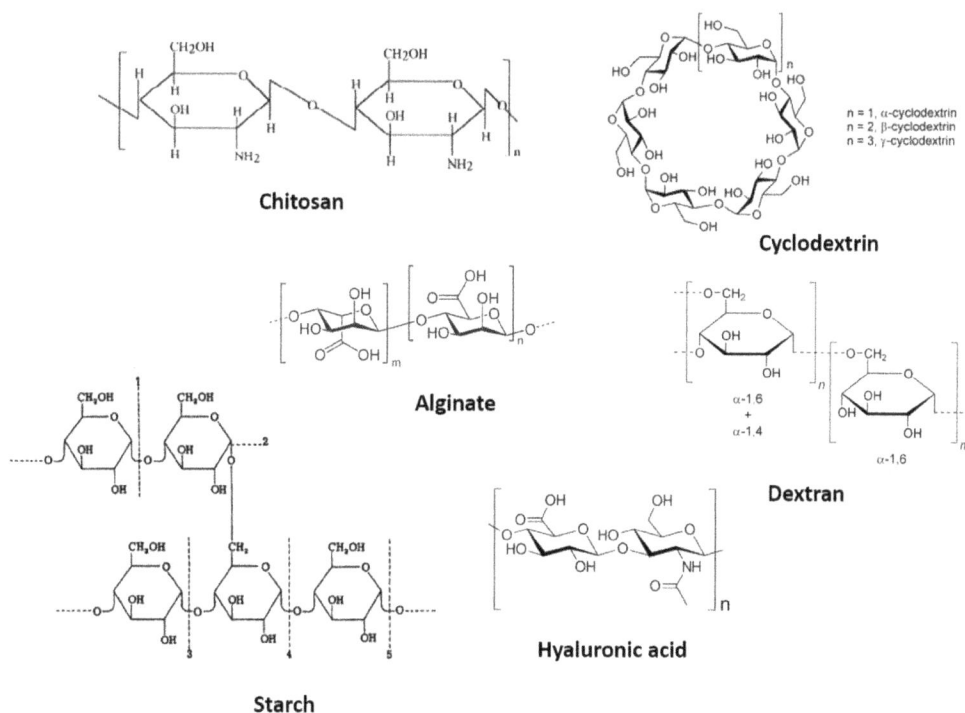

FIGURE 2.12 The chemical structures of the bio-polysaccharide-based polymers used in cancer therapy applications.

TABLE 2.4

Examples of Some Recent Applications of HA-Based Drug Release Systems in Cancer Therapy

PNPs Complex	Tumor	Ref.
HA NPs/β-cyclodextrin/Poly-L-lysine	Hepatocellular carcinoma	[168]
HA/Fucoidan/Zein NPs	Oral cancer	[169]
L-glutamate, HA NPs	Lung cancer	[170]
HA NPs/Physicochemical Entrapped Indocyanine Green	Pancreatic cancer	[171]
HA NPs/PEG	MDA-MB-231 and MCF-7 Cell lines	[172]
HA NPs/Alendronate sodium	A549 Cell line	[173]

leads to the uncontrollable cellular growth (cancer progression, migration, proliferation, and aggressiveness of solid tumors). On the other hand, the high molecular weight polymer (>200 kDa) exhibits water-holding capacity (hydration) and has a lower affinity for CD44 contribution; therefore, its external administration counteracts the oligo-HA in cancer cell growth [126]. Several antitumor applications were reported using HANPs as frameworks of standalone carriers and copolymers for the targeted delivery of antitumor drugs, metallic NPs, and the co-delivery of genes and small xenobiotics in tumor cells (Table 2.4).

2.7.2.3 Alginates

The monovalent salts of alginic acid (alginates) are natural polymers obtained from seaweeds, which might entrap impurities such as heavy metals and protein molecules. They are water-soluble linear anionic polymers of irregularly repeated units from -L-guluronic and -d-mannuronic acids linked with 1,4 glycosidic linkages. Alginates, particularly hydrogel forms, have a great capacity for constructing drug delivery systems for anticancer drugs. They have the advantage of being gel under mild temperature and pH conditions without toxic solvents, and the water molecules can penetrate this gel matrix [127].

Alginates are smart, mucoadhesive, nontoxic, biocompatible, biodegradable, and eco-friendly natural polymers. They are pH-sensitive, making them ideal candidates for the delivery of delicate bioactive agents such as drugs, proteins, hormones, peptides, and genes to their target sites within the host's cells (Figure 2.12) [174].

Recently, graphene oxide with sodium alginate NPs was modified as a carrier system for loading and targeting 5-FU for colon cancer treatment [175]. Also, alginate NPs were reported for loading and targeting other anticancer drugs, such as DOX [176–178]. Besides, alginate NPs can be combined with other polymers, such as CS, for curcumin delivery [179] of DOX [180,181], and conjugated with cyclodextrin for delivery of 5-fluorouracil (5-FU) [182], and with synthetic biodegradable polymers such as PLGA for the delivery of DOX [183].

2.7.2.4 Dextrans

They are biopolymers of high molecular weight and a linear backbone of α-linked D-glucopyranosyl repeating units [184,185]. In addition, their structure contains many hydroxyl groups (-OH) capable of conjugating bioactive agents (Figure 2.12). Dextrans have different carbohydrate sources and, therefore, their molecular structure varies depending on their source [127,186]. They have been used to form various nanosystems by themselves or as a coating agent, for controlled drug delivery and targeting of several anticancer reagents such as 5-FU [187], DOX [188,189], and curcumin [190]. Dextrans are flexible for several functional modifications owing to their suitable aqueous solubility; however, their molecular weight and branching control their biological efficacy. They and their derivatives can develop drug carriers such as micelles and hydrogels that improve cargo bioavailability without considerable toxicity.

2.7.2.5 Starch

Starch is a mixture of amylose (with linear chains of glucose molecules) and amylopectin (with branched chains of glucose [191] (Figure 2.12). The final starch properties (stability, viscosity, etc) were affected and controlled by the ratio of amylose/amylopectin [192]. Because of its availability at a low cost and biocompatibility, starch is widely used in pharmaceutical and biomedical applications [193]. Starch NPs can be prepared by different methods, including nano-precipitation [194], enzymatic hydrolysis using certain enzymes, such as amylase [195], chemical hydrolysis [196], or physical methods, such as ultrasonication and high-pressure homogenization [197]. Recently, starch NPs were used as a nanodevice for the delivery of different anticancer agents, such as gefitinib [198] paclitaxel [199], and curcumin [200,201].

2.7.2.6 Cyclodextrin-Based PNPs (CDs-PNPs)

CD is a cyclic oligosaccharide that combines different numbers of glucose units (6, 7, and 8 units for α-CD, β-CD, and γ-CD, respectively) linked by α-1,4-glycosidic linkage (Figure 2.12), forming a hollow cone-shaped structure. They are characterized by a hydrophobic cavity and hydrophilic outer surface, enabling them to include various hydrophobic bioactive agents via hydrophobic–hydrophobic interaction (inclusion complexation) [202].

The applications of native CDs are limited by their poor aqueous solubility. Therefore, modifying such molecules by polymerization or derivatization through the secondary -OH groups helped to overcome this challenge. CDs and their derivatives are active excipients in different pharmaceutical applications, including NP formulation with loading capacity, providing efficient targeting and delivery [203,204]. For example, β-CD-based NPs were reported for loading and release of anticancer drugs such as Tamoxifen [205] since the cargo can be loaded inside CD cavities or entangled on the aliphatic chains, allowing its controlled release. The solubility issue of β-CD can be solved by structural modification of native CDs via coating of AuNPs to produce β-CD-AuNPs. The modified CD-based AuNPs system was used for the loading and delivery of violacein (a hydrophobic antitumor drug) [206]. The results indicated that the constructed nano-system maintained the violacein cytotoxicity on HL60 leukemia cells (increased drug bioavailability) on the one hand and increased the selectivity and targeting ability towards cancer cells (less cytotoxic to normal cells) [206].

Polyrotaxans (PRs) are necklace-like structures constructed from CD molecules (host), and PEG or PPG (guests) is considered attractive drug carriers for anticancer drugs. For example, camptothecin (a poorly soluble anticancer) was loaded via ester bonds on the preconstructed PRs. The results showed higher efficacy toward the cancer cells and reduced toxicity toward the normal ones [205]. A similar finding was obtained with other anti-cancer drugs [204,207–209].

2.8 THE NATURAL RESOURCES OF ANTICANCER-BASED BIOPOLYMERS

The natural resources of polymers are diverse in all environments; however, the aquatic environment, especially the marine, is the most predominant in providing an enormous variation of bioactive and bio-safe natural therapeutic polymeric and non-polymeric reagents [210,211]. Despite their limited diversity, as they construct < 10% of the underwater phytogenic species, the higher plants in aquatic environments are a considerable resource of anticancer reagents. Marine plant extracts contain bioactive alkaloids, terpenes, polyphenols, and xanthones that work on the onset of apoptosis. Therefore, plant extracts are promising for anticancer drug fabrication, working on pro- and anti-apoptotic protein regulation [212]. Recently, therapeutics with antimicrobial, anti-inflammatory, and apoptotic efficacy have been extracted from the predominant marine plants, Mangroves [213].

The marine-based reagents can initiate cell apoptosis and cause DNA fragmentation, which makes them effective against cancer cell growth and proliferation. Most of these biopolymers were isolated from nature and validated in the laboratory for future clinical application in cancer therapy or anticancer mediation. Therefore, the FDA approved several marine-derived anticancer reagents isolated from microorganism resources, including bacteria and fungi, phytogenic resources of algae (micro- and macro-algae "seaweeds"), or animal resources of sponges, snails, squids, tunicates, and fish [211,214,215].

Several anti-cancer peptides were extracted from the marine sponges and symbiotic or commensal bacteria and fungi [216]. Similarly, sponges, crustaceans, and tunicates are a natural source of several kinds of anticancer polysaccharides [217]. However, fish are the most prominent in providing almost all kinds of anticancer-mediated biopolymers and are advanced by their high popularity, safety, and specificity. Therefore, the recent research trend focuses on extracting fish-derived biopolymers and their validation for cancer therapy.

2.8.1 Fish-Derived Biopolymers for Cancer Therapy

Fish, particularly marine, have recently been considered a therapeutic resource against various diseases, including cancer. Their tissues and secretions construct bioactive polymeric complexes widely used as biodegradable polymeric platforms for anticancer drug delivery and/or treatment [218]. On the one hand, fish tissues such as scales, skin, and bones are employed as advanced medical tools for topical and tissue engineering. On the other hand, they are bio-sources for several bio-polymeric frameworks of wide medical applications. However, employing fish for human medicine should be under the care of expert specialists to ensure their healthy status, as fish are susceptible to several kinds of stress, infections, pollutants, and toxins in their environment, which might affect the efficacy and/or safety of their products [219–223].

Regarding fish tissues, the biodegradable fish scales showed improved cellular adhesion and were incorporated into the fabrication of tissue engineering and regeneration scaffolds, particularly bones [224–226]. In addition, fish skin has recently been widely considered for advanced medical and pharmaceutical applications. Lyophilized or xenograft forms of fish skin have shown effectiveness for the topical biomedical application of treating partial thickness burns [227–230]. Moreover, fish bones are rich in functional amino acids and phenols that provide anti-inflammatory, antioxidant, and anticarcinogenic properties [231]. Furthermore, bone derivatives are bioactive for the fabrication of blended composites of drug delivery systems [232].

Regarding the derived biopolymers, fish tissues are a valuable source of various bioactive polymers usable therapeutically. Chitin and CS are active bioingredients extracted from fish scales and show characteristics competing with those from crustacean sources [233–235]. Low-molecular-weight CS modified from fish scales was integrated into banana/starch films to construct effective antibacterial films against *Escherichia coli* and *Staphylococcus aureus* growth [236]. The antimicrobial peptides abundant in fish skin mucus secretions and protein hydrolysates of fish muscles elicit antitumor proliferative properties [237,238]. Therefore, the 33-amino acid chain sequence of the naturally secreted peptide, pardaxin, by the Red Sea Moses sole, *Pardachirus mamoratus*, was synthesized and applied for antitumor activities *in vitro* and *in vivo* [239,240]. Collagen derived from fish scales or skin is characterized by its moisture, reparative, and super-absorbing capacity of polymers, which makes it incorporated in the modification of high-quality personal care products [241,242]. In addition, it formulated biocomplexes suitable for the topical treatment of pathogens and wound healing applications [243]. Gelatin isolated from fish skin and bones or modified from fish collagen is bioactive for drug delivery applications [244,245]. Recently, an antibacterial and antioxidant hydrogel was formulated from fish skin gelatin nanofibers [246]. Fish oil extracted from fatty fish meat, liver, or fish by-products has also been reported for anticancer applications [247]. It constructed an effective topical bi-gel colloidal system for the controlled release of Imiquimod against epidermoid carcinoma [248]. Interestingly, several bioactive anticarcinogenic ingredients of amino acids, peptides, oils, collagen, gelatin, chitin, CS, and so forth are isolated from fish waste obtained during fish catch and processing activities [249,250]. In addition, fish by-products are biosources for active anticarcinogenic peptides [251]. Examples of recent studies that employed fish tissues for anticancer activities are collected in Table 2.5.

TABLE 2.5

Examples of the Different Fish Tissues and Polymers Used for Cancer Treatment

Fish Tissue/Polymer		Complex	Tumor	Ref.
Liver	Oil	Topical bi-gel drug delivery vehicle/Imiquimod	Skin cancer	[248]
Skin	Collagen	Nanoemulsion, nanoliposome, and nanogold & their folic acid conjugates with Taiwan tilapia skin collagen	Lung cancer cell line	[252]
		Low molecular weight collagen peptides from Unicorn Leatherjacket (*Aluterus Monoceros*)	COLO320 Cancer cell line	[253]
		Bluefin tuna abdominal skin collagen and collagen peptides (Type I Collagen)	HepG2 and HeLa cell lines	[254]
	Gelatin	Hydrogel from methacrylate fish gelatin microparticles encapsulating curcumin	Gastric cancer cell line and postoperative treatment *in vivo*	[255]
		Hydrogel of 3D-printed patches of fish gelatin/PEGylated Liposomal Doxorubicin	Drug release efficacy	[256]

(Continued)

TABLE 2.5 (*Continued*)

Examples of the Different Fish Tissues and Polymers Used for Cancer Treatment

Fish Tissue/ Polymer	Complex	Tumor	Ref.
Epidermal Mucus	Crude skin mucus of *Mugil cephalus*	Laryngeal cancer cell lines	[257]
	Crude skin mucus of *Anabas testudineus*	Human breast cancer and melanoma cell lines (MCF7, and A375.S2)	[237]
	Acetic extracts of the skin mucus of 3 freshwater fish (*Channa punctatus, Channa striatus*, and *Heteropneustes fossilis*)	Lung adenocarcinoma	[258]
Bones	Fermented milkfish (*Chanos chanos*) Bone	Human colorectal cancer (HCT-116) cell line	[231]
Protein hydrolysates	Skin hydrolysate of rainbow trout (*Oncorhynchus mykiss*)	Human colorectal cancer (HCT-116) cell line	[259]
	Muscle hydrolysates (18 proteins) of several European fis	Human breast cancer cell lines (MCF-7/6 and MDA-MB-231)	[238]
Antimicrobial peptides (AMPs) in fish secretions	**Synthetic Pardaxin**	MN-11 cells of Murine Fibrosarcoma *in vitro* and *in vivo*	[260]
		Human fibrosarcoma (HT-1080) and epithelial carcinoma (HeLa) cell lines	[239]
	Synthetic Tilapia Hepcidin TH2-3	Human fibrosarcoma cell line	[261]
	Synthetic AMP5 of *Anabas testudineus* fish	Breast cancer cell line (MDA-MB-231)	[262]
	AMPs Skin mucus peptides extracted from *Anabas testudineus* fish secretions	Breast cancer cell lines (MCF7 and MDA-MB-231)	[263]
	Oxysterols A skin mucus peptide extracted from Catfish (*Arius bilineatus*) secretions	Breast cancer cell line (K-562 (CML, MDAMB-231, and MCF-7)	[264]
	Epinecidin-1 A skin mucus peptide extracted from grouper fish (*Epinephelus coioides*) secretions	HT1080 Fibrosarcoma cell line	[265]
	Pituitary adenylate cyclase-activating polypeptide A skin mucus polypeptide extracted from North African catfish (*Clarias gariepinus*) secretions	Lung cancer cell line H460 (ATCC, HTB-117)	[266]

2.8.2 Synthetic Polymeric-Based NPs

In this section, many synthetic polymeric NPs have been mentioned and discussed as nanocarriers for both delivery and targeting anticancer agents due to their unique physicochemical properties. Smart nano-systems of specific functionalities, such as biodegradable biopolymers, amphiphilic block copolymers, and stimuli-responsive polymers, have recently gained great attention. Compared to natural polymers, synthetic polymers have attracted more interest because of the possibility of their physicochemical characteristics modification to match a wide range of applications.

In addition, these polymers can encapsulate a variety of bioactive agents, either small compounds or those of very high molecular weight, such as proteins, peptides, and DNA [267]. The ideal synthetic polymeric NPs should be stable in blood with minimized immunogenicity and toxicity. These systems should also protect the encapsulated medicaments from degradation before reaching the site of action (cancer tissues). The properties and applications of diverse synthetic polymers are mentioned in the following sections.

2.8.2.1 Hydrophilic/Hydrophobic and Block Copolymers

Many hydrophobic polymers can be used for PNP fabrication, including PLA, PLGA, polyglycolic acid (PGA), PCL, and polypropylene oxide. These polymers have been approved by the FDA as vehicles for drug delivery applications. On the other hand, different hydrophilic polymers are used for NPs preparation, most frequently, polyethylene glycol, polyethyleneimine, and hydroxypropyl methacrylic acid.

2.8.2.2 PLGA-Based NPs (PLGA-NPs)

PLGA is a biodegradable copolymer, composed of lactic acid and glycolic acid, and it has been approved by the FDA for drug delivery [268] (see Figure 2.13). PLGA is one of the most frequently used synthetic drug vehicles due to its biocompatibility, high bioavailability, and less systemic toxicity [269]. Its biodegradability by hydrolysis inside the body ensures its biocompatibility and minimizes systemic toxicity [270]. Therefore, PLGA was used by many research groups worldwide as a suitable base for different nano-pharmaceutical systems.

Several anticancer drugs have been successfully loaded onto constructed vehicles of PLGA NPs. In addition, numerous macromolecules of genes, peptides, growth factors, and so forth have been effectively delivered to tumor tissues on PLGA-NPs vehicles. He et al. [271] fabricated PLGA-NPs loaded with palmitic acid (saturated fatty acid) and DOX through the double emulsion solvent evaporation method. A diminution of tumor growth and metastasis was reported by this system. *Gahtani et al.* [272] used PLGA as a base for PLGA-NPs preparation, which were loaded with 5-fluorouracil (5-FU) as an anticancer drug. The results showed that the modified NPs had higher cytotoxicity (higher activity) when compared with the pure drug at the same dose. Regarding cancer diagnosis, the radiolabeled PLGA is an effective tool for cancer imaging. Ekinci and his coworkers [273] modified a radiolabeled lamivudine-loaded PLGA-NP system using radioactive technetium (99mTc) and reported it for lung cancer diagnosis. This modified system was a promising diagnostic tool for lung cancer imaging.

As a limitation, the immune system might partially refuse the PLGA-based NPs systems, as they are recognized as foreign bodies. This issue can be overcome by further conjugation of these NPs with other biomimetic NPs [274]. In this line, a novel fabrication of macrophage membrane-coated NPs involved with gemcitabine and encapsulated within PLGA-NPs was reported. This approach helped to enhance drug accumulation within tumors [275]. Table 2.6 shows some applications of PLGA-NPs for the delivery of anticancer drugs.

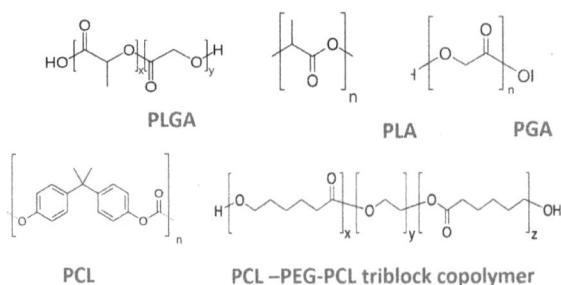

FIGURE 2.13 Chemical structures of the synthetic polymers and triblock copolymer used in cancer therapy applications including PLA, PGA, PLGA, PCL, and a triblock copolymer composed of PCL-PEG-PCL.

TABLE 2.6
PLGA-Based NP Systems for the Delivery of Anticancer Drugs

Polymer	Cargo	Technique	Cancer Type	Ref.
PLGA-NPs	Afatinib	Emulsification and Solvent Evaporation	Lung cancer cell line	[276]
PEG-PLGA	Curcumin	Solvent evaporation	Breast cancer	[277]
Cs-FA-PLGA	Docetaxel	Nanoprecipitation	RPMI 2650, Calu-3, and A549	[278]
HA-PLF127-PLGA	Irinotecan	Nanoprecipitation	U87-MG Glioblastoma and A549 Lung cancer (NSCLC)	[279]
CD56 antibody conjugated PLGA	Irinotecan	Double emulsion and solvent evaporation	Lung cancer	[280]
PLGA – NPs	Raloxifene HCl	Emulsification, solvent diffusion, and evaporation	MCF-7 Breast Cancer Cell Line	[281]
PLGA	Paclitaxel	Nanoprecipitation	MCF-7 Breast Cancer Cell Line	[282]

TABLE 2.7
PLA-Based NP Systems for the Delivery of Anticancer Drugs

Polymer	Cargo	Technique	Cancer Type	Ref.
PLA-PEG-FA-guided redox-responsive NPs	Pirarubicin and Salinomycin	Nanoprecipitation	Breast Cancer	[287]
Hybrid blockm of PEG-PLA and PLA-pluronic L-61 copolymer	Salinomycin and DOX	Emulsion solvent evaporation	*In vitro* Release	[288]
PLA-mPEG/PLA-L61-PLA	Navitoclax/Decitabine	Solvent evaporation		[289]
PLA-Cs	5-fluorouracil and Irinotecan	Ultrasound emulsification	Colon Cancer	[290]
Poly (lactic acid)-hyperbranched polyglycerol (PLA-HPG)	Camptothecin (CPT)	Single emulsion	Skin Cancer	[291]
mPEG-PLA	DTX	Emulsion solvent diffusion	Sarcoma	[292]

Alternatively, PEG-conjugated PLGA polymeric nanocarriers improved drug delivery issues and were characterized by accepted tolerance by the immune system [283]. Besides, PEGylated lipid-PLGA polymeric nanocarriers are characterized by their enhanced NPs stability significantly, which improves the drug's cellular uptake, offering higher efficacy for *in vivo* therapy [284]. Additionally, the linkage of PLGA with PEG stabilizes the formulations by inhibiting the binding of the nonspecific plasma proteins. It also prevents NP aggregation and macrophage clearance, and consequently, enhances the persistence of blood circulation [284].

2.8.2.3 Polylactic Acid

PLA is a lactic acid derivative that is extracted from renewable agricultural sources such as straw, wheat, and corn, and agricultural wastes such as olive pits and sugar cane bagasse [285,286]. Due to its tunable mechanical properties and biodegradability, PLA has been widely used for drug delivery applications [97]. Table 2.7 shows the utilization of PLA-based polymeric NPs for anti-cancer delivery.

Qin et al. [290] modified PLA-CS NPs by using an ultrasound emulsification technique. The constructed NPs were loaded with two anticancer drugs named 5-FU and irinotecan. Like PLGA and PLA, when used alone for NP formation, the drawbacks of poor water solubility and the possible recognition by reticuloendothelial cells of the immune system were observed. Therefore, PEG addition will be the solution to such a problem [293]. Recently, mPEG-PLA NPs were designed by *Chen et al.* [292] and used for docetaxel (DTX) loading, an anticancer drug designed for sarcoma treatment. The results indicated that the activity of the formulated NPs is 1.24 times higher than the free injectable drug at the same dose.

Improved hydrophobicity of the di-block copolymer core and denser NPs were reported by *Anees et al.* [288], who formulated a hybrid system of penta-block copolymer modified from PLA-pluronic L-61-PLA conjugated with mPEG-PLA. Loading the fabricated system with PIRA and DOX revealed a higher antitumor cytotoxic efficacy compared to the free drugs at the same doses [288]. Similar findings were obtained by *Mehrotra et al.* [289], who modified a new drug delivery system of hybrid-block copolymer NPs constructed from PLA-mPEG/PLA-L61-PLA NPs. This system offered the co-delivery of two chemotherapeutic drugs (navitoclax and decitabine).

2.8.2.4   PCL-Based NPs

PCL is a linear aliphatic polyester with semi-crystallinity [294]. It has biodegradable, biocompatible, and nontoxic attributes; therefore, it is applicable for biological, pharmaceutical, and medical applications [295,296]. The formulation of PCL-NPs by the nanoprecipitation technique was reported by Jan et al., and then the formulated NPs were loaded with Arabinosylcytosine (anticancer agent). The results indicated that the activity of the loaded drug against KG-1 leukemia cells increased twofold compared with the free drug injection [297]. The drug release was sustained for 48 hours, offering a controlled release mechanism. *Khan et al.* [298] fabricated a hybrid NPs system from 5-fluorouracil loaded with lipid polymer using the nano-precipitation technique. Higher activity of NPs compared to the free drug was reported, where the half-maximal inhibitory concentration (IC_{50}) of MCF-7 cells by the free drug was 58.35 µg/mL and that by the drug-loaded NPs was 43.33 µg/mL. Additionally, *Hasanbegloo et al.* formulated a PCL/C nanofiber system for paclitaxel/liposomes loading and release. An enhanced drug delivery pattern was observed on the modified nanofibers, reaching up to 30 days of sustained release [299].

Notably, the PNPs of singular polymers are poorly soluble and susceptible to being recognized and eliminated by the reticuloendothelial immune system, which limits their bio-applications. Therefore, this problem can be overcome by linking a hydrophilic polymer to PCL, forming amphiphilic block copolymers, to enhance NP properties and exhibit a controlled release behavior and long-term therapeutic effects [283]. One of these trials is the PCL linking with PEG as a hydrophilic polymer to form PEG-PCL.

Nowadays, the modified copolymeric NP drug delivery systems have become essential for decreasing non-specific drug uptake and increasing specific drug targeting and accumulation [300]. Recently, methoxy polyethylene glycol-block-poly(ε-caprolactone) copolymer loaded with DOX was modified and elicited superior cytotoxic efficacy on the HCT116 cells [301]. Table 2.8 shows the utilization of PCL-based polymeric NPs for anti-cancer delivery.

TABLE 2.8
PCL-Based PNP Systems for the Delivery of Anticancer Drugs

Polymer	Cargo	Cancer Type	Ref.
mPEG-b-PCL NPs	DOX	Colorectal Cancer	[302]
PEG-PCL	Camptothecin (CPT)	Breast Cancer	[303]
mPEG-b-PCL NPs	Camptothecin (CPT)	Liver and Lung Cancer	[304]
PCL-CS	PTX	Breast Cancer	[299]
mPEG-b-PCL NPs	PTX	Ovarian Cancer	[305]

2.8.2.5   Vinyl-Based Polymers

Despite their non-biodegradability, several vinyl-based polymers were used effectively as carriers for anticancer drugs *in vivo,* including divinyl ether-maleic anhydride copolymer-methotrexate (MTX), PVP, ethylene-vinyl alcohol copolymer-fluorouracil (5-FU) conjugate, poly (styrene-co-maleic acid) copolymer neocarzinstatin, and N- (2-hydroxypropyl) methacrylamide-DOX. Modifying several administration formulations (tablets, capsules, gels, and solutions) made the vinyl polymers suitable for oral, topical, and intravascular administration. In *in vivo* applications, they are water-soluble, biocompatible, non-carcinogenic, and non-immunogenic and exhibit minimal inflammatory reactions with tissues. Features of the vinyl-based drug delivery polymers and administration routes were reviewed by Pereira et al. [306].

2.8.2.6   Lipid-Based PNPs

Lipid-based PNPs (LP-NPs) are a category of synthetic copolymeric NPs that mimic the natural amphiphilic structure of cell membranes with further improvement to form membranous vesicles usable for wide diagnostic and therapeutic applications [307]. They provide a platform of drug vehicles, and three forms were reported in cancer chemotherapy, including niosomes, liposomes, and solid lipid NPs [308,309]. Drug carriers of LP-NPs improve drug localization and internalization into the tumor cells, which augments the anticancer efficacy and minimizes the systemic adverse effects of conventional drug administrations [310]. They were frequently reported for the targeted delivery of anticancer drugs in breast cancer cells [311,312]. In addition, LP-NPs offered platforms for colorectal cancer therapy [313].

Carriers of LPNPs stabilize therapeutic cargo, keep their homogenous distribution, and help with cellular absorption; therefore, they have recently been employed as active vehicles for cancer targeting and therapy, particularly lung cancer [314]. More recently, LP-NPs, including liposomes and solid-lipid NPs, have offered prominent strategies for brain cancer diagnosis and treatment [315]. In addition, liposomes worked with other polymers in constructing hybrid carrier platforms of lipid/PNPs or lipid-coating PNPs in the form of core/shell type NPs that are very suitable for drug delivery and tumor targeting [316]. Moreover, LP-NPs provide agonists targeting gene stimulation and gene transduction for some cellular transmembrane proteins, such as interferon, which provide cancer immunotherapy via agonist therapy [317,318].

2.8.2.7   Stimuli-Responsive PNPs

The recent trend of therapeutic delivery was enhanced using NPs of stimuli-responsive polymers that employ numerous internal and external triggers to stimulate the release of their drug cargo. Therefore, stimuli-responsive polymers were effectively applied to comprise nanoplatforms of stimuli-responsive nanoparticle carriers for anticancer drugs [14,28]. Two categories of responsive PNPs were fabricated: physical and biological.

2.8.2.7.1   Physically Responsive PNPs

Physical stimuli such as magnetic, enzymes, changes in pH, or redox potential are frequently used for the drug cargo release. Acidified vesicles allow for the internalization of PNPs into cells; therefore, pH fluctuation has sparked interest as a delivery motive for the stimuli-responsive PNPs, particularly to cancer cells with a lower pH than the surrounding tissue. The pH-responsive PNP design, based on their charge shifting and H^+ labile linkages, was reviewed recently [319]. Furthermore, several dual- and multi-stimuli-responsive PNPs were established from multiple polymers and used efficiently for the programmed site-specific drug delivery. Some selective example studies on the dual- and multi-stimuli-responsive PNPs systems for the controlled-release of anticancer drugs are collected in Table 2.9. The dual-stimuli-responsive PNPs include temperature/magnetic-, pH/redox-, pH/temperature-, pH/magnetic-, and double pH-responsive NPs. The multi-stimuli-responsive polymeric NPs include temperature/pH/magnetic-, temperature/redox/guest molecule-, and temperature/pH/guest molecule-responsive NPs [320].

TABLE 2.9

Dual Stimuli-Responsive NP-Based Systems for the Delivery of Anticancer Drugs

System	Polymeric NPs	Cargo	Ref.
	A. Dual-Stimuli-Responsive NPs		
1. Temperature and Magnetic	Pluronic F127 (F127) matrix (Pluronic-Magnetic Fe_3O_4 NPs)	DOX	[321]
2. pH and Redox	Poly(2-(pyridin-2-yldisulfanyl) ethyl acrylate) NPs-Polyethylene glycol and cyclo (Arg-Gly-Asp-d-Phe-Cys) (cRGD) peptide	DOX	[322]
3. pH and Temperature	Poly (d,l-lactide)-graft-poly(N-isopropyl acrylamide-co-methacrylic acid	5-fluorouracil	[323]
4. pH and Magnetic	Magnetic Fe_3O_4 NPs	DOX	[324]
5. Double pH	PPC-Hyd-DA NPs	DOX	[325]
	B. Multi-Stimuli-Responsive NPs		
1. Temperature, pH, and redox	A copolymer of Monomethyl oligo (ethylene glycol) acrylate (OEGA) and an ortho ester-containing acrylic monomer, 2- (5,5-dimethyl-1,3-dioxan-2-yloxy) ethyl acrylate (DMDEA), with bis (2-acryloyloxyethyl) disulfide (BADS)	Nile Red, Paclitaxel, and DOX	[326]
2. Temperature, pH, and magnetic	Poly (N-isopropylacrylamide-co-methacrylic acid) (P(NIPAM-co-MAA)) coated magnetic mesoporous silica NPs (M-MSN)	DOX Hydrochloride	[327]
3. Temperature, redox, and guest molecule	Vesicles based on host-guest complex formation between C4AS & MVC12	DOX Hydrochloride	[328]
4. Temperature, pH, and guest molecule	Micelles of double hydrophilic block copolymer (DHBC) held together by cucurbit [8] uril (CB [8]) ternary complexation and poly (dimethylaminoethylmethacrylate) (PDMAE-MA)	DOX	[329]

2.8.2.7.2   Biologically Responsive PNPs

Nanobiotechnology has improved various techniques, achieving specific targeting of drug delivery issues. Interestingly, some responsive PNPs can release their drug cargo in response to the hosting tissue's inconstant pH and oxidative stress conditions. This opened new horizons for the achievement of the specific drug delivery to a particular targeted location during a specific biological or disease status in what is known as "functionally targeted" drug delivery [330]. Moreover, simultaneous drug loading for targeted delivery was achieved by using a supramolecular self-assembly approach. In this regard, polyacrylate-based supramolecular PNPs were fabricated by host–guest complexation between adamantane and β-cyclodextrin self-assembled for the simultaneous targeted delivery of the anticancer drug DOX [27]. These supramolecular PNPs could augment the targeted delivery and the therapeutic efficacy of DOX against tumor growth *in vitro* and *in vivo* [27]. By a self-assembly approach, effective bio-reducible anticancer siRNA-based nanomedicine replaced the ineffective DNA-delivery polymers for tumor targeting. A bio-reducible delivery nanomaterial (poly(β-amino ester)s) self-assembled with siRNA and produced an environmentally-triggered drug release system that is triggered by the reducing cytoplasm for siRNA release in the cytoplasm of human brain cancer cells [331]. Furthermore, vehicles of charged PNPs can offer physical protection and targeted delivery of the delicate therapeutic proteins characterized by their intrinsic sensitivity to the surrounding conditions. According to the tissue pH, the electrostatic interactions with the charged polymers of the PNPs provide gentle protection to the protein cargo against the surroundings, increase its cellular uptake, and provide a long-term therapeutic effect [332].

2.9 CONCLUSION

The use of PNPs for cancer treatment has grown in popularity within the last ten years. Better efficacy and less toxicity in cancer treatment may result from PNP fabrication that can transport medications directly to cancer cells at a steady and regulated rate. The current chapter includes the different polymers usable for PNP preparation that are synthetic or natural. Also, the methods of PNP preparation were discussed, including polymerization, nanoprecipitation, emulsification, solvent evaporation, freeze-drying, and spray-drying techniques. Moreover, the analytical methods for PNP characterization were highlighted, including the microscopy for the morphological characterization and the DLS analysis for particle size and zeta potential determinations. In addition, the reported natural resources of the anticancer-based biopolymers were surveyed. As fish is a wealth that can naturally construct most anticancer biopolymers, a survey of the recent applications of fish-based polymers was included.

Although the PNPs are essential in contemporary medicine, their use in clinical settings necessitates close observation because of the possible toxicity of their constituent parts. Even though most polymers are biodegradable, and normal metabolic processes can readily remove their oligomers, caution is still required during their applications. Therefore, the toxicity, release, and stability of different PNPs were highlighted in this chapter.

REFERENCES

1. El-Readi MZ, Althubiti MA. Cancer nanomedicine: A new era of successful targeted therapy. *J Nanomater.* 2019;2019:1–13.
2. Begines B, Ortiz T, Perez-Aranda M, et al. Polymeric nanoparticles for drug delivery: Recent developments and future prospects. *Nanomaterials.* 2020;10(7):1403.
3. Chen Y, Li N, Xu B, et al. Polymer-based nanoparticles for chemo/gene-therapy: Evaluation its therapeutic efficacy and toxicity against colorectal carcinoma. *Biomed Pharmacother.* 2019;118:109257.
4. Alven S, Aderibigbe, BA. Efficacy of polymer-based nanocarriers for co-delivery of curcumin and selected anticancer drugs. *Nanomaterials.* 2020;10:1556.
5. Li Y, Thambi T, Lee DS. Co-delivery of drugs and genes using polymeric nanoparticles for synergistic cancer therapeutic effects. *Adv Healthc Mater.* 2018;7(1):1700886.
6. Nasery MM, Abadi B, Poormoghadam D, et al. Curcumin delivery mediated by bio-based nanoparticles: A review. *Molecules.* 2020;25:689.
7. Gregoriou Y, Gregoriou G, Manoli A, et al. Photophysical and biological assessment of coumarin-6 loaded polymeric nanoparticles as a cancer imaging agent. *Sensors Diagnostics.* 2023;2(5):1277–1285.
8. Caraway CA, Gaitsch H, Wicks EE, et al. Polymeric nanoparticles in brain cancer therapy: A review of current approaches. *Polymers.* 2022;14(14):2963.
9. Chen S, Cheng SX, Zhuo RX. Self-assembly strategy for the preparation of polymer-based nanoparticles for drug and gene delivery. *Macromol Biosci.* 2011;11(5):576–589.
10. Eckmann DM, Composto RJ, Tsourkas A, et al. Nanogel carrier design for targeted drug delivery. *J Mater Chem B.* 2014;2(46):8085–8097.
11. Vlerken LEV, Amiji MM. Multi-functional polymeric nanoparticles for tumour-targeted drug delivery. *Expert Opin Drug Deliv.* 2006;3(2):205–216.
12. Sun L, Wu Q, Peng F, et al. Strategies of polymeric nanoparticles for enhanced internalization in cancer therapy. *Colloids Surf B Biointerfaces.* 2015;135:56–72.
13. Gagliardi A, Giuliano E, Venkateswararao E, et al. Biodegradable polymeric nanoparticles for drug delivery to solid tumors. *Front Pharmacol.* 2021;12:601626.
14. El-Say KM, El-Sawy HS. Polymeric nanoparticles: Promising platform for drug delivery. *Int J Pharm.* 2017;528(1–2):675–691.
15. Bal-Öztürk A, Tietilu SD, Yücel O, et al. Hyperbranched polymer-based nanoparticle drug delivery platform for the nucleus-targeting in cancer therapy. *J Drug Deliv Sci Technol.* 2023;81:104195.
16. Ferrari R, Sponchioni M, Morbidelli M, et al. Polymer nanoparticles for the intravenous delivery of anticancer drugs: The checkpoints on the road from the synthesis to clinical translation. *Nanoscale.* 2018;10(48):22701–22719.
17. Mainini F, Eccles MR. Lipid and polymer-based nanoparticle siRNA delivery systems for cancer therapy. *Molecules.* 2020;25:2692.

18. Prajapati SK, Jain A, Jain A, et al. Biodegradable polymers and constructs: A novel approach in drug delivery. *Eur Polym J.* 2019;120:109191.

19. Bhardwaj H, Jangde RK. Current updated review on preparation of polymeric nanoparticles for drug delivery and biomedical applications. *Next Nanotechnol.* 2023;2:100013.

20. Mishra P, Dey RK. Development of docetaxel loaded PEG-PLA nano-particles using surfactant Free method for controlled release studies. *Int J Polym Mater Polym Biomater.* 2017;67:535–542.

21. Pandey SK, Ghosh S, Maiti P, et al. Therapeutic efficacy and toxicity of tamoxifen loaded PLA nanoparticles for breast cancer. *Int J Biol Macromol.* 2015;72:309–319.

22. Babos G, Biro E, Meiczinger M, et al. Dual drug delivery of sorafenib and doxorubicin from PLGA and PEG-PLGA polymeric nanoparticles. *Polymers.* 2018;10(8):895.

23. Li Y, Wu M, Zhang N, et al. Mechanisms of enhanced antiglioma efficacy of polysorbate 80-modified paclitaxel-loaded PLGA nanoparticles by focused ultrasound. *J Cell Mol Med.* 2018;22(9):4171–4182.

24. Wu ST, Fowler AJ, Garmon CB, et al. Treatment of pancreatic ductal adenocarcinoma with tumor antigen specific-targeted delivery of paclitaxel loaded PLGA nanoparticles. *BMC Cancer.* 2018;18(1):457.

25. Huang D, Zhou Y, Xiang Y, et al. Polyurethane/doxorubicin nanoparticles based on electrostatic interactions as pH-sensitive drug delivery carriers. *Polym Int.* 2018;67(9):1186–1193.

26. Tian G, Sun X, Bai J, et al. Doxorubicin-loaded dual-functional hyaluronic acid nanoparticles: Preparation, characterization and antitumor efficacy in vitro and in vivo. *Mol Med Rep.* 2019;19(1):133–142.

27. Ang CY, Tan SY, Wang X, et al. Supramolecular nanoparticle carriers self-assembled from cyclodextrin- and adamantane-functionalized polyacrylates for tumor-targeted drug delivery. *J Mater Chem B.* 2014;2(13):1879–1890.

28. Guo X, Wang L, Wei X, et al. Polymer-based drug delivery systems for cancer treatment. *J Polym Sci A: Poly Chem.* 2016;54(22):3525–3550.

29. Yong HW, Kakkar A. Nanoengineering branched star polymer-based formulations: Scope, strategies, and advances. *Macromol Biosci.* 2021;21(8):e2100105.

30. Wang Y, Li P, Truong-Dinh Tran T, et al. Manufacturing techniques and surface engineering of polymer based nanoparticles for targeted drug delivery to cancer. *Nanomaterials.* 2016;6(2):26.

31. Daniel RI, Fumio W, Uo M. Microparticle formation and its mechanism in single and double emulsion solvent evaporation. *J Control Release.* 2004;99(2):271–280.

32. Mora-Huertas CE, Fessi H, Elaissari A. Polymer-based nanocapsules for drug delivery. *Int J Pharm.* 2010;385:113–142.

33. Rawat M, Saraf S. Formulation optimization of double emulsification method for preparation of enzyme-loaded Eudragit S100 microspheres. *J Microencapsul.* 2009;26:306–314.

34. Sahana DK, Mittal G, Bhardwaj V, et al. PLGA nanoparticles for oral delivery of hydrophobic drugs: Influence of organic solvent on nanoparticle formation and release behavior in vitro and in vivo using estradiol as a model drug. *J Pharm Sci.* 2008;97(4):1530–1542.

35. Hariharan S, Bhardwaj V, Bala I, et al. Design of estradiol loaded PLGA nanoparticulate formulations: A potential oral delivery system for hormone therapy. *Pharm Res.* 2006;23(1):184–195.

36. Astete CE, Kumar CSSR, Sabliov CM. Size control of poly(d,l-lactide-co-glycolide) and poly(d,l-lactide-co-glycolide)-magnetite nanoparticles synthesized by emulsion evaporation technique. *Colloids Surf A Physicochem Eng Aspects.* 2007;299(1–3):209–216.

37. Hood E, Simone E, Wattamwar P, et al. Nanocarriers for vascular delivery of antioxidants. *Nanomedicine.* 2011;6:1257–1272.

38. Sahoo SK, Panyam J, Prabha S, et al. Residual polyvinyl alcohol associated with poly (D,L-lactide-co-glycolide) nanoparticles affects their physical properties and cellular uptake. *J Control Release.* 2002;82:105–114.

39. Xie S, Wang S, Zhao B, et al. Effect of PLGA as a polymeric emulsifier on preparation of hydrophilic protein-loaded solid lipid nanoparticles. *Colloids Surf B Biointerfaces.* 2008;67(2):199–204.

40. Fessi H, Puisieux F, Devissaguet JP, et al. Nanocapsule formation by interfacial polymer deposition following solvent displacement. *Int J Pharm.* 1989;55:R1–R4.

41. Zielinska A, Carreiro F, Oliveira AM, et al. Polymeric nanoparticles: Production, characterization, toxicology and ecotoxicology. *Molecules.* 2020;25(16):3731.

42. Betancourt T, Brown B, Brannon-peppas L. Doxorubicin-loaded PLGA nanoparticles by nanoprecipitation: Preparation, characterization and in vitro evaluation. *Nanomedicine.* 2007;2:219–232.

43. Bilati U, Allemann E, Doelker E. Development of a nanoprecipitation method intended for the entrapment of hydrophilic drugs into nanoparticles. *Eur J Pharm Sci.* 2005;24(1):67–75.

44. Pisania E, Fattala E, Parisa J, et al. Surfactant dependent morphology of polymeric capsules of perfluo-rooctyl bromide: Influence of polymer adsorption at the dichloromethane–water interface. *J Colloid Interface Sci.* 2008;326:66–71.
45. Hoa LTM, Chi NT, Nguyen LH, et al. Preparation and characterisation of nanoparticles containing ketoprofen and acrylic polymers prepared by emulsion solvent evaporation method. *J Exp Nanosci.* 2012;7(2):189–197.
46. Jaiswal J, Gupta SK, Kreuter J. Preparation of biodegradable cyclosporine nanoparticles by high-pressure emulsification-solvent evaporation process. *J Control Release.* 2004;96(1):169–178.
47. Wang Y, Li P, Peng Z, et al. Microencapsulation of nanoparticles with enhanced drug loading for pH-sensitive oral drug delivery for the treatment of colon cancer. *J Appl Poly Sci.* 2013;129(2):714–720.
48. Xu Y, Du Y, Huang R, et al. Preparation and modification of N-(2-hydroxyl) propyl-3-trimethyl ammonium chitosan chloride nanoparticle as a protein carrier. *Biomaterials.* 2003;24(27):5015–5022.
49. Calvo P, Vila-Jato JL, Alonso Ma J. Evaluation of cationic polymer-coated nanocapsules as ocular drug carriers. *Int J Pharm.* 1997;153:41–50.
50. Gan Q, Wang T. Chitosan nanoparticle as protein delivery carrier--systematic examination of fabrication conditions for efficient loading and release. *Colloids Surf B Biointerfaces.* 2007;59(1):24–34.
51. Agnihotri SA, Mallikarjuna NN, Aminabhavi TM. Recent advances on chitosan-based micro- and nanoparticles in drug delivery. *J Control Release.* 2004;100(1):5–28.
52. Yanat M, Schroën K. Preparation methods and applications of chitosan nanoparticles; with an outlook toward reinforcement of biodegradable packaging. *React Funct Polym.* 2021;161:104849.
53. Vauthier C, Bouchemal K. Methods for the preparation and manufacture of polymeric nanoparticles. *Pharm Res.* 2009;26(5):1025–1058.
54. Rezazadeh M, Davatsaz Z, Emami J, et al. Preparation and characterization of spray-dried inhalable powders containing polymeric micelles for pulmonary delivery of Paclitaxel in lung cancer. *J Pharm Pharm Sci.* 2018;21:200s–214s.
55. Abdelwahed W, Degobert G, Stainmesse S, et al. Freeze-drying of nanoparticles: Formulation, process and storage considerations. *Adv Drug Deliv Rev.* 2006;58(15):1688–1713.
56. Ho HN, Do TT, Nguyen TC, et al. Preparation, characterisation and in vitro/in vivo anticancer activity of lyophilised artesunate-loaded nanoparticles. *J Drug Deliv Sci Technol.* 2020;58:101801.
57. De Marco I. Supercritical fluids and nanoparticles in cancer therapy. *Micromachines.* 2022;13(9):1449.
58. Cooper AI. Polymer synthesis and processing using supercritical carbon dioxide. *J Mater Chem.* 2000;10:207–234.
59. Prieto C, Calvo L, editors. Supercritical fluid extraction of emulsions for the production of vitamin E nanocapsulates. *XXI Int Conf Bioencapsul*, Berlin, Germany; 2013.
60. Ajiboye AL, Trivedi V, Mitchell JC. Preparation of polycaprolactone nanoparticles via supercritical carbon dioxide extraction of emulsions. *Drug Deliv Transl Res.* 2018;8(6):1790–1796.
61. Xiang S-T, Chen B-Q, Kankala RK, et al. Solubility measurement and RESOLV-assisted nanonization of gambogic acid in supercritical carbon dioxide for cancer therapy. *J Supercrit Fluids.* 2019;150:147–155.
62. Sharma SK, Al Hosani S, Kalmouni M, et al. Supercritical CO_2 processing generates aqueous cisplatin solutions with enhanced cancer specificity. *ACS Omega.* 2020;5(9):4558–4567.
63. Pathak P, Prasad GL, Meziani MJ, et al. Nanosized paclitaxel particles from supercritical carbon diox-ide processing and their biological evaluation. *Langmuir.* 2007;23:2674–2679.
64. Román JV, Rodríguez-Rodríguez JA, del Valle EMM, et al. Synthesis of a new nanoparticle system based on electrostatic alginate-piperazine interactions. *Polym Adv Technol.* 2016;27:623–629.
65. Maciel VBV, Yoshida CMP, Pereira S, et al. Electrostatic self-assembled chitosan-pectin nano- and microparticles for insulin delivery. *Molecules.* 2017;22(10):1707.
66. Pulingam T, Foroozandeh P, Chuah J Λ, et al. Exploring various techniques for the chemical and biological synthesis of polymeric nanoparticles. *Nanomaterials.* 2022;12:576.
67. Gharieh A, Khoee S, Mahdavian AR. Emulsion and miniemulsion techniques in preparation of polymer nanoparticles with versatile characteristics. *Adv Colloid Interface Sci.* 2019;269:152–186.
68. Khan MU, Reddy KR, Snguanwongchai T, et al. Polymer brush synthesis on surface modified carbon nanotubes via in situ emulsion polymerization. *Colloid Polym Sci.* 2016;294(10):1599–1610.
69. Garay-Jimenez JC, Gergeres D, Young A, et al. Physical properties and biological activity of poly(butyl acrylate-styrene) nanoparticle emulsions prepared with conventional and polymerizable surfactants. *Nanomedicine.* 2009;5(4):443–451.
70. Gao J, Wu C. Modified structural model for predicting particle size in the microemulsion and emulsion polymerization of styrene under microwave irradiation. *Langmuir.* 2005;21(2):782–785.

71. Ghayempour S, Montazer M. A modified microemulsion method for fabrication of hydrogel Tragacanth nanofibers. *Int J Biol Macromol.* 2018;115:317–323.

72. Hermanson KD, Kaler EW. Kinetics and mechanism of the multiple addition microemulsion polymerization of hexyl methacrylate. *Macromolecules.* 2003;36:1836–1842.

73. Ramírez AG, López RG, Tauer K. Studies on semibatch microemulsion polymerization of butyl acrylate: Influence of the potassium peroxodisulfate concentration. *Macromolecules.* 2004;37:2738–2747.

74. Nauman N, Zaquen N, Junkers T, et al. Particle size control in miniemulsion polymerization via membrane emulsification. *Macromolecules.* 2019;52:4492–4499.

75. Li WSJ, Negrell C, Ladmiral V, et al. Cardanol-based polymer latex by radical aqueous miniemulsion polymerization. *Polym Chem.* 2018;9(18):2468–2477.

76. Ham HT, Choi YS, Chee MG, et al. Singlewall carbon nanotubes covered with polystyrene nanoparticles by in-situ miniemulsion polymerization. *J Polym Sci A: Polym Chem.* 2006;44(1):573–584.

77. Ziegler A, Landfester K, Musyanovych A. Synthesis of phosphonate-functionalized polystyrene and poly(methyl methacrylate) particles and their kinetic behavior in miniemulsion polymerization. *Colloid Polym Sci.* 2009;287(11):1261–1271.

78. Errezma M, Mabrouk AB, Magnin A, et al. Surfactant-free emulsion Pickering polymerization stabilized by aldehyde-functionalized cellulose nanocrystals. *Carbohydr Polym.* 2018;202:621–630.

79. Heo HJ, Park IJ, Lee SG, et al. Surfactant-free preparation of poly(vinylidene fluoride) nanoparticle dispersions and their use as surface coating agents. *Green Chem.* 2018;20(2):502–505.

80. Kassim S, Zahari SB, Tahrin RAA, et al. Co-polymerization of methyl methacrylate and styrene via surfactant-free emulsion polymerization, as a potential material for photonic crystal application. 2017;1885(1).

81. Ulery BD, Nair LS, Laurencin CT. Biomedical applications of biodegradable polymers. *J Polym Sci B Polym Phys.* 2011;49(12):832–864.

82. Bishnoi M, Jain A, Hurkat P, et al. Chondroitin sulphate: A focus on osteoarthritis. *Glycoconj J.* 2016;33(5):693–705.

83. Guerrero-Cázares H, Tzeng SY, Young NP, et al. Biodegradable polymeric nanoparticles show high efficacy and specificity at DNA delivery to human glioblastoma in vitro and in vivo. *ACS Nano.* 2014;8:5141–5153.

84. Kim J, Wilson DR, Zamboni CG, et al. Targeted polymeric nanoparticles for cancer gene therapy. *J Drug Target.* 2015;23(7–8):627–641.

85. Poon Z, Lee JB, Morton SW, et al. Controlling in vivo stability and biodistribution in electrostatically assembled nanoparticles for systemic delivery. *Nano Lett.* 2011;11:2096–2103.

86. Kaur S, Manhas P, Swami A, et al. Bioengineered PLGA-chitosan nanoparticles for brain targeted intranasal delivery of antiepileptic TRH analogues. *Chem Eng J.* 2018;346:630–639.

87. Robertson JD, Rizzello L, Avila-Olias M, et al. Purification of nanoparticles by size and shape. *Sci Rep.* 2016;6:27494.

88. Molpeceres J, Aberturas MR, Guzman M. Biodegrable nanoparticles as a delivery system for cyclosporine: Preparation and characterization. *J Microencapsul.* 2000;17:599–614.

89. Zhou W, Wang ZL, (eds.). *Scanning Microscopy for Nanotechnology: Techniques and Applications.* Springer Science & Business Media; 2007.

90. Morris VJ, Kirby AR, Gunning PA. *Atomic Force Microscopy for Biologists.* World Scientific; 2009.

91. Hu L, Hach D, Chaumont D, et al. One step grafting of monomethoxy poly(ethylene glycol) during synthesis of maghemite nanoparticles in aqueous medium. *Colloids Surf A: Physicochem Eng Aspects.* 2008;330(1):1–7.

92. Ural MS, Dartois E, Mathurin J, et al. Quantification of drug loading in polymeric nanoparticles using AFM-IR technique: A novel method to map and evaluate drug distribution in drug nanocarriers. *Analyst.* 2022;147(23):5564–5578.

93. Koopmans C, Ritter H. Formation of physical hydrogels via host–guest interactions of β-cyclodextrin polymers and copolymers bearing adamantyl groups. *Macromolecules.* 2008;41:7418–7422.

94. Bootz A, Vogel V, Schubert D, et al. Comparison of scanning electron microscopy, dynamic light scattering and analytical ultracentrifugation for the sizing of poly(butyl cyanoacrylate) nanoparticles. *Eur J Pharm Biopharm.* 2004;57(2):369–375.

95. Clogston JD, Patri AK. Zeta Potential Measurement. In Characterization of Nanoparticles Intended for Drug Delivery. Methods in Molecular Biology, Vol. 697, Humana Press, 2011:63–70.

96. Ho HN, Tran TH, Tran TB, et al. Optimization and characterization of artesunate-loaded chitosan-decorated poly(D,L-lactide-co-glycolide) acid nanoparticles. *J Nanomater.* 2015;2015(1):674175.

97. Lee JH, Yeo Y. Controlled drug release from pharmaceutical nanocarriers. *Chem Eng Sci.* 2015; 125:75–84.

98. Fu Q, Sun J, Zhang W, et al. Nanoparticle albumin-bound (NAB) technology is a promising method for anti-cancer drug delivery. *Recent Pat Anticancer Drug Discov.* 2009;4:262–272.

99. González AE. Colloidal aggregation coupled with sedimentation: A comprehensive overview. *Adv Colloid Sci.* 2016;211:65699.

100. Kamiya H, Gotoh K, Shimada M, et al. Characteristics and behavior of nanoparticles and its dispersion systems. In Naito M, Yokoyama T, Hosokawa K, et al. (eds.). *Nanoparticle Technology Handbook.* Elsevier; 2008. pp. 109–168; 113–176.

101. Heinz H, Pramanik C, Heinz O, et al. Nanoparticle decoration with surfactants: Molecular interactions, assembly, and applications. *Surf Sci Rep.* 2017;72(1):1–58.

102. Wanning S, Suverkrup R, Lamprecht A. Pharmaceutical spray freeze drying. *Int J Pharm.* 2015; 488(1–2):136–153.

103. Ziaee A, Albadarin AB, Padrela L, et al. Spray drying of pharmaceuticals and biopharmaceuticals: Critical parameters and experimental process optimization approaches. *Eur J Pharm Sci.* 2019;127:300–318.

104. Shen J, Burgess DJ. In vitro dissolution testing strategies for nanoparticulate drug delivery systems: Recent developments and challenges. *Drug Deliv Transl Res.* 2013;3(5):409–415.

105. Soppimath KS, Aminabhavi TM, Kulkarni AR, et al. Biodegradable polymeric nanoparticles as drug delivery devices. *J Control Release.* 2001;70:1–20.

106. Bohrey S, Chourasiya V, Pandey A. Polymeric nanoparticles containing diazepam: Preparation, optimization, characterization, in-vitro drug release and release kinetic study. *Nano Converg.* 2016;3(1):3.

107. Kamaly N, Yameen B, Wu J, et al. Degradable controlled-release polymers and polymeric nanoparticles: Mechanisms of controlling drug release. *Chem Rev.* 2016;116(4):2602–2663.

108. Kwolek-Mirek M, Zadrag-Tecza R. Comparison of methods used for assessing the viability and vitality of yeast cells. *FEMS Yeast Res.* 2014;14(7):1068–1079.

109. Ishiyama M, Tominaga H, Shiga M, et al. A combined assay off cell viability and in vitro cytotoxicity with a highly water-soluble tetrazolium salt,neutral red and crystal violet. *Biol Pharm Bull.* 1996;19:1518–1520.

110. Ahmed F, Soliman FM, Adly MA, et al. Recent progress in biomedical applications of chitosan and its nanocomposites in aquaculture: A review. *Res Vet Sci.* 2019;126:68–82.

111. Rostami M, Mazaheri H, Joshaghani AH, et al. Using experimental design to optimize the photodegradation of P-Nitro toluene by nano-TiO_2 in synthetic wastewater. *Int J Eng.* 2019;32:1074–1081.

112. Strober W. Trypan blue exclusion test of cell viability. *Curr Protoc Immunol.* 2015;111:A3 B 1–A3 B 3.

113. Duellman SJ, Zhou W, Meisenheimer P, et al. Bioluminescent, nonlytic, real-time cell viability assay and use in inhibitor screening. *Assay Drug Dev Technol.* 2015;13(8):456–465.

114. Mueller H, Kassack MU, Wiese M. Comparison of the usefulness of the MTT, ATP, and calcein assays to predict the potency of cytotoxic agents in various human cancer cell lines. *SLAS-Discov.* 2004;9:506–515.

115. Mosmann T. Rapid colorimetric assay for cellular growth and survival: Application to proliferation and cytotoxicity assays. *J lmmunol Methods.* 1983;65:55–63.

116. Tada H, Shiho O, Kuroshima Ki, et al. An improved colorimetric assay for interleukin 2. *J Immunol Methods.* 1986;93:157–165.

117. Almawash S, El Hamd MA, Osman SK. Polymerized beta-cyclodextrin-based injectable hydrogel for sustained release of 5-fluorouracil/methotrexate mixture in breast cancer management: In vitro and in vivo analytical validations. *Pharmaceutics.* 2022;14(4):817.

118. Rubinstein LV, Shoemaker RH, Paull KD, et al. Comparison of in vitro anticancer-drug-screening data generated with a tetrazolium assay versus a protein assay against a diverse panel of human tumor cell lines. *J Natl Cancer Inst.* 1990;82:1113–1117.

119. van Tonder A, Joubert AM, Cromarty AD. Limitations of the 3-(4,5-dimethylthiazol-2-yl)-2,5-diphenyl-2H-tetrazolium bromide (MTT) assay when compared to three commonly used cell enumeration assays. *BMC Res Notes.* 2015;8:47.

120. Vichai V, Kirtikara K. Sulforhodamine B colorimetric assay for cytotoxicity screening. *Nat Protoc.* 2006;1(3):1112–1116.

121. Rai Y, Pathak R, Kumari N, et al. Mitochondrial biogenesis and metabolic hyperactivation limits the application of MTT assay in the estimation of radiation induced growth inhibition. *Sci Rep.* 2018;8:1531.

122. Repetto G, del Peso A, Zurita JL. Neutral red uptake assay for the estimation of cell viability/cytotoxicity. *Nat Protocols.* 2008;3(7):1125–1131.

123. Lavogina D, Lust H, Tahk MJ, et al. Revisiting the resazurin-based sensing of cellular viability: Widening the application horizon. *Biosensors*. 2022;12(4):196.

124. Stockert JC, Blazquez-Castro A, Canete M, et al. MTT assay for cell viability: Intracellular localization of the formazan product is in lipid droplets. *Acta Histochem*. 2012;114(8):785–796.

125. DeFrates K, Markiewicz T, Gallo P, et al. Protein polymer-based nanoparticles: Fabrication and medical applications. *Int J Mol Sci*. 2018;19(6):1717.

126. Ding L, Agrawal P, Singh SK, et al. Polymer-based drug delivery systems for cancer therapeutics. *Polymers*. 2024;16(6):843.

127. Wong KH, Lu A, Chen X, et al. Natural ingredient-based polymeric nanoparticles for cancer treatment. *Molecules*. 2020;25(16):3620.

128. Lin T, Zhao P, Jiang Y, et al. Blood-brain-barrier-penetrating albumin nanoparticles for biomimetic drug delivery via albumin-binding protein pathways for antiglioma therapy. *ACS Nano*. 2016;10(11):9999–10012.

129. Khandelia R, Bhandari S, Pan UN, et al. Gold nanocluster embedded albumin nanoparticles for two-photon imaging of cancer cells accompanying drug delivery. *Small*. 2015;11(33):4075–4081.

130. Cho H, Jeon SI, Ahn CH, et al. Emerging albumin-binding anticancer urugs for tumor-targeted drug delivery: Current understandings and clinical translation. *Pharmaceutics*. 2022;14(4):728.

131. Mad-Ali S, Benjakul S, Prodpran T, et al. Characteristics and gelling properties of gelatin from goat skin as affected by drying methods. *J Food Sci Technol*. 2017;54:1646–1654.

132. Kushwah V, Katiyar SS, Dora CP, et al. Co-delivery of docetaxel and gemcitabine by anacardic acid modified self-assembled albumin nanoparticles for effective breast cancer management. *Acta Biomater*. 2018;73:424–436.

133. Watanabe K, Nishio Y, Makiura R, et al. Paclitaxel-loaded hydroxyapatite/collagen hybrid gels as drug delivery systems for metastatic cancer cells. *Int J Pharm*. 2013;446(1–2):81–86.

134. Anandhakumar S, Krishnamoorthy G, Ramkumar KM, et al. Preparation of collagen peptide functionalized chitosan nanoparticles by ionic gelation method: An effective carrier system for encapsulation and release of doxorubicin for cancer drug delivery. *Mater Sci Eng C Mater Biol Appl*. 2017;70 (Pt 1):378–385.

135. Hu G, Wang Y, He Q, et al. Multistage drug delivery system based on microenvironment-responsive dendrimer-gelatin nanoparticles for deep tumor penetration. *RSC Adv*. 2015;5:85933–85937.

136. Wang X, Wei B, Cheng X, et al. Phenylboronic acid-decorated gelatin nanoparticles for enhanced tumor targeting and penetration. *Nanotechnology*. 2016;27(38):385101.

137. Xu Y, Zhang J, Liu X, et al. MMP-2-responsive gelatin nanoparticles for synergistic tumor therapy. *Pharm Dev Technol*. 2019;24(8):1002–1013.

138. Ahmed F, Soliman FM, Adly MA, et al. In vitro assessment of the antimicrobial efficacy of chitosan nanoparticles against major fish pathogens and their cytotoxicity to fish cell lines. *J Fish Dis*. 2020;43(9):1049–1063.

139. Al-Zahrani SS, Bora RS, Al-Garni SM. Antimicrobial activity of chitosan nanoparticles. *Biotechnol Biotechnol Equip*. 2022;35(1):1874–1880.

140. Chandrasekaran M, Kim K, Chun S. Antibacterial activity of chitosan nanoparticles: A review. *Processes*. 2020;8(9):1173.

141. Ahmed F, Soliman FM, Adly MA, et al. Dietary chitosan nanoparticles: Potential role in modulation of rainbow trout (*Oncorhynchus mykiss*) antibacterial defense and intestinal immunity against enteric redmouth disease. *Mar Drugs*. 2021;19(2):72.

142. Saleh M, Essawy E, Shaalan M, et al. Therapeutic intervention with dietary chitosan nanoparticles alleviates fish pathological and molecular systemic inflammatory responses against infections. *Mar Drugs*. 2022;20(7):425.

143. Zhang C, Zhang P-Q, Guo S, et al. Dual-targeting polymer nanoparticles efficiently deliver DNA vaccine and induce robust prophylactic immunity against spring viremia of carp virus infection. *Microbiol Spectr*. 2022;10(5):e03085–22.

144. Tong X, Pan W, Su T, et al. Recent advances in natural polymer-based drug delivery systems. *React Funct Polym*. 2020;148:104501.

145. Abdellatif AA, Mohammed AM, Saleem I, et al. Smart injectable chitosan hydrogels loaded with 5-fluorouracil for the treatment of breast cancer. *Pharmaceutics*. 2022;14(3):661.

146. Herdiana Y, Wathoni N, Shamsuddin S, et al. Chitosan-based nanoparticles of targeted drug delivery system in breast cancer treatment. *Polymers*. 2021;13(11):1717.

147. Niu S, Williams GR, Wu J, et al. A chitosan-based cascade-responsive drug delivery system for triple-negative breast cancer therapy. 2019;17(1):95.

148. Vivek R, Nipun Babu V, Thangam R, et al. pH-responsive drug delivery of chitosan nanoparticles as Tamoxifen carriers for effective anti-tumor activity in breast cancer cells. *Colloids Surf B Biointerfaces.* 2013;111:117–123.

149. Tian B, Hua S, Liu J. Multi-functional chitosan-based nanoparticles for drug delivery: Recent advanced insight into cancer therapy. *Carbohydr Polym.* 2023;315:120972.

150. Yu Z, Shen X, Yu H, et al. Smart polymeric nanoparticles in cancer immunotherapy. *Pharmaceutics.* 2023;15(3):775.

151. Esmaeili Y, Khavani M, Bigham A, et al. Mesoporous silica@chitosan@gold nanoparticles as "on/off" optical biosensor and pH-sensitive theranostic platform against cancer. *Int J Biol Macromol.* 2022;202:241–255.

152. Amin H, Amin MA, Osman SK, et al. Chitosan nanoparticles as a smart nanocarrier for gefitinib for tackling lung cancer: Design of experiment and in vitro cytotoxicity study. *Int J Biol Macromol.* 2023;246:125638.

153. Mahmoud MA, El-bana MA, Morsy SM, et al. Synthesis and characterization of berberine-loaded chitosan nanoparticles for the protection of urethane-induced lung cancer. *Int J Pharm.* 2022;618(March):121652.

154. Zhu X, Yu Z, Feng L, et al. Chitosan-based nanoparticle co-delivery of docetaxel and curcumin ameliorates anti-tumor chemoimmunotherapy in lung cancer. *Carbohydr Polym.* 2021;268:118237.

155. Ghobadi-Oghaz N, Asoodeh A, Mohammadi M. Fabrication, characterization and in vitro cell exposure study of zein-chitosan nanoparticles for co-delivery of curcumin and berberine. *Int J Biol Macromol.* 2022;204:576–586.

156. Yadav AS, Radharani NNV, Gorain M, et al. RGD functionalized chitosan nanoparticle mediated targeted delivery of raloxifene selectively suppresses angiogenesis and tumor growth in breast cancer. *Nanoscale.* 2020;12(19):10664–10684.

157. Esfandiarpour-Boroujeni S, Bagheri-Khoulenjani S, Mirzadeh H, et al. Fabrication and study of curcumin loaded nanoparticles based on folate-chitosan for breast cancer therapy application. *Carbohydr Polym.* 2017;168:14–21.

158. Liang Y, Wang Y, Wang L, et al. Self-crosslinkable chitosan-hyaluronic acid dialdehyde nanoparticles for CD44-targeted siRNA delivery to treat bladder cancer. *Bioact Mater.* 2021;6(2):433–446.

159. Zhang E, Xing R, Liu S, et al. Vascular targeted chitosan-derived nanoparticles as docetaxel carriers for gastric cancer therapy. *Int J Biol Macromol.* 2019;126:662–672.

160. Faraji Dizaji B, Hasani Azerbaijan M, Sheisi N, et al. Synthesis of PLGA/chitosan/zeolites and PLGA/chitosan/metal organic frameworks nanofibers for targeted delivery of Paclitaxel toward prostate cancer cells death. *Int J Biol Macromol.* 2020;164:1461–1474.

161. Fathi M, Barar J, Erfan-Niya H, et al. Methotrexate-conjugated chitosan-grafted pH- and thermo-responsive magnetic nanoparticles for targeted therapy of ovarian cancer. *Int J Biol Macromol.* 2020;154:1175–1184.

162. Bhattacharya S. Fabrication and characterization of chitosan-based polymeric nanoparticles of Imatinib for colorectal cancer targeting application. *Int J Biol Macromol.* 2020;151:104–115.

163. Agrawal P, Singh RP, Sonali, et al. TPGS-chitosan cross-linked targeted nanoparticles for effective brain cancer therapy. *Mater Sci Eng C Mater Biol Appl.* 2017;74:167–176.

164. hevtsov M, Nikolaev B, Marchenko Y, et al. Targeting experimental orthotopic glioblastoma with chitosan-based superparamagnetic iron oxide nanoparticles (CS-DX-SPIONs). *Int J Nanomedicine.* 2018;13:1471–1482.

165. Balan P, Indrakumar J, Murali P, et al. Bi-faceted delivery of phytochemicals through chitosan nanoparticles impregnated nanofibers for cancer therapeutics. *Int J Biol Macromol.* 2020;142:201–211.

166. Sekar V, Rajendran K, Vallinayagam S, et al. Synthesis and characterization of chitosan ascorbate nanoparticles for therapeutic inhibition for cervical cancer and their in silico modeling. J Ind Eng Chem. 2018;62:239–249.

167. Nascimento et al., Mad2 checkpoint gene silencing using epidermal growth factor receptor-targeted chitosan nanoparticles in non-small cell lung cancer model. *Mol Pharm.* 2014;11:3515–3527.

168. Xiong Q, Cui M, Bai Y, et al. A supramolecular nanoparticle system based on beta-cyclodextrin-conjugated poly-l-lysine and hyaluronic acid for co-delivery of gene and chemotherapy agent targeting hepatocellular carcinoma. *Colloids Surf B Biointerfaces.* 2017;155:93–103.

169. Moustafa MA, El-refaie WM, Elnaggar YSR, et al. Fucoidan/hyaluronic acid cross-linked zein nanoparticles Loaded with fisetin as a novel targeted nanotherapy for oral cancer. *Int J Biol Macromol.* 2023;241(April):124528.

170. Jeannot V, Gauche C, Mazzaferro S, et al. Anti-tumor efficacy of hyaluronan-based nanoparticles for the co-delivery of drugs in lung cancer. *J Control Release.* 2018;275:117–128.

171. Qi B, Crawford AJ, Wojtynek NE, et al. Indocyanine green loaded hyaluronan-derived nanoparticles for fluorescence-enhanced surgical imaging of pancreatic cancer. *Nanomed Nanotechnol Biol Med.* 2018;14(3):769–780.

172. Sargazi A, Shiri F, Keikha S, et al. Hyaluronan magnetic nanoparticle for mitoxantrone delivery toward CD44-positive cancer cells. *Colloids Surf B Biointerfaces.* 2018;171:150–158.

173. Gao C, Wang M, Zhu P, et al. Preparation, characterization and in vitro antitumor activity evaluation of hyaluronic acid-alendronate-methotrexate nanoparticles. *Int J Biol Macromol.* 2021;166:71–79.

174. Lakkakula J, Roy A, Krishnamoorthy K, et al. Alginate-based nanosystems for therapeutic applications. *J Nanomater.* 2022;2022(1):6182815.

175. Zhang B, Yan Y, Shen Q, et al. A colon targeted drug delivery system based on alginate modificated graphene oxide for colorectal liver metastasis. *Mater Sci Eng C Mater Biol Appl.* 2017;79:185–190.

176. Baghbani F, Chegeni M, Moztarzadeh F, et al. Ultrasonic nanotherapy of breast cancer using novel ultrasound-responsive alginate-shelled perfluorohexane nanodroplets: In vitro and in vivo evaluation. *Mater Sci Eng C Mater Biol Appl.* 2017;77:698–707.

177. Baghbani F, Moztarzadeh F. Bypassing multidrug resistant ovarian cancer using ultrasound responsive doxorubicin/curcumin co-deliver alginate nanodroplets. *Colloids Surf B Biointerfaces.* 2017;153:132–140.

178. Zhang C, Wu Y, Liu T, et al. Antitumor activity of drug loaded glycyrrhetinic acid modified alginate nanoparticles on mice bearing orthotopic liver tumor. *J. Control. Release.* 2011;152:e111–e113.

179. Sorasitthiyanukarn FN, Muangnoi C, Ratnatilaka Na Bhuket P, et al. Chitosan/alginate nanoparticles as a promising approach for oral delivery of curcumin diglutaric acid for cancer treatment. *Mater Sci Eng C Mater Biol Appl.* 2018;93:178–190.

180. Di Martino A, Trusova ME, Postnikov PS, et al. Folic acid-chitosan-alginate nanocomplexes for multiple delivery of chemotherapeutic agents. *J Drug Deliv Sci Technol.* 2018;47:67–76.

181. Rosch JG, Winter H, DuRoss AN, et al. Inverse-micelle synthesis of doxorubicin-loaded alginate/chitosan nanoparticles and in vitro assessment of breast cancer cytotoxicity. *Colloid Interface Sci Commun.* 2019;28:69–74.

182. Hosseinifar T, Sheybani S, Abdouss M, et al. Pressure responsive nanogel base on alginate-cyclodextrin with enhanced apoptosis mechanism for colon cancer delivery. *J Biomed Mater Res A.* 2018;106(2):349–359.

183. Chai F, Sun L, He X, et al. Doxorubicin-loaded poly (lactic-co-glycolic acid) nanoparticles coated with chitosan/alginate by layer by layer technology for antitumor applications. *Int J Nanomedicine.* 2017;12:1791–1802.

184. Campos Fd S, Cassimiro DL, Crespi MS, et al. Preparation and characterisation of Dextran-70 hydrogel for controlled release of praziquantel. *Braz J Pharm Sci.* 2013;49:75–83.

185. Petrovici AR, Pinteala M, Simionescu NJM. Dextran formulations as effective delivery systems of therapeutic agents. *Molecules.* 2023;28(3):1086.

186. Díaz-Montes E. Dextran: Sources, structures, and properties. *Polysaccharides.* 2021;2(3):554–565.

187. Abid M, Naveed M, Azeem I, et al. Colon specific enzyme responsive oligoester crosslinked dextran nanoparticles for controlled release of 5-fluorouracil. *Int J Pharm.* 2020;586:119605.

188. Wu L, Zhang L, Shi G, et al. Zwitterionic pH/redox nanoparticles based on dextran as drug carriers for enhancing tumor intercellular uptake of doxorubicin. *Mater Sci Eng C Mater Biol Appl.* 2016;61:278–285.

189. Zhang M, Liu J, Kuang Y, et al. Ingenious pH-sensitive dextran/mesoporous silica nanoparticles based drug delivery systems for controlled intracellular drug release. *Int J Biol Macromol.* 2017;98:691–700.

190. Sampath M, Pichaimani A, Kumpati P, et al. The remarkable role of emulsifier and chitosan, dextran and PEG as capping agents in the enhanced delivery of curcumin by nanoparticles in breast cancer cells. *Int J Biol Macromol.* 2020;162:748–761.

191. Ashfaq A, Khursheed N, Fatima S, et al. Application of nanotechnology in food packaging: Pros and cons. *J Agric Food Res.* 2022;7:100270.

192. Marta H, Rizki DI, Mardawati E, et al. Starch nanoparticles: Preparation, properties and applications. *Polymers* 2023;15(5):1167.

193. Tagliapietra BL, de Melo BG, Sanches EA, et al. From micro to nanoscale: A critical review on the concept, production, characterization, and application of starch nanostructure. *Starch – Stärke.* 2021;73(11–12):2100079.

194. Wu X, Chang Y, Fu Y, et al. Effects of non-solvent and starch solution on formation of starch nanoparticles by nanoprecipitation. *Starch – Stärke.* 2016;68(3–4):258–263.

195. Qiu C, Hu Y, Jin Z, et al. A review of green techniques for the synthesis of size-controlled starch-based nanoparticles and their applications as nanodelivery systems. *Trends in Food Science & Technology* 2019;92:138–151.

196. Wang S, Copeland L. Effect of acid hydrolysis on starch structure and functionality: A review. *Crit Rev Food Sci Nutr.* 2015;55(8):1081–1097.
197. BeMiller JN. Physical modification of starch. In Sjöö M, Nilsson L (eds.), *Starch in Food.* Elsevier; 2018. pp. 223–253.
198. Amin H, Osman SK, Mohammed AM, et al. Gefitinib-loaded starch nanoparticles for battling lung cancer: Optimization by full factorial design and in vitro cytotoxicity evaluation. *Saudi Pharm J.* 2023;31(1):29–54.
199. Wang L, Zhao X, Yang F, et al. Loading paclitaxel into porous starch in the form of nanoparticles to improve its dissolution and bioavailability. *Int J Biol Macromol.* 2019;138:207–214.
200. Acevedo-Guevara L, Nieto-Suaza L, Sanchez LT, et al. Development of native and modified banana starch nanoparticles as vehicles for curcumin. *Int J Biol Macromol.* 2018;111:498–504.
201. Nieto-Suaza L, Acevedo-Guevara L, Sánchez LT, et al. Characterization of *Aloe vera*-banana starch composite films reinforced with curcumin-loaded starch nanoparticles. *Food Struct.* 2019;22:100131.
202. Abdellatif AAH, Ahmed F, Mohammed AM, et al. Recent advances in the pharmaceutical and biomedical applications of cyclodextrin-capped gold nanoparticles. *Int J Nanomedicine.* 2023;18:3247–3281.
203. Czarniecki M. Small molecule modulators of toll-like receptors. *Journal of Medicinal Chemistry.* 2008;51(21):6621–6626.
204. Xiong Q, Zhang M, Zhang Z, et al. Anti-tumor drug delivery system based on cyclodextrin-containing pH-responsive star polymer: In vitro and in vivo evaluation. *Int J Pharm.* 2014;474(1–2):232–240.
205. Zhang M, Xiong Q, Chen J, et al. A novel cyclodextrin-containing pH-responsive star polymer for nano-structure fabrication and drug delivery. *Polym Chem.* 2013;4(19): 5086–5095.
206. Hu J, Zhang G, Ge Z, et al. Stimuli-responsive tertiary amine methacrylate-based block copolymers: Synthesis, supramolecular self-assembly and functional applications. *Prog Polym Sci.* 2014;39(6):1096–1143.
207. Lin W, Yao N, Li H, et al. Co-delivery of imiquimod and plasmid DNA via an amphiphilic pH-responsive star polymer that forms unimolecular micelles in water. *Polymers.* 2016;8(11):397.
208. Shi X, Ma X, Hou M, et al. pH-Responsive unimolecular micelles based on amphiphilic star-like copolymers with high drug loading for effective drug delivery and cellular imaging. *J Mater Chem B.* 2017;5(33):6847–6859.
209. Wang Y, Zhang S, Li H, et al. Small-molecule modulators of toll-like receptors. *Acc Chem. Res.* 2020;53(5):1046–1055.
210. Ahuja R, Panwar N, Meena J, et al. Natural products and polymeric nanocarriers for cancer treatment: A review. *Environ Chem Lett.* 2020;18(6):2021–2030.
211. Saeed AF, Su J, Ouyang SJB, et al. Marine-derived drugs: Recent advances in cancer therapy and immune signaling. *Biomed Pharmacother.* 2021;134:111091.
212. Chaudhry GE, Md Akim A, Sung YY, et al. Cancer and apoptosis: The apoptotic activity of plant and marine natural products and their potential as targeted cancer therapeutics. *Front Pharmacol.* 2022;13:842376.
213. Khalifa SAM, Elias N, Farag MA, et al. Marine natural products: A source of novel anticancer drugs. *Mar Drugs.* 2019;17(9):491.
214. Barreca M, Spano V, Montalbano A, et al. Marine anticancer agents: An overview with a particular focus on their chemical classes. *Mar Drugs.* 2020;18(12):619.
215. Ruiz-Torres V, Encinar JA, Herranz-Lopez M, et al. An updated review on marine anticancer compounds: The use of virtual screening for the discovery of small-molecule cancer drugs. *Molecules.* 2017;22(7):1037.
216. Sable R, Parajuli P, Jois SJMD. Peptides, peptidomimetics, and polypeptides from marine sources: A wealth of natural sources for pharmaceutical applications. *Mar Drugs.* 2017;15(4):124.
217. Fedorov SN, Ermakova SP, Zvyagintseva TN, et al. Anticancer and cancer preventive properties of marine polysaccharides: Some results and prospects. *Mar Drugs.* 2013;11(12):4876–4901.
218. Mirunalini S, Maruthanila VL. The impact of bioactive compounds derived from marine fish on cancer. *Anticancer Agents Med Chem.* 2022;22(15):2757–2765.
219. Abd El-Galil MAEAA, Abd-Elaal Hassan HAA, Abd Alhamed Ahmed FE, et al. Impact of transportation in freshwater and brackish water on Nile tilapia (*Oreochromis niloticus*) resistance. *BMC Vet Res.* 2024;20(1):396.
220. Ahmed F, Ali H, Bakheet Y, et al. Fish's crustacean parasites: Types, prevalence, clinical signs, and control. *Sohag J Sci.* 2022;7(2):123–129.
221. Ahmed F, Mahrous A, Yasser A, et al. Manipulative strategies of aquatic arthropods for fish parasitism: A review. *Sohag J Sci.* 2024;9(3):357–366.
222. El-Sayed A, Soliman HA, Ahmed F, et al. Nephrotoxicity of single or combined exposure to micro-plastics (MPs) and sodium Lauryl sulfate (SLS) in African catfish (*Clarias gariepinus*). *Sohag J Sci.* 2024;9(3):342–349.

223. Zahran E, Ahmed F, Hassan Z, et al. Toxicity evaluation, oxidative, and immune responses of mercury on Nile tilapia: Modulatory role of dietary *Nannochloropsis oculata. Biol Trace Element Res.* 2024;202(4):1752–1766.

224. Kara A, Gunes OC, Albayrak AZ, et al. Fish scale/poly(3-hydroxybutyrate-co-3-hydroxyvalerate) nanofibrous composite scaffolds for bone regeneration. *J Biomater Appl.* 2020;34(9):1201–1215.

225. Kara A, Tamburaci S, Tihminlioglu F, et al. Bioactive fish scale incorporated chitosan biocomposite scaffolds for bone tissue engineering. *Int J Biol Macromol.* 2019;130:266–279.

226. Wu W, Zhou Z, Sun G, et al. Construction and characterization of degradable fish scales for enhancing cellular adhesion and potential using as tissue engineering scaffolds. *Mater Sci Eng C Mater Biol Appl.* 2021;122:111919.

227. Lima Júnior EM, de Moraes Filho MO, Costa BA, et al. A randomized comparison study of lyophilized Nile tilapia skin and silver-impregnated sodium carboxymethylcellulose for the treatment of superficial partial-thickness burns. *J Burn Care Res.* 2021;42(1):41–48.

228. Lima Junior EM, de Moraes Filho MO, Costa BA, et al. Innovative burn treatment using tilapia skin as a xenograft. *J Burn Care Res.* 2020;41:585–592.

229. Lima-Junior EM, de Moraes Filho MO, Costa BA, et al. Innovative treatment using tilapia skin as a xenograft for partial thickness burns after a gunpowder explosion. *J Surg Case Rep.* 2019;2019(6):rjz181.

230. Song W-K, Liu D, Sun L-L, et al. Physicochemical and biocompatibility properties of type I collagen from the skin of *Nile tilapia* (*Oreochromis niloticus*) for biomedical applications. *Mar Drugs.* 2019;17(3):137.

231. Chen Y-T, Chen S-J, Hu C-Y, et al. Exploring the anti-cancer effects of fish bone fermented using *Monascus Purpureus*: Induction of apoptosis and autophagy in human colorectal cancer cells. *Molecules.* 2023;28(15):5679.

232. Ahamed AF, Manimohan M, Kalaivasan N. Fabrication of biologically active fish bone derived hydroxyapatite and montmorillonite blended sodium alginate composite for in-vitro drug delivery studies. *J Inorg Organomet Polym Mater.* 2022;32(10):3902–3922.

233. Aboudamia FZ, Kharroubi M, Neffa M, et al. Potential of discarded sardine scales (*Sardina pilchardus*) as chitosan sources. *J Air Waste Manag Assoc.* 2020;70(11):1186–1197.

234. Kumari S, Kumar Annamareddy SH, Abanti S, et al. Physicochemical properties and characterization of chitosan synthesized from fish scales, crab and shrimp shells. *Int J Biol Macromol.* 2017;104 (Pt B):1697–1705.

235. Putri DKT, Wijayanti D, Oktiani BW, et al. Synthesis and characteristics of Chitosan from Haruan (*Channa striata*) fish scales. *Syst Rev Pharm.* 2021;11(001).

236. Molina-Ramírez C, Mazo P, Zuluaga R, et al. Characterization of chitosan extracted from fish scales of the Colombian endemic species *Prochilodus magdalenae* as a novel source for antibacterial starch-based films. *Polymers.* 2021;13(13):2079.

237. Najm AA, Azfaralarriff A, Dyari HRE, et al. A systematic review of antimicrobial peptides from fish with anticancer properties. *Pertanika J Sci Technol.* 2022;30(2):1171–1196.

238. Picot L, Bordenave S, Didelot S, et al. Antiproliferative activity of fish protein hydrolysates on human breast cancer cell lines. *Process Biochem.* 2006;41(5):1217–1222.

239. Hsu JC, Lin LC, Tzen JT, et al. Pardaxin-induced apoptosis enhances antitumor activity in HeLa cells. *Peptides.* 2011;32(6):1110–1116.

240. Huang TC, Lee JF, Chen JY. Pardaxin, an antimicrobial peptide, triggers caspase-dependent and ROS-mediated apoptosis in HT-1080 cells. *Mar Drugs.* 2011;9(10):1995–2009.

241. Antipova L, Storublevtsev S, Titov S, et al. A study of the use of modified collagen of freshwater fish as. In Ahmad M (ed.) *Wound Healing.* Rijeka: IntechOpen; 2019. p. 17.

242. Chinh NT, Huynh MD, Lien LTN, et al. Preparation and characterization of materials based on fish scale collagen and polyphenols extracted from *Camellia chrysantha. Vietnam J Sci Technol.* 2023;61(1):72–83.

243. Tayel AA, Ghanem RA, Al-Saggaf MS, et al. Application of fish collagen-nanochitosan-henna extract composites for the control of skin pathogens and accelerating wound healing. *Int J Polym Sci.* 2021;2021(1):1907914.

244. Alipal J, Mohd Pu'ad NAS, Lee TC, et al. A review of gelatin: Properties, sources, process, applications, and commercialisation. *Mater Today: Proc.* 2021;42:240–250.

245. Kang MG, Lee MY, Cha JM, et al. Nanogels derived from fish gelatin: Application to drug delivery system. *Mar Drugs.* 2019;17(4):246.

246. Liu Y, Xia X, Li X, et al. Design and characterization of edible chitooligosaccharide/fish skin gelatin nanofiber-based hydrogel with antibacterial and antioxidant characteristics. *Int J Biol Macromol.* 2024;262:130033.

247. Eshari F, Keley MT, Habibi-Rezaei M, et al. A review of the fish oil extraction methods and omega 3 concentration technologies technologies. *Food Process Preserv.* 2022;14(3):101–124.

248. Rehman K, Zulfakar MH. Novel fish oil-based bigel system for controlled drug delivery and its influence on immunomodulatory activity of imiquimod against skin cancer. *J Pharm Res.* 2017;34:36–48.

249. Behera A, Das R, Patnaik P, et al. A review on fish peptides isolated from fish waste with their potent bioactivities. *J. Appl. Biol. Biotechnol.* 2022;10(3):195–209.

250. Roy S, Chaudhuri S, Mukherjee P, et al. Biomedical applications of chitin, chitosan, their derivatives, and processing by-products from fish waste. In Maqsood S, Naseer MN, Benjakul S, Zaidi AA (eds.), *Fish Waste to Valuable Products.* Springer; 2024. pp. 279–300.

251. Nurdiani R, Vasiljevic T, Singh T, et al. Bioactive peptides from fish by-products with anticarcinogenic potential. *Int Food Res J.* 2017;24(5):1840–1849.

252. Inbaraj BS, Lai Y-W, Chen B-H. A comparative study on inhibition of lung cancer cells by nanoemulsion, nanoliposome, nanogold and their folic acid conjugates prepared with collagen peptides from Taiwan tilapia skin. *Int J Biol Macromol.* 2024;261:129722.

253. Kumar LV, Shakila RJ, Jeyasekaran G. In Vitro anti-cancer, anti-diabetic, anti-inflammation and wound healing properties of collagen peptides derived from unicorn leatherjacket (Aluterus Monoceros) at different hydrolysis. *Turk J Fish Aquat Sci.* 2018;19(7):551–560.

254. Han S-H, Uzawa Y, Moriyama T, et al. Effect of collagen and collagen peptides from bluefin tuna abdominal skin on cancer cells. *Health.* 2011;3(3):129–134.

255. Zhu T, Liang D, Zhang Q, et al. Curcumin-encapsulated fish gelatin-based microparticles from microfluidic electrospray for postoperative gastric cancer treatment. *Int J Biol Macromol.* 2024;254:127763.

256. Liu J, Tagami T, Ozeki T. Fabrication of 3D-printed fish-gelatin-based polymer hydrogel patches for local delivery of PEGylated liposomal doxorubicin. *Mar Drugs.* 2020 20;18(6):325.

257. Balasubramanian S, Revathi A, Gunasekaran G. Studies on anticancer, haemolytic activity and chemical composition of crude epidermal mucus of fish *Mugil cephalus. Int J Fish Aquat Stud.* 2016;4(5):438–443.

258. Jameel F, Agarwal P, Ahmad R, et al. Skin mucus extract derived from *Channa punctatus,Channa striatus*, and *Heteropneustes fossilisinduces* apoptosis and suppresses proliferation in human adenocarcinoma cells via ROS mediated pathway and cell cycle arrest. 2024.

259. Yaghoubzadeh Z, Ghadikolaii FP, Kaboosi H, et al. Antioxidant activity and anticancer effect of bioactive peptides from rainbow trout (Oncorhynchus Mykiss) skin hydrolysate. *Int J Peptide Res Ther.* 2020;26(1):625–632.

260. Wu SP, Huang TC, Lin CC, et al. Pardaxin, a fish antimicrobial peptide, exhibits antitumor activity toward murine fibrosarcoma in vitro and in vivo. *Mar Drugs.* 2012;10(8):1852–1872.

261. Chen J-Y, Lin W-J, Lin T-L. A fish antimicrobial peptide, tilapia hepcidin TH2–3, shows potent antitumor activity against human fibrosarcoma cells. *Peptides.* 2009;30:1636–1642.

262. Law D, Najm AA, Chong JX, et al. In silico identification and in vitro assessment of a potential anti-breast cancer activity of antimicrobial peptide retrieved from the ATMP1 anabas testudineus fish peptide. *PeerJ.* 2023;11:e15651.

263. Najm AAK, Azfaralariff A, Dyari HRE, et al. Anti-breast cancer synthetic peptides derived from the anabas testudineus skin mucus fractions. *Sci Rep.* 2021;11:23182.

264. Al-Hassan JM, Afzal M, Oommen S, et al. Oxysterols in catfish skin secretions (*Arius bilineatus*, Val.) exhibit anti-cancer properties. *Front Pharmacol.* 2022;13:1001067.

265. Lin WJ, Chien YL, Pan CY, et al. Epinecidin-1, an antimicrobial peptide from fish (*Epinephelus coioides*) which has an antitumor effect like lytic peptides in human fibrosarcoma cells. *Peptides.* 2009;30(2):283–290.

266. Lugo JM, Tafalla C, Oliva A, et al. Evidence for antimicrobial and anticancer activity of pituitary adenylate cyclase-activating polypeptide (PACAP) from North African catfish (*Clarias gariepinus*): Its potential use as novel therapeutic agent in fish and humans. *Fish Shellfish Immunol.* 2019;86:559–570.

267. Avramovic N, Mandic B, Savic-Radojevic A, et al. Polymeric nanocarriers of drug delivery systems in cancer therapy. *Pharmaceutics.* 2020;12(4):298.

268. Makadia HK, Siegel SJJP. Poly lactic-co-glycolic acid (PLGA) as biodegradable controlled drug delivery carrier. *Polymers.* 2011;3(3):1377–1397.

269. Molavi F, Barzegar-Jalali M, Hamishehkar CR. Polyester based polymeric nano and microparticles for pharmaceutical purposes: A review on formulation approaches. *J Control Release.* 2020;320:265–282.

270. Zhu S, Xing H, Gordiichuk P, et al. PLGA spherical nucleic acids. *Adv Mater.* 2018;30(22):e1707113.

271. He Y, de Araujo Junior RF, Cavalcante RS, et al. Effective breast cancer therapy based on palmitic acid-loaded PLGA nanoparticles. *Biomater Adv.* 2023;145:213270.

272. Gahtani RM, Alqahtani A, Alqahtani T, et al. Retracted: 5-fluorouracil-loaded PLGA nanoparticles: Formulation, physicochemical characterisation, and in vitroanti-cancer activity. *Bioinorg Chem Appl.* 2023;2023:9846184.

273. Ekinci M, Öztürk AA, Santos-Oliveira R, et al. The use of Lamivudine-loaded PLGA nanoparticles in the diagnosis of lung cancer: Preparation, characterization, radiolabeling with 99mTc and cell binding. *J Drug Deliv Sci Technol.* 2022;69:103139.

274. He Y, Li X, Jia D, et al. A transcriptomics-based analysis of the toxicity mechanisms of gabapentin to zebrafish embryos at realistic environmental concentrations. *J Environ Pollut.* 2019;251:746–755.

275. Cai H, Wang R, Guo X, et al. Combining gemcitabine-loaded macrophage-like nanoparticles and erlotinib for pancreatic cancer therapy. *Mol Pharm.* 2021;18(7):2495–2506.

276. Elbatanony RS, Parvathaneni V, Kulkarni NS, et al. Afatinib-loaded inhalable PLGA nanoparticles for localized therapy of non-small cell lung cancer (NSCLC)-development and in-vitro efficacy. *Drug Deliv Transl Res.* 2021;11(3):927–943.

277. Prabhuraj RS, Bomb K, Srivastava R, et al. Selection of superior targeting ligands using PEGylated PLGA nanoparticles for delivery of curcumin in the treatment of triple-negative breast cancer cells. *J Drug Deliv Sci Technol.* 2020;57(December 2019):101722.

278. Al-Nemrawi NK, Altawabeyeh RM, Darweesh RS. Preparation and characterization of docetaxel-PLGA nanoparticles coated with folic acid-chitosan conjugate for cancer treatmen. *J Pharm Sci.* 2022;111(2):485–494.

279. Fabozzi A, Barretta M, Valente T, et al. Preparation and optimization of hyaluronic acid decorated irinotecan-loaded poly (Lactic-Co-Glycolic Acid) nanoparticles by microfluidics for cancer therapy applications. *Colloids Surf A Physicochem Eng Aspects.* 2023;674(April):131790.

280. Arslan FB, Ozturk K, Tavukcuoglu E, et al. A novel combination for the treatment of small cell lung cancer: Active targeted irinotecan and stattic co-loaded PLGA nanoparticles. *Int J Pharm.* 2023;632:122573.

281. Maddiboyina B, Roy H, Nakkala RK, et al. Formulation, optimization and characterization of raloxifene hydrochloride loaded PLGA nanoparticles by using taguchi design for breast cancer application. *Chem Biol Drug Des.* 2023;102(3):457–470.

282. Al-Humaidi RB, Fayed B, Shakartalla SB, et al. Optimum inhibition of MCF-7 breast cancer cells by efficient targeting of the macropinocytosis using optimized paclitaxel-loaded nanoparticles. *Life Sci.* 2022;305:120778.

283. Zhang D, Liu L, Wang J, et al. Drug-loaded PEG-PLGA nanoparticles for cancer treatment. *Front Pharmacol.* 2022;13:1–12.

284. Suk JS, Xu Q, Kim N, et al. PEGylation as a strategy for improving nanoparticle-based drug and gene delivery. *Adv Drug Deliv Rev.* 2016;99(Pt A):28–51.

285. Cox R, Narisetty V, Castro E, et al. Fermentative valorisation of xylose-rich hemicellulosic hydrolysates from agricultural waste residues for lactic acid production under non-sterile conditions. *Waste Manag.* 2023;166:336–345.

286. Rydz J, Sikorska W, Kyulavska M, et al. Polyester-based (Bio) degradable polymers as environmentally friendly materials for sustainable development. *Int J Mol Sci.* 2015;16:564–596.

287. Gupta P, Bansal A, Kaur H, et al. Folic acid-targeted redox responsive polylactic acid-based nanoparticles co-delivering pirarubicin and salinomycin suppress breast cancer tumor growth in vivo. *Nanoscale.* 2024;16(43):20131–20146.

288. Anees M, Tiwari S, Mehrotra N, et al. Development and evaluation of PLA based hybrid block copolymeric nanoparticles for systemic delivery of pirarubicin as an anti-cancer agent. *Int J Pharm.* 2022;620:121761.

289. Mehrotra N, Anees M, Tiwari S, et al. Polylactic acid based polymeric nanoparticle mediated co-delivery of navitoclax and decitabine for cancer therapy. *Nanomed Nanotechnol Biol Med.* 2023;47:102627.

290. Qin C, Shen Y, Wang B, et al. An acellular tissue matrix-based drug carriers with dual chemo-agents for colon cancer growth suppression. *Biomed Pharmacother.* 2019;117:109048.

291. Hu JK, Suh HW, Qureshi M, et al. Nonsurgical treatment of skin cancer with local delivery of bioadhesive nanoparticles. *Proc Natl Acad Sci U S A.* 2021;118(7):e2020575118.

292. Chen J, Ning E, Wang Z, et al. Docetaxel loaded mPEG-PLA nanoparticles for sarcoma therapy: Preparation, characterization, pharmacokinetics, and anti-tumor efficacy. *Drug Deliv.* 2021;28(1):1389–1396.

293. Im SH, Im DH, Park SJ, et al. Stereocomplex polylactide for drug delivery and biomedical applications: A review. *Molecules.* 2021;26(10):2846.

294. Khan I, Ray Dutta J, Ganesan R. Lactobacillus sps. lipase mediated poly (epsilon-caprolactone) degradation. *Int J Biol Macromol.* 2017;95:126–131.

295. Bhadran A, Shah T, Babanyinah GK, et al. Recent advances in polycaprolactones for anticancer drug delivery. *Pharmaceutics*. 2023;15(7).
296. Murab S, Herold S, Hawk T, et al. Advances in additive manufacturing of polycaprolactone based scaffolds for bone regeneration. *J Mater Chem B*. 2023;11:7250–7279.
297. Jan N, Madni A, Rahim MA, et al. In vitro anti-leukemic assessment and sustained release behaviour of cytarabine loaded biodegradable polymer based nanoparticles. *Life Sci*. 2021;267:118971.
298. Khan S, Aamir MN, Madni A, et al. Lipid poly (varepsilon-caprolactone) hybrid nanoparticles of 5-fluorouracil for sustained release and enhanced anticancer efficacy. *Life Sci*. 2021;284:119909.
299. Hasanbegloo K, Banihashem S, Faraji Dizaji B, et al. Paclitaxel-loaded liposome-incorporated chitosan (core)/poly(epsilon-caprolactone)/chitosan (shell) nanofibers for the treatment of breast cancer. *Int J Biol Macromol*. 2023;230:123380.
300. Grossen P, Witzigmann D, Sieber S, et al. PEG-PCL-based nanomedicines: A biodegradable drug delivery system and its application. *J Control Release*. 2017;260:46–60.
301. Chang D, Ma Y, Xu X, et al. Stimuli-responsive polymeric nanoplatforms for cancer therapy. *Front Bioeng Biotechnol*. 2021;9:707319.
302. Shen H, Liu Q, Liu D, et al. Fabrication of doxorubicin conjugated methoxy poly(ethylene glycol)-block-poly(epsilon-caprolactone) nanoparticles and study on their in vitro antitumor activities. *J Biomater Sci Polym Ed*. 2021;32(13):1703–1717.
303. Behl A, Sarwalia P, Kumar S, et al. Codelivery of gemcitabine and MUC1 inhibitor using PEG-PCL nanoparticles for breast cancer therapy. *Mol Pharm*. 2022;19(7):2429–2440.
304. Yang DC, Yang XZ, Luo CM, et al. A promising strategy for synergistic cancer therapy by integrating a photosensitizer into a hypoxia-activated prodrug. *Eur J Med Chem*. 2022;243:114749.
305. Jin CE, Yoon MS, Jo MJ, et al. Synergistic encapsulation of paclitaxel and sorafenib by methoxy poly(ethylene glycol)-b-poly(caprolactone) polymeric micelles for ovarian cancer therapy. *Pharmaceutics*. 2023;15(4):1206.
306. Pereira P, Serra AC, Coelho JFJ. Vinyl polymer-based technologies towards the efficient delivery of chemotherapeutic drugs. *Prog Polym Sci*. 2021;121:101432.
307. Feldman D. Polymers and polymer nanocomposites for cancer therapy. *Appl Sci*. 2019;9(18):3899.
308. Kim MW, Kwon SH, Choi JH, et al. A promising biocompatible platform: Lipid-based and bio-inspired smart drug delivery systems for cancer therapy. *Int J Mol Sci*. 2018;19(12):3859.
309. Yingchoncharoen P, Kalinowski DS, Richardson DR. Lipid-based drug delivery systems in cancer therapy: What is available and what is yet to come. *Pharmacol Rev*. 2016;68(3):701–787.
310. Arias JL, Clares B, Morales ME, et al. Lipid-based drug delivery systems for cancer treatment. *Curr Drug Targets*. 2011;12(8):1151–1165.
311. Marcial SPDS, Carneiro G, Leite EA. Lipid-based nanoparticles as drug delivery system for paclitaxel in breast cancer treatment. *J Nanoparticle Res*. 2017;19:340.
312. Rethi L, Mutalik C, Anurogo D, et al. Lipid-based nanomaterials for drug delivery systems in breast cancer therapy. *Nanomaterials*. 2022;12:2948.
313. Yang C, Merlin D. Lipid-based drug delivery nanoplatforms for colorectal cancer therapy. *Nanomaterials*. 2020;10(7):1424.
314. Kim SJ, Puranik N, Yadav D, et al. Lipid nanocarrier-based drug delivery systems: Therapeutic advances in the treatment of lung cancer. *Int J Nanomedicine*. 2023;18:2659–2676.
315. Duwa R, Emami F, Lee S, et al. Polymeric and lipid-based drug delivery systems for treatment of glioblastoma multiforme. *J Ind Eng Chem*. 2019;79:261–273.
316. Krishnamurthy S, Vaiyapuri R, Zhang L, et al. Lipid-coated polymeric nanoparticles for cancer drug delivery. *Biomater Sci*. 2015;3(7):923–936.
317. Namiki Y, Fuchigami T, Tada N, et al. Nanomedicine for cancer: Lipid-based nanostructures for drug delivery and monitoring. *Acc Chem Res*. 2011;44:1080–1093.
318. Hao Y, Ji Z, Zhou H, et al. Lipid-based nanoparticles as drug delivery systems for cancer immunotherapy. *MedComm (2020)*. 2023;4(4):e339.
319. Deirram N, Zhang C, Kermaniyan SS, et al. pH-responsive polymer nanoparticles for drug delivery. *Macromol Rapid Commun*. 2019;40(10):e1800917.
320. Cheng R, Meng F, Deng C, et al. Dual and multi-stimuli responsive polymeric nanoparticles for programmed site-specific drug delivery. *Biomaterials*. 2013;34(14):3647–3657.
321. Liu T-Y, Hu S-H, Liu K-H, et al. Instantaneous drug delivery of magnetic/thermally sensitive nanospheres by a high-frequency magnetic field. *Langmuir*. 2008;24:13306–13311.
322. Remant Bahadur KC, Thapa B, Xu P. PH and redox dual responsive nanoparticle for nuclear targeted drug delivery. *Mol Pharm*. 2012;9(9):2719–2729.

323. Lo CL, Lin KM, Hsiue GH. Preparation and characterization of intelligent core-shell nanoparticles based on poly(D,L-lactide)-g-poly(N-isopropyl acrylamide-co-methacrylic acid). *J Control Release.* 2005;104(3):477–488.

324. Zhao Z, Huang D, Yin Z, et al. Magnetite nanoparticles as smart carriers to manipulate the cytotoxicity of anticancer drugs: Magnetic control and pH-responsive release. *J Mater Chem.* 2012;22(31):15717–15725.

325. Du J-Z, Du X-J, Mao C-Q, et al. Tailor-made dual PH-sensitive polymer–doxorubicin nanoparticles for efficient anticancer drug delivery. *J Am Chem Soc.* 2011;133(44):17560–17563.

326. Qiao Z-Y, Zhang R, Du F-S, et al. Multi-responsive nanogels containing motifs of ortho ester, oligo(ethylene glycol) and disulfide linkage as carriers of hydrophobic anti-cancer drugs. *J Control Release.* 2011;152(1):57–66.

327. Chang B, Sha X, Guo J, et al. Thermo and PH dual responsive, polymer shell coated, magnetic mesoporous silica nanoparticles for controlled drug release. *J Mater Chem.* 2011;21(25):9239–9247.

328. Wang K, Guo D-S, Wang X, et al. Multistimuli responsive supramolecular vesicles based on the recognition of p-sulfonatocalixarene and its controllable release of doxorubicin. *ACS Nano.* 2011;5(4):2880–2894.

329. Loh XJ, del Barrio J, Toh PP, et al. Triply triggered doxorubicin release from supramolecular nanocontainers. *Biomacromolecules.* 2012;13(1):84–91.

330. Colson YL, Grinstaff MW. Biologically responsive polymeric nanoparticles for drug delivery. *Adv Mater.* 2012;24(28):3878–3886.

331. Kozielski KL, Tzeng SY, Mendoza BAHD, et al. Bioreducible cationic polymer-based environmentally triggered cytoplasmic siRNA delivery to primary human brain cancer cells. *ACS Nano.* 2014;8:3232–3241.

332. Zhao H, Lin ZY, Yildirimer L, et al. Polymer-based nanoparticles for protein delivery: Design, strategies and applications. *J Mater Chem B.* 2016;4(23):4060–4071.

3 Lipid Nanoparticle Formulations for Targeted Anticancer Drug Delivery

Masoud Ojarudi, Asal Golchin,
Yusef Rasmi, and Imran Saleem

3.1 INTRODUCTION

Organic nanoparticles (NPs), made of proteins, lipids, carbohydrates, polymers, or other organic compounds, are non-toxic and biodegradable and exhibit sensitivity to thermal energy and electromagnetic radiation. They are formed through non-covalent intermolecular interactions, resulting in less stability and increased excretion. Their potential uses include targeted drug delivery and cancer treatment. These nanoparticles are currently used in the biomedical field due to their composition, surface shape, stability, and carrying capacity [1–3].

3.2 LIPID NPs

Lipids include various organic compounds such as hormones, fats, co-solvents, surfactants, waxes, phospholipids, and steroids. They are mainly used to facilitate the penetration of water-insoluble drugs into the body [4].

In recent decades, numerous lipid-based nanoparticles have been rapidly produced [3,5–9]. The main categories comprise solid lipid nanoparticles (SLNs), non-structured lipid nanoparticles (NLCs), lipid–polymer hybrid nanoparticles (LPHNs), lipid nanocapsules (LNCs), liposomes, niosomes, stereoisomers, and liposomes (Figure 3.1).

Lipid particle systems have been engineered for precise localization, enhanced bioavailability, and stability. The notion of liposomal drug delivery systems, initially expressed by Alec Bangham in 1961, has influenced the pharmaceutical field. Liposomes are spherical vesicles composed of a lipid bilayer that encases an aqueous core, making them appropriate for encapsulating hydrophobic and hydrophilic substances. Phospholipids or synthetic amphiphiles, in conjunction with sterols such as cholesterol, affect membrane permeability. Niosomes comprising non-ionic surfactants are non-toxic and adaptable carriers for many pharmaceuticals. Bilosomes, vesicles composed of bile salts, have superior stability and improve transmembrane drug absorption relative to niosomes and liposomes. They enhance the translocation of polypeptides, proteins, and polysaccharides over the mucosal interface. Vesicular nanocarriers are classified into multilamellar vesicles (MLV), large unilamellar vesicles (LUV), and small unilamellar vesicles (SUV) according to their size and layer count [10–14]. The vesicles are shown in Figure 3.2. Sterosomes are formed from non-phospholipid, single-chain amphiphilic molecules characterized by a high sterol concentration. They create massive unilamellar liposomes at the nanoscale, providing benefits such as reduced bilayer permeability and uniform dispersion. The inclusion of sterols improves the stability of stereoisomers, mitigating the disintegration and restricted shelf-life concerns associated with traditional liposomes. Sterosomes enhance encapsulation efficiency (EE) and regulate the release of therapeutic substances, positioning them as a promising platform for sophisticated drug delivery applications [15–17].

DOI: 10.1201/9781003517870-3

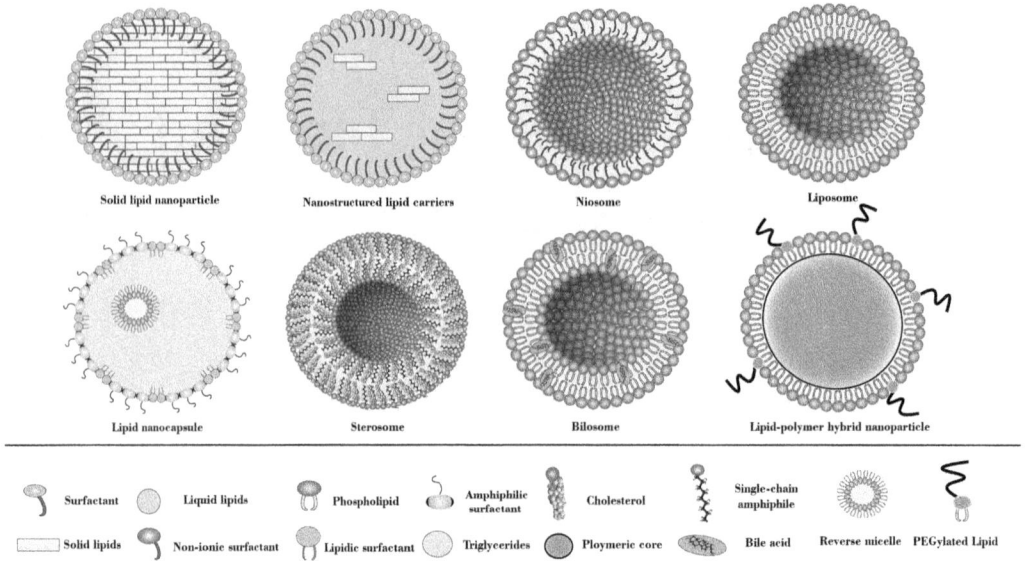

FIGURE 3.1 Different lipid-based nanoparticles. Solid lipid nanoparticles consist of a single layer of surfactant with solid lipids inside. NLCs are the second generation of solid lipid nanoparticles that use liquid lipids instead of solid lipids. Niosomes are lipid bilayer vesicles formed from nonionic surfactants. Liposomes are a classical form of lipid nanoparticles formed from phospholipids. The use of bile acids in the composition of liposomes results in the formation of bilosomes. Sterosomes are vesicles of non-phospholipid, single-chain amphiphilic molecules. Lipid–polymer hybrid nanoparticles consist of an inner polymer matrix (core), a lipid layer, and an outer polymer shell. LNCs are an oily core enclosed in a shell of solid lipids and emulsifying agents.

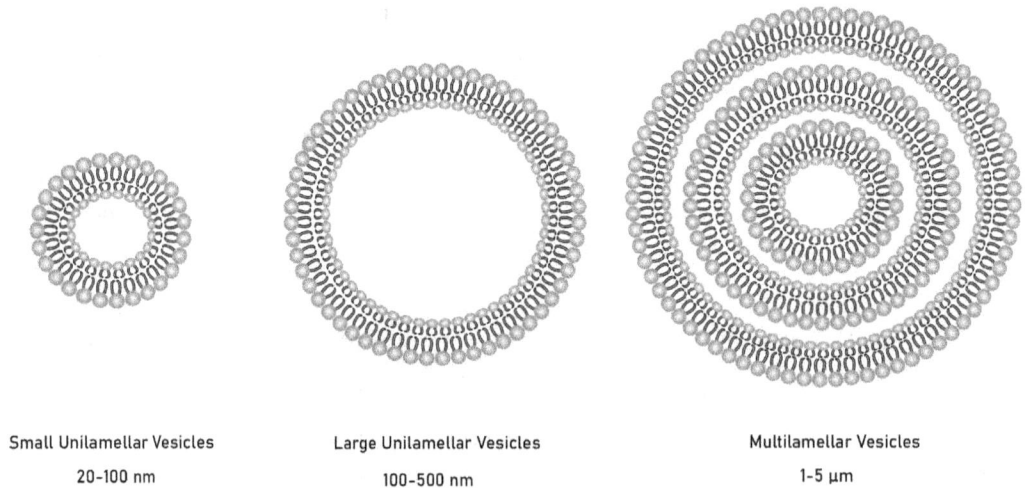

Small Unilamellar Vesicles
20–100 nm

Large Unilamellar Vesicles
100–500 nm

Multilamellar Vesicles
1–5 μm

FIGURE 3.2 Classification of vesicular nanocarriers. Vesicular nanoparticles are divided into three categories based on the number of layers and size: MLV, LUV, and SUV.

In the 1990s, Müller and collaborators introduced SLNs and NLCs. SLNs integrate the advantages of nanoparticles, mostly metallic and polymeric, with lipid-based emulsions composed of non-toxic, biodegradable lipids. These nanoparticles are considered safer alternatives to other nano-systems, possessing a solid matrix that facilitates regulated drug release, improved stability, and cost-effectiveness relative to phospholipid-based liposomes [18,19].

SLNs encounter obstacles like inadequate long-term drug retention and restricted drug loading capacity. The lipid matrix experiences polymorphic changes throughout storage, leading to a more structured crystalline lattice and the progressive release of encapsulated drugs. To enhance stability, liquid lipids or solubilizers were included, resulting in the formulation of NLCs. NLCs, the second generation of SLNs, substitute solid lipids with liquid lipids, thereby improving storage stability and drug loading capacity [20,21].

NLCs are classified into three categories: defective amorphous type, multiple type, and crystal type. The defective crystal structure possesses a markedly uneven matrix, leading to elevated drug loading capacity but diminished EE. Numerous NLCs utilize elevated oil content combined with solid lipids to facilitate phase separation, resulting in drug-encapsulated oily nanocompartments that improve drug solubility and enhance EE. Amorphous NLCs possess a robust, unstructured matrix that inhibits drug leakage resulting from crystallization [9,22,23].

LPHNs comprise three functional components: an inner polymer matrix (core), a lipid layer, and an outer polymer shell. The polymer core efficiently encapsulates hydrophobic or poorly water-soluble medicinal agents with excellent loading efficacy. The core is encased in a lipid layer encircled by a polymer shell that improves biocompatibility and facilitates medication retention. The outermost layer, consisting of PEG lipids, enhances stability and extends circulation time in the body via steric stabilization. LPHNs can encapsulate, adsorb, or covalently attach many bioactive therapeutic molecules, such as genes, drugs, proteins, vaccines, peptides, targeting ligands, and diagnostic imaging agents [24].

LNCs are hybrid entities that integrate characteristics of liposomes and polymeric nanoparticles. They comprise an oily nucleus encased in a shell of solid lipids and emulsifying agents. LNCs provide substantial benefits, such as biodegradability, biocompatibility, and elevated drug entrapment efficiency, especially for lipophilic pharmaceuticals. They can encapsulate many pharmaceuticals, protecting them from breakdown and minimizing toxicity to healthy cells [7,8].

3.3 MECHANISMS OF TARGETING: NAVIGATING BIOLOGICAL BARRIERS

Numerous initial versions of NPs failed to overcome the biological obstacles related to delivery. Recent developments in NPs design have markedly enhanced their capacity to surmount biological delivery obstacles. The enhancements include regulated synthesis methods that facilitate the integration of complex structures and targeting agents. Notwithstanding considerable progress, lipid nanocarriers encounter various challenges in overcoming all biological barriers. The biodistribution and delivery of NPs encounter difficulties due to physical and biological obstacles, such as rapid absorption and elimination. These obstacles restrict the percentage of NPs reaching the intended therapeutic location. Also, different biological obstacles among diseases and within individual patients prevent the implementation of a standardized method. Understanding these biological obstacles is crucial for designing perfect LNPs, as they vary among diseases and manifest variably in individual patients at systemic, microenvironmental, and cellular levels. Administration routes and patient disease progression influence biological barriers. Local delivery methods can bypass systemic barriers but require more invasive procedures and complex techniques. These methods often require more complicated techniques [5,25].

Multiple parameters during circulation, including excretion, blood flow, and phagocytic cells, can substantially influence the stability and distribution effectiveness of NPs. The physicochemical characteristics of NPs affect these variables. The kidneys rapidly excrete nanoparticles with sizes under 10 nm; however, those over 200 nm may activate the complement system. Various nanoparticle formulations include polyethylene glycol (PEG) as a stealth coating to reduce fast excretion. PEGylation extends circulation time by changing the shape and solubility of NPs, while protecting the surface against enzymatic and antibody destruction (Figure 3.3). Nonetheless, this physical barrier does not entirely inhibit detection by macrophages or components of the immune system [26]. NPs in the bloodstream encounter mechanical stress from variable flow rates, possibly damaging

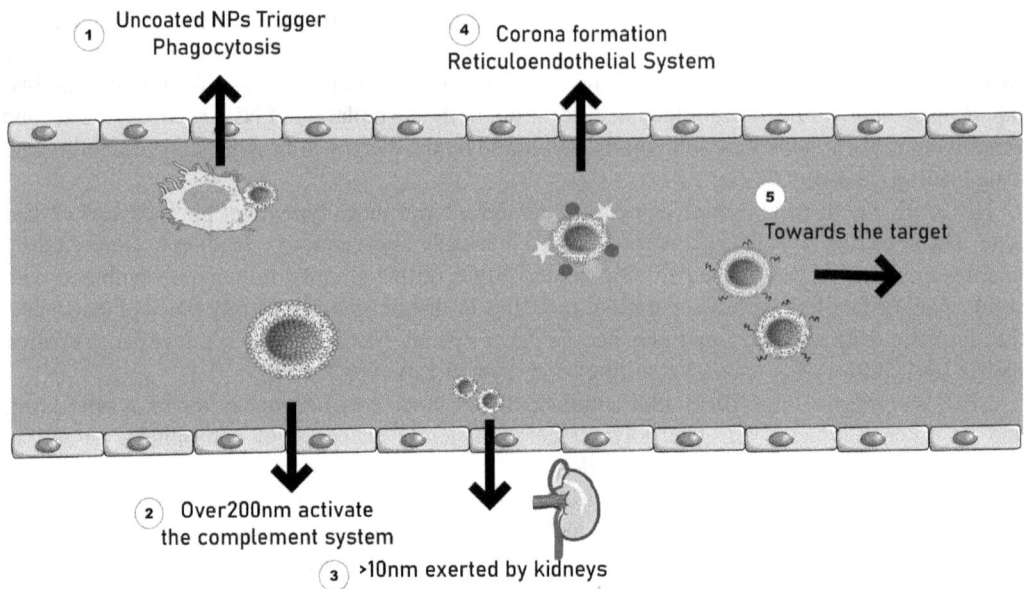

FIGURE 3.3 Biological barriers of delivery systems. 1: In the bloodstream, uncoated nanoparticles may be phagocytosed by the immune system. 2: Nanoparticles larger than 200 nm are removed from the bloodstream by the complement system. 3: Nanoparticles smaller than 10 nm are rapidly excreted through the kidneys. 4: Most nanoparticles, especially those that are not coated, tend to form a protein corona upon entering biological systems, which subsequently leads to their removal by the reticuloendothelial system. 5: Ultimately, a small number of nanoparticles that have the right size and the right coating can reach the target.

their integrity and hindering their escape from circulation. Non-spherical particles, including ellipsoids, disc-shaped nanoparticles, and nanorods with increased aspect ratios, show enhanced localization to blood vessels. The nonspecific attachment of serum proteins and lipids leads to the development of a protein corona on the NP's surface, impacted by compounds in the bloodstream and the characteristics of the NP's surface. The protein corona can affect nanoparticle biodistribution and stability. Integrating auxiliary lipids, cholesterol, and PEGylated lipids into lipid-based formulations can enhance their stability and prevent corona [15].

Vascular escape is an initial step, allowing NPs to reach target tissues in the bloodstream, notably dependent on NP size. Small NPs penetrate capillary walls more quickly. Pathological situations, such as the structure of tumor blood vessels, can modify this distribution by promoting the exit of larger NPs. NP systems have targeted moieties such as antibodies, glucose, transferrin, folate, transporters, and integrin ligands to reduce random distribution. These moieties precisely bind with target cell surface molecules, improving targeted delivery. Also, to make targeted delivery systems work better, it's essential to ensure that the NPs have the correct number of targeting moieties [27].

Another challenge in delivering NPs to the intended areas is the variability of microenvironments, including chemical conditions and physical barriers. Tumor microenvironments (TMEs) are generally acidic and possess extensive extracellular matrices (ECM) that obstruct NP infiltration. Specific features of NPs, including shape, size, composition, and surface charge, can increase their accumulation in malignancies. Upon reaching target cells, NPs face obstacles to uptake, such as the protein corona and membrane characteristics. Charge influences the interactions between NPs and cell surfaces, rejecting anionic particles and damaging membranes with excessively cationic ones. NPs often utilize endocytosis to cross cell membranes. NPs are surrounded by vesicles that turn into lysosomes during this process. The low pH and enzymes inside these vesicles could make the NPs less stable. So, making NPs that can get past these microenvironmental and intracellular barriers is essential for quickly getting them to the right organelles [1,27].

3.4 PERSONALIZED MEDICINE: TAILORING LIPID NANOPARTICLES FOR PATIENT-SPECIFIC THERAPIES

Personalized medicine is a customized management strategy designed to administer the appropriate medication at the ideal dosage for each patient.

Nanolipid particles must vary in size according to the specific tissue targets. LNPs, including liposomes, are generated via the self-assembly of phospholipids, with their size strictly adjusted by methods such as extrusion. The size of the NPs is accurately controlled using techniques such as extrusion. These NPs usually have a size range of 40 nm–1 μm and can be produced in different sizes according to application requirements. Surface modification/surface binding of molecular recognition elements such as peptides, antibodies, and aptamers enhances the performance of personalized medicine by facilitating the selective targeting of particular biomolecules [28–30].

Modifying lipid nanocarriers can be accomplished by altering their components, such as integrating surfactants into liposomes to control drug release, leading to personalized therapy. For example, in a study, the release characteristics of liposomal ciprofloxacin for inhalation were enhanced by incorporating surfactants [31]. Stimuli-responsive liposomal formulations are lipid-based NPs in personalized medicine that may release their payload regarding certain stimuli, such as temperature, light, pH, or biological signals. This targeted release mechanism optimizes drug delivery by ensuring therapeutic chemicals are released at specific sites or times, thus enhancing efficacy and minimizing side effects. pH-sensitive liposomes release their contents in reaction to pH fluctuations, facilitating targeted delivery to locations such as tumors, which exhibit a lower pH than normal tissues [32,33]. Mild hyperthermia (HT), which raises tissue temperature to 43°C, has been commonly used with chemotherapy and radiation therapy to enhance the efficiency of cancer therapy. When utilized with thermosensitive liposomes, HT enhances drug therapeutic efficacy through multiple mechanisms such as (1) regulating drug release into the tumor's vascular, (2) improving liposome accumulation in tumor tissue, (3) enhancing permeability of cell membranes and drug sensitivity, and (4) inducing direct cytotoxic effects on tumor cells [28,34].

Another advancement in the field of LNPs for personalized medicine is theranostics. This emerging science utilizes imaging technologies to determine the real-time efficacy and safety of treatments for intricate disorders, including cancer. Nano-theranostics encompasses both diagnostic and therapeutic applications. Theranostics attempts to improve targeted and individualized disease management by delivering therapies at the right moment and dosage, thus increasing efficacy and reducing hazards. In this regard, NPs function as a complex structure that can concurrently incorporate diverse substances, including therapeutic agents (e.g., chemotherapeutic agents, etc.) and various imaging agents (e.g., fluorescent polymers and fluorescent dyes). LNPs possess significant clinical importance and potential influence on theranostics by integrating diagnostic and therapeutic capabilities. This dual feature enables the accurate targeting of cancer cells, while reducing damage to whole tissues. LNPs may improve the distribution and efficacy of chemotherapeutic agents, reduce adverse effects, and increase imaging capacity, thus providing an integrated approach to cancer therapy. Implementing these NPs into clinical practice might alter cancer treatments by presenting patients with more effective and less invasive treatment options. Liposomes, which typically have a size of 50–250 nm and are either negatively charged or neutrally, are primarily employed in theranostics because of their exceptional stability. Many studies concentrate on the synthesis and enhancement of liposomes for theranostics. One study formulated tailored PEGylated liposome-indocyanine green to encapsulate doxorubicin (DOX) for both therapeutic and diagnostic applications [34–36]. Lipid nanotheranostics for cancer therapy encounter numerous problems, including needing individualized chemotherapy, compliance with specific size parameters (≤100 nm), and factors such as protein interaction, biodistribution, and toxicity. A significant challenge is the large-scale fabrication of intricate nanoplatforms, impeded by inconsistencies in physical and chemical attributes, low yields, and reproducibility of final characteristics. Nanotheranostics include diagnostic and therapeutic elements, which may require different delivery methods, necessitating a careful equilibrium to attain synergistic benefits. A critical issue is guaranteeing the safety profile of nanotheranostics in humans, necessitating extensive long-term patient monitoring during clinical trials. It is anticipated that these issues will be resolved in the future.

3.5 SYNERGISTIC STRATEGIES: COMBINING LNPs WITH OTHER NANOCARRIERS

The current version of NPs integrates the biomimetic characteristics of lipid materials with the biomechanical benefits of polymeric materials. These systems incorporate different lipids and polymers, each possessing unique attributes like structure, size, and charge, thereby enabling the encapsulation of a broad spectrum of active chemicals. Lipid–polymer nanoparticles (LPNs), such as polymer-modified liposomes, solid surface-modified lipid nanoparticles, polymer micelles, LPHNs, and various lipid composites, exhibit distinctive properties that improve their performance. Hybrid NPs exhibit considerable biomedical potential, especially regarding therapeutic efficacy, safety, targeted therapy, stimulus-responsive release, and reduced toxicity and adverse effects [37]. The reason for using hybrid nanoparticles originates from the unique characteristics of the polymers. PNs are biodegradable. They can use many structures, including core-shell structures, solid nanoparticles, polymeric micelles, and polyplexes. PNs have better stability than LNPs, providing prolonged drug delivery and exhibiting stimuli-responsive capabilities (e.g., temperature and pH). These features offer precise control over the intracellular release of pharmaceutical compounds. Nonetheless, PNs often demonstrate poor biocompatibility and lower affinity for cell membranes, leading to impaired drug delivery effectiveness [38]. Recent attempts have focused on integrating the benefits of lipid materials and polymeric into a unified, innovative carrier system. This has been performed using suitable methods and precisely selecting biocompatible lipid–polymer combinations. These combinations guarantee strong affinity with cell membranes, provide regulated drug discharge overextended durations, and permit the co-encapsulation of drugs with different properties. These carriers can load and carry various functional molecules, including anticancer drugs, peptides, genetic material, vitamins, metallic inclusions, and other therapies [39]. Lipid–polymeric technology refers to the adjustment of specific lipids, like polyethyleneimine stearate and PEG lipid, as well as the modification of whole lipid/polymeric particles. This leads to the development of hybrid systems that, based on their structure and manufacturing method, can be categorized into three primary types:

1. Monolithic matrix lipid polymeric NPs: These comprise a lipid phase with a uniformly distributed polymeric drug complex or a polymeric core enveloped by a lipid shell.
2. Core–shell lipid polymeric NPs are defined as a polymeric core enveloped by a lipid shell or a lipid core encased in a polymer shell.
3. Polymer-decorated liposomes: These liposomes possess surfaces embellished with polymeric substances [40].

LPHNPs are attractive carriers for cancer treatments due to their considerable load capacity for different drugs, exceptional stability in circulation, and efficient in vivo drug delivery characteristics. These NPs provide programmable release kinetics while reducing off-target effects. Their unique structure encapsulates many drugs, allowing the co-encapsulation of many drugs to benefit from synergistic effects. Optimizing loading techniques can enhance the co-entrapment efficiency of hydrophilic and hydrophobic drugs, which is necessary for co-administration strategies to battle complex disorders such as cancer. Improvements in manufacturing processes are required for the industrial-scale application of LPN-functionalized delivery systems. Existing production constraints stem from traditional methods that fail to satisfy industrial requirements. Conventional batch procedures are limited in scale and labor-intensive and insufficiently regulate the end product's characteristics. Consequently, there is an immediate need for more adaptable and economical methods that facilitate ongoing, large-scale production [37,38,41–43].

3.6 LIPID NANOPARTICLES IN ACTION: CASE STUDIES AND CLINICAL APPLICATIONS

More than 50 nano-pharmaceuticals have successfully entered the market, with LNPs establishing themselves as the predominant approach. Despite more than 30 years of investigation on liposomes as drug delivery systems, the first FDA-approved liposomal formulation, Doxil®, was not realized until the 1990s. Doxil is a stealth liposome that encapsulates DOX and is used for treating ovarian cancer, metastatic breast cancer, and many myelomas. Encapsulation in PEGylated liposomes markedly diminishes the detrimental effects of free DOX, including congestive heart failure and persistent cardiomyopathy. Doxil exhibits less cardiotoxicity and higher anticancer activity relative to free DOX due to the improved retention effect and permeability [44,45].

NeXstar Pharmaceuticals USA introduced DaunoXome®, a liposomal version of daunorubicin (DNR) citrate, which received U.S. FDA approval in 1996 to treat advanced HIV-associated Kaposi's sarcoma. DaunoXome® is an intravenous infusion solution that is free of preservatives, germs, and pyrogens, containing 50 mg of DNR base encapsulated in liposomes made up of cholesterol and distearoylphosphatidylcholine (DSPC) in a 2:1 molar ratio, with an average particle size of about 45 nm. This formulation enhances the stability of encapsulated DNR and reduces its rapid metabolism and corona development.

Other liposomal formulations have been developed for oncological treatment, including Onivyde™, approved for metastatic pancreatic adenocarcinoma. Liposomal formulations have been created for multiple diseases, such as Abelcet® and Ambisome® for fungal infections, as well as liposomal vaccinations like Epaxal® and Inflexal® V for the prevention of hepatitis and influenza. Moreover, liposomes have been used to synthesize a range of medications, such as cytarabine, verteporfin, morphine, and irinotecan. Important examples comprise Curosurf®, a liposomal formulation for the management of respiratory distress syndrome in premature newborns, and Vyxeos®, which combines cytarabine and DNR inside liposomes for the therapeutic management of acute myeloid leukemia. These enhancements emphasize the adaptability of liposomal technology in optimizing drug delivery and improving clinical outcomes across different diseases [11].

Nucleic acid medications, such as small activating RNAs (saRNA), small interfering RNAs (siRNA), and messenger RNA (mRNA), have attracted considerable interest for their possible uses in disease treatment. These oligonucleotides enhance therapeutic values relative to traditional chemotherapeutics, especially when designed for focused tissue delivery. The vulnerability of naked nucleic acids is a problem for gene delivery. Formulations providing in vivo stability and tissue-specific targeting have been designed to overcome this challenge.

The FDA approved Onpattro® in 2018, a medication that Alnylam Pharmaceuticals and Sanofi Genzyme produced. Onpattro comprises siRNA encapsulated in LNP and is authorized for treating polyneuropathy in individuals with hereditary transthyretin-mediated amyloidosis. The LNP formulation comprises DLin-MC3-DMA lipid, DSPC, cholesterol, and PEG-DMG (1,2-dimyristoyl-rac-glycero-3-methoxypolyethylene glycol), which guides the LNPs to liver hepatocytes, hence augmenting siRNA targeting efficacy [46].

Several NPs have been produced and employed in preclinical and clinical research. With the advancement of research, these lipid-based nanocarriers are expected to advance to clinical trials and eventually be utilized in clinical practice.

3.7 OVERCOMING CANCER DRUG RESISTANCE WITH LNPs

LNPs represent a significant advancement in nanomedicine, offering improved treatment outcomes for resistant types of cancer [47,48]. One of the outstanding attributes of LNPs is the improvement in the solubility and stability of the associated drugs, which makes LNPs more appropriate for drug-resistant cancer cells.

A commonly reported drug-resistant mechanism is the hyperactivity of efflux transporters, such as P-glycoprotein (P-gp), which pump chemotherapeutic drugs out of cancer cells [49]. Through the design of LNPs, it is possible to co-deliver P-gp inhibitors and chemotherapeutic agents, enhancing the intracellular concentration of the drugs and thus overcoming P-gp-based resistance. For example, a study showed that DOX and Edelfosine conjugated LPHNs displayed synergistic anticancer activity against drug-resistant osteosarcoma [47]. This approach not only increases the deposition of drugs within the tumor but can also provide long-lasting release of the drugs, increasing their effectiveness and reducing side effects.

Besides their capacity to encapsulate and deliver both the drug and the targeting moiety, LNPs can also be designed to exhibit a response to a particular trigger within the TME. For example, another study recently described chemoimmunotherapy with thermosensitive exosome-liposome hybrid NPs for enhancing the loading efficiency of metastatic peritoneal cancer [50]. Combination therapy uses nanoparticles to improve drug delivery by utilizing specific properties from the tumor-specific microenvironment, such as temperature and pH, thereby increasing effectiveness against resistant tumors.

Also, LNPs could be effectively used to deliver mRNAs to induce the cancer immune response, pointing out their capacity to boost the action against tumors [51]. LNPs containing mRNA that encodes tumor antigens stimulate the immune system and can bypass the immunosuppressive environment characteristic of drug-resistant malignancies [51,52]. Figure 3.4 illustrates the mechanism of LNPs in overcoming cancer drug resistance.

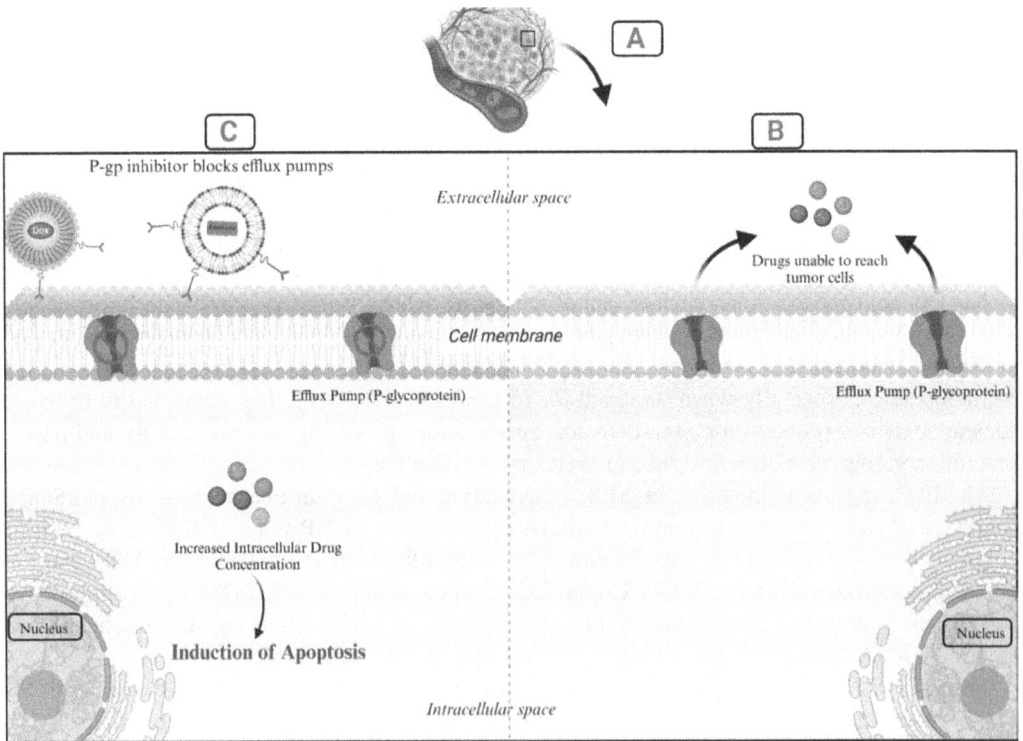

FIGURE 3.4 Mechanism of LNPs in overcoming cancer drug resistance. (a) Schematic representation of cancer cells in EMT. (b) Cancer cells with overactive efflux pumps (P-gp) on the membrane expel chemotherapeutic drugs. (c) LNPs deliver chemotherapeutic drugs and P-gp inhibitors, blocking efflux pumps, increasing intracellular drug concentration and accumulation, leading to apoptosis in cancer cells, overcoming drug resistance, and improving therapeutic outcomes. LNPs are engineered to overcome drug resistance in cancer cells by encapsulating chemotherapeutic drugs (e.g., DOX) and efflux pump inhibitors (e.g., edelfosine) within their lipid bilayer. The LNPs fuse with the cancer cell membrane, releasing their payload into the cell. The efflux pump inhibitor blocks P-gp, a key efflux transporter overexpressed in drug-resistant cancer cells, preventing the expulsion of the chemotherapeutic drug. This results in increased intracellular drug concentration, enhanced cytotoxicity, and induction of apoptosis. (Created with BioRender.)

3.7.1 siRNA Delivery to Target Drug Resistance

siRNA can silence specific genes responsible for drug resistance, such as those encoding efflux transporters like P-gp. By reducing the expression of these transporters, LNP-mediated siRNA delivery elevates the intracellular concentration of chemotherapeutic drugs, enhancing their cytotoxicity. Preclinical models have demonstrated significant efficacy using this approach, suggesting its potential for clinical translation. A study reported that siRNA-loaded LNP targeting P-gp achieved a marked reduction in drug resistance, with increased apoptosis in resistant cancer cells [51].

3.7.2 Stimuli-Responsive LNPs

Stimuli-responsive LNPs exhibit biocompatibility, maintain stability during circulation, and selectively release therapeutic agents only in tumor regions. This precise delivery minimizes adverse effects and enhances the therapeutic efficacy of chemotherapy [52,53].

It also underlies a novel stimuli-responsive lipid nanoparticle (SRLNP), which is considered a powerful strategy to reverse drug resistance in chemotherapy [54,55]. The SRLNP's responsiveness to particular elements of the TME or inside cancer cells provides specific drug delivery, which increases treatment effectiveness and eliminates or diminishes toxicity side effects [56].

SRLNPs can skip efflux pumps, increase drug permeation, and increase drug accumulation in tumor sites, which are reasonable solutions to obstacles associated with cancer drug resistance [57]. For example, LPHNs with magnetic beads are shown to have near-infrared photothermal agent-triggered drug release and a marked inhibition of cancer cell growth [58]. This targeted approach enhances nanomedicine drug delivery systems' pharmacokinetics and therapeutic utility. However, there are limitations tied to SRLNPs when considering the application of these systems in clinical practice, such as stability, reproducibility, and production costs.

3.8 IMMUNOMODULATION WITH LNPs: ENHANCING VACCINE EFFICACY

This vaccine delivery technique, known as LNPs, has gained prevalence due to its ability to enhance vaccination efficacy through compelling administration. They can effectively encapsulate various medicinal chemicals and mRNA, an essential element in modern vaccines. LNPs primarily enhance vaccination effectiveness in multiple ways. The main role of LNPs is to transport mRNA and other nucleic acids across cell membranes, while simultaneously ensuring mRNA stability to prevent degradation by cellular enzymes. Ion-conducting lipids, neutral lipids, and PEG lipids help make stable nanoparticles that can get through cell membranes [54,55].

Since LNPs have immunostimulatory properties, they can be seriously considered adjuvants for vaccines. Medical research has also demonstrated that LNP formulations can stimulate T follicular helper (Tfh) cells and also improve humoral immunity by stimulating the formation of long-lived plasma cells and germinal center B cells. The ionizable lipid ingredient found in LNPs is essential for this adjuvant effect as it triggers pro-inflammatory cytokines like IL-6 and further amplifies immune responses [59,60].

Besides infectious diseases, LNPs are also considered in cancer immunotherapy. They can directly translocate mRNA encoding tumor-associated antigens into dendritic cells and significantly improve the presentation of these antigens and the subsequent immune response against tumors [61–63]. This approach seems to have the potential to create specific cancer vaccines based on patient characteristics.

3.8.1 Mechanisms of Immune Activation by LNPs

In several methods, LNPs have served as a critical component of mRNA vaccines and immune response modulation. This subsection explains four significant ways LNPs promote and modulate immune responses, including mRNA protection and delivery, cellular uptake, and endosomal release, activation of innate immunity, and the role of ionizable lipids.

3.8.1.1   Protection and Delivery of mRNA

The LNPs function as active components, preventing extracellular nucleases from breaking down mRNA. They need to target specific cell types, like dendritic cells and macrophages. The incorporation of mRNA into LNPs not only protects it from degradation by nucleases and enables it to cross physiological barriers [64,65]. The relationship between the ionizable lipids in LNPs and the negatively charged phosphate backbone of mRNA is critical for this process because it stabilizes stable complexes that effectively transport mRNA into target cells [66].

3.8.1.2   Cellular Uptake and Endosomal Escape

Following the internalization of LNPs by immune cells, they must be released from endosomes containing the mRNA payload to facilitate the delivery of the appropriate protein. The ionizable lipids in LNPs are especially significant in this context. The optimized pKa values enhance protonation in the acidic environment of the endosome, resulting in membrane disruption and the release of mRNA into the cytoplasm [66,67].

3.8.1.3   Stimulation of Innate Immunity

LNPs act not merely as vectors for nucleic acid delivery but also enhance innate immune signaling. Following exposure to immune cells, LNPs can trigger the generation of pro-inflammatory cytokines like as TNF-α and IL-6, essential for an elevated immune response [68,69]. This activation is necessary to initiate specific immunity, T cell, and B cell activation. The nature of lipid components in the formulation of LNPs can also influence these immune responses and, thus, improve the adjuvanticity [70].

3.8.1.4   Role of Ionizable Lipids

Among the elements of LNP structure, ionizable lipids play an essential role and are responsible for adjuvant effects. These lipids also contribute to the stabilization of the NPs and improve immune responses as they integrate with cellular membranes [64,66]. The fact that ionizable lipids can be protonated and deprotonated depending on the pH value is central to further enhancing mRNA delivery and the immune response. For instance, a higher pKa of ionizable lipids indicates better endosomal escape and enhanced immunogenicity of modern vaccines [66,67]. This characteristic enables the LNPs to have a remarkable ability in immunomodulation, and they make a perfect interaction with immune cells, vital to the efficiency of the mRNA vaccines.

3.9   REGULATORY AND SAFETY CONSIDERATIONS: NAVIGATING THE APPROVAL LANDSCAPE

The issue of regulation becomes much more critical when new forms of drug delivery, such as LNPs, are developed, especially in the context of mRNA vaccines. This section will focus on three crucial areas: manufacturing processes are highly standardized, the need for preclinical and clinical trials, and regulation by authorities, including the FDA.

3.9.1   Standardization of Manufacturing Processes

One of the critical challenges in developing LNPs is the need to attain the requisite stability and reproducibility of the production procedures. The FDA and similar regulatory agencies demand a comprehensive description of LNPs to know their quality and efficacy. These include determining their physicochemical characteristics, stability, and interactions with biological systems [71,72]. The production of LNPs requires high manufacturing standards to minimize batch variability, which can significantly affect the effectiveness and safety of the medications. The lipid formulation method, preparation temperature, and mixing techniques must be optimized to produce nanoparticles with specified characteristics [73,74]. Microfluidics is one of the crucial developments in manufacturing technologies recently adopted to improve the reproducibility of LNP manufacturing. The repeated microfluidic operations enable the well-defined lipid–aqueous phase emulsification procedures to be conveniently managed, which resulted in more uniform particle size and better scalable manufacturing [74,75]. Microfluidic technology can exert control over the lipid and aqueous phase diffusion, which is a critical step in forming well-defined LNPs. The systemic differences in mixing and processing conditions prevalent in lengthy traditional processes impede the ability to standardize particle size in emulsions, as exemplified by lipid film hydration and extrusion [75,76]. Microfluidics solves this problem by allowing the continuous flow of fluids in microscale channels where laminar flow characteristics make mixing at a much smaller scale than bulk operations possible [73,77]. This leads to LNPs with regulated sizes, typically less than 100 nm, and suitable for drug delivery and sterile filtration [78,79]. The flexibility of controlling other process parameters, including flow rate, lipid concentration, and different mixtures, enables precise tuning of the LNP properties, including the entrapment efficiency and stability [80].

Incorporating microfluidics in the production of LNP complies with the regulatory requirements of Good Manufacturing Practices (GMP) because it involves high degrees of process standardization and quality controls. GMP guidelines argue that the manufacturing processes should be adequately described and controlled with the purpose of products meeting the required quality standard [81]. Microfluidics naturally provides for these requirements because LNP synthesis requires a more controlled environment than traditional batch processes [82]. Furthermore, the design of the microfluidic systems can be integrated with specific features for monitoring flow rate and temperature, which are effective process parameters, as well as controlling them in situ during the LNP production process [83]. Additionally, process analytical technologies (PAT) integration to microfluidic LNP manufacturing increases the capability for real-time measurement of critical quality attributes (CQAs). PAT tools allow for measuring parameters such as zeta potential, particle size, and EE in real time during product manufacturing [84].

3.9.2 Preclinical and Clinical Testing

Extensive preparatory work is required in preclinical studies to assess the efficacy and safety of LNP-based systems before advancing them to clinical trials. This subsection includes various in vitro and in vivo investigations to explain essential variables such as toxicity, immunogenicity, and the primary distribution of the drug. These studies are crucial to determining the safety profile of LNPs and avoiding adverse immune reactions.

3.9.2.1 Preclinical Testing: In Vitro and In Vivo Studies

In vitro experiments are crucial to preclinical testing, examining the interactions between LNPs and cellular biology. These tests determine the composition's cytotoxic and immunogenic properties. For instance, cytotoxicity assessment involves exposing an LNP to different concentrations in different cell lines and assessing cell survival [85,86]. In vivo investigations are essential for understanding the pharmacokinetics and biodistribution of LNPs, as they involve animal models to observe the dissemination of drugs and their cellular binding affinities [87,88].

3.9.2.2 Clinical Trials Involving LNPs

LNPs are also being explored in terms of the role of cancer treatment. For instance, a liposomal RNA, RNA-LPX formulation, called Lipo-MERIT, is planned for a phase 1 clinical trial in patients diagnosed with stage III/IV melanoma [89]. The current investigational study seeks to evaluate the safety profile and therapeutic outcomes of utilizing LNPs to transfect RNA-based therapies to tumorigenic cells. Since LNPs can enhance the aggregation of therapeutic agents at the tumor site, they could be a valuable resource in cancer treatment.

3.9.3   Safety and Biocompatibility of Lipid Nanoparticles

3.9.3.1   Biocompatibility and Toxicity

Lipid composition, surface charge, and particle size significantly influence the biocompatibility of LNPs. Different lipid components, such as cationic lipids, can dramatically affect their toxicity range, which can induce cytotoxicity due to the interaction between charged lipids and cell membranes [90]. Particle size considerably influences the biocompatibility of LNPs, with smaller particles being more immunogenic and more internalized by cells than larger particles of 100 nm [91]. Figure 3.5 outlines the nanoparticle storage and stability strategies and the quality control measures involved in maintaining the stability of LNPs.

3.9.3.2   Biodistribution and Off-Target Effects

The biodistribution of LNPs is crucial for improving formulations and reducing off-target delivery effects. Therapeutic effects depend on cell types and tissue distribution. Surface charge and hydrophilicity affect LNPs' interaction with serum proteins, in turn affecting their half-life and buildup in organisms [87]. According to some researchers, LNP biodistribution can lead to their deposition in non-target organs, potentially causing organ toxicity, such as the liver, spleen, or lungs [92].

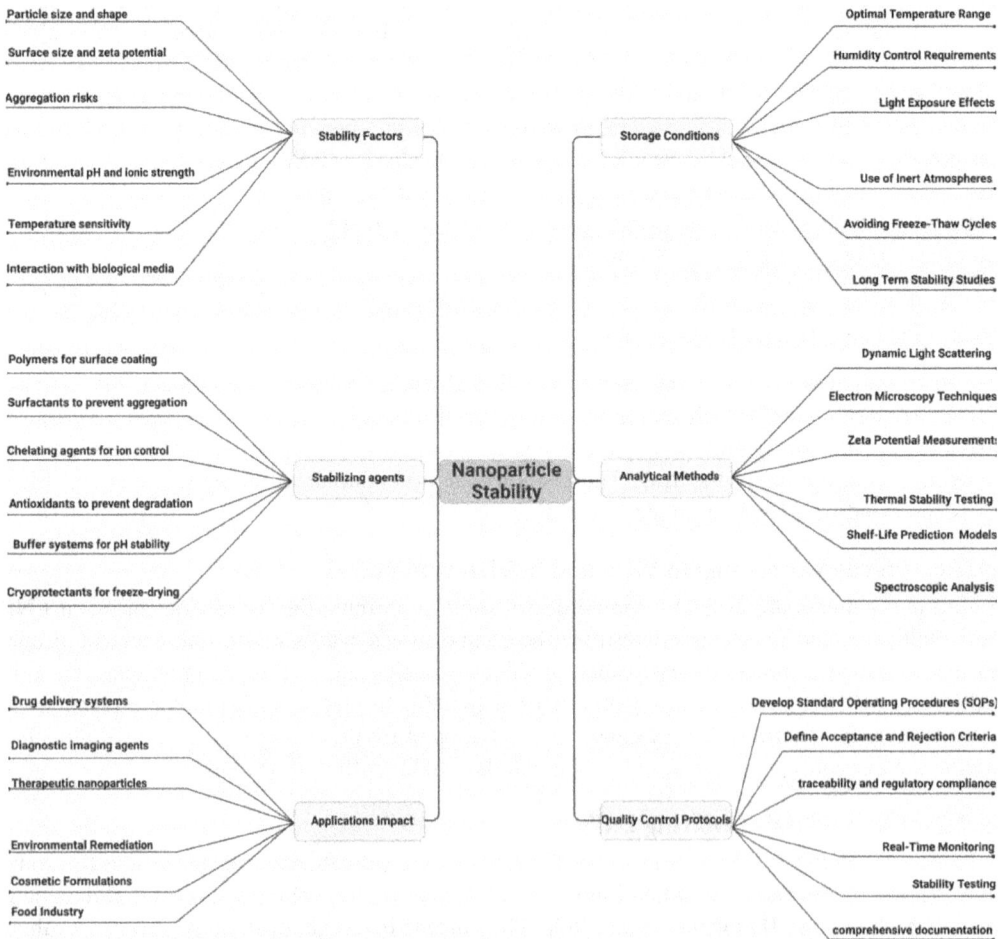

FIGURE 3.5   This flowchart outlines the nanoparticle stability and storage, strategies, and quality control measures involved in maintaining the stability of LNPs (Created with BioRender.)

3.9.3.3 Long-term Safety

Extended toxicity assessments are required for LNPs in drug delivery to determine chronic toxicity, immunogenicity, and other effects. Short-term investigations reveal acute impacts, multiple doses causing anti-PEG antibodies, and long-term safety issues [93].

3.10 INNOVATIONS ON THE HORIZON FOR LNPs

LNPs represent a disruptive technology in developing epigenetic therapies by offering new opportunities for gene expression modulation and correction of epigenetic alterations linked to disease. The following section shows the future of LNPs in epigenetic modulator delivery, RNA-based epigenetic editing, combination therapies, challenges, and opportunities in this fast-moving area.

3.10.1 Delivery of Epigenetic Modulators

LNPs may serve as a delivery vehicle for epigenetic drugs, including DNA methyltransferase inhibitors (Azacitidine) and histone deacetylase inhibitors (Vorinostat). The drugs correct gene expression, reversing the aberrant epigenetic modifications usually associated with many diseases, including cancers, neurological disorders, and autoimmune diseases [94]. LNPs may deliver histone deacetylase inhibitors, leading to the restoration of typical patterns of histone acetylation and making the way to re-expressing genes that are important in cellular functions and maintaining homeostasis [95].

3.10.2 Combination Therapies

The versatility of LNPs, like other nanoparticles, originates from their capacity to co-deliver several therapeutic agents: epigenetic modulators, chemotherapeutic medicines, and immunotherapeutics. A combinatorial approach will be considered to enhance treatment efficacy, therefore eliminating resistance mechanisms prevalent in cancers and other complex disorders. Consequently, LNPs can be engineered to encapsulate only chemotherapeutic drugs alongside an epigenetic modifier, thereby enhancing synergy and improving therapeutic efficacy [96,97].

Although LNPs hold great potential for epigenetic therapy, several challenges must be addressed regarding their optimization for efficacy and safety. A better understanding of the underlying epigenetic mechanisms will be essential to optimize LNP-based therapies, including how various epigenetic regulators play their roles and interact with LNP information that can be used to design more effective therapeutic strategies [98]. LNPs remain one of the most powerful tools for developing epigenetic therapies, including delivery approaches based on epigenetic modulators, RNA-based editing, and even combination therapies.

3.11 NANOPARTICLE STABILITY AND STORAGE: ENSURING EFFICACY OVER TIME

Maintaining the stability of the LNPs during storage and transport will be essential for their successful application in drug delivery systems, which include mRNA vaccines and other therapeutic agents. Some of the main problems at this step include degradation, aggregation, and alteration of some physicochemical properties.

3.11.1 Stability Challenges

LNPs are degraded by environmental conditions, including temperature, light, and oxygen exposure. The lipids, especially unsaturated lipids, are easily oxidized; this may result in the formation of reactive oxygen species that may affect the integrity of nanoparticles and therapeutic agents encapsulated within the particles [99]. This degradation can result in decreased activity and even produce toxicity; thus, efforts are now underway to enhance the stability of lipid nanoparticle formulations.

3.11.1.1 Aggregation

Another serious problem that may arise with storage is the aggregation of NPs. That might, in turn, impact changes in size distribution and particle size, affecting the pharmacokinetics and biodistribution of LNPs [100]. Ionic strength, pH, and the presence of surfactants can also affect LNP stability; thus, they are essential parameters to be optimized during formulation development to minimize aggregation [101].

3.11.2 STRATEGIES TO ENHANCE STABILITY

3.11.2.1 Refining Formulation Ingredients

While enhancing the stability of lipid nanoparticles depends primarily on the composition, this will mainly entail a proper choice of adequate lipids and surfactants that are typically included to create an environment of excipients, helping to stabilize these NPs. More particularly, when incorporated into LNPs, the formulation of antioxidants strongly impedes oxidized-induced particle deterioration [102]. Similarly, adding stabilizers such as polyethylene glycol may contribute to increased steric stability, reducing the aggregation of LNPs [103].

3.11.2.2 Regulated Environmental Parameters

LNPs require specific conditions for stability. A significant reduction in degradation rate at around -80°C helps preserve the efficacy of LNPs for an extended duration [104]. However, this requires handling logistics concerning the distribution and administration of such particles. To that end, various efforts are underway to produce modified versions that can stabilize at higher temperatures [105].

3.11.2.3 Application of Cryoprotectants

Cryoprotectants can enhance the stability of LNPs during freeze-drying or during long-term storage. Such materials may protect NPs from the action of freeze-thaw that may damage them, thereby preserving their structure and function [106]. It has been noticed that adding cryoprotectants improves both the recovery and stability of LNPs after being stored, which means that the therapeutic activity of such LNPs is preserved upon their reconstitution [107].

3.11.2.4 Characterization and Quality Control

Dynamic light scattering (DLS) measures the size distribution of nanoparticles suspended in a solution. DLS uses the variation of scattered light caused by Brownian motion fluctuations of particles in a liquid. This information is essential to predict the hydrodynamic diameter, a critical parameter of LNPs concerning their behavior in biologics. DLS also offers the possibility to determine agglomeration or size changes necessary for the stability assessment of LNP formulations [108]. Also, transmission electron microscopy (TEM) has a high resolution for imaging LNPs for in-depth morphological and structural information on their size. It will show nanoparticles' shape, size, and internal structure, which has an available purpose to determine their interaction with biological systems. This approach was especially beneficial in observing the homogeneity of LNP formulations and quantifying morphological artifacts that may impact their function [109]. The zeta potential defines the surface charge of NPs. A more considerable absolute zeta potential (negative or positive) indicates stronger electrostatic repulsion between particles, which can be used to prevent particles from aggregating. Measuring zeta potential over time can inform the stability of LNP formulations and their interactions with biological materials (e.g., proteins and cells) [110].

3.12 THE NEED FOR QUALITY CONTROL PROTOCOLS

There are substantial quality control protocols necessary for ensuring LNP quality throughout its shelf life. Implementation of Standard Operating Procedures (SOPs) for characterization and quality assessment of LNPs will promote consistency and reproducibility in the tests. These protocols and procedures must specify analytical methods along with the way of sample preparation and include a set of acceptance or rejection criteria by CQAs [111]. Periodic stability testing is important to understand the long-term behavior of LNPs from the perspective of pharmaceutical formulation. It studies the role of storage environments, such as temperature and light exposure, during the degradation or development of the chemical and physical properties of the LNP over time. Stability studies must adhere to the regulatory guidance established by the International Council for Harmonisation (ICH) [112]. Proper and complete documentation on all activities carried out in characterization and quality control may guarantee traceability and use in compliance with regulatory agencies. Relevant points include analytical tests carried out, variations to SOPs, and corrective actions taken to ensure the quality and safety of LNP compositions visible to regulatory agencies [113].

It will be important to establish robust quality control protocols that can ensure LNPs meet specifications throughout shelf life. This presents several challenges with respect to ensuring the stability and efficacy of LNP during storage and transportation, including degradation, aggregation, and phase transitions. However, these are overcome by optimized formulation components, controlled storage conditions, cryoprotectants, and rigorous characterization, together with quality control. Since this technology is in the process of ongoing development, securing the stability of the proposed systems will be relevant for successful clinical use related to targeted drug delivery and personalized medicine.

REFERENCES

1. Joudeh N, Linke D. Nanoparticle Classification, Physicochemical Properties, Characterization, and Applications: A Comprehensive Review for Biologists. *Journal of Nanobiotechnology*. 2022;20(1):262.
2. Mitchell MJ, Billingsley MM, Haley RM, et al. Engineering Precision Nanoparticles for Drug Delivery. *Nature Reviews Drug Discovery*. 2021;20(2):101–124.
3. Xu Y, Fourniols T, Labrak Y, et al. Surface Modification of Lipid-Based Nanoparticles. *ACS Nano*. 2022;16(5):7168–7196.
4. Hald Albertsen C, Kulkarni JA, Witzigmann D, et al. The Role of Lipid Components in Lipid Nanoparticles for Vaccines and Gene Therapy. *Advanced Drug Delivery Reviews*. 2022;188:114416.
5. Andhari S, Gupta R, Khandare J. Drug Delivery Systems: Lipid Nanoparticles Technology in Clinic. In: Jagadeesh G, Balakumar P, Senatore F, editors. *The Quintessence of Basic and Clinical Research and Scientific Publishing*. Singapore: Springer Nature Singapore; 2023. pp. 181–200.
6. Battaglia L, Gallarate M. Lipid Nanoparticles: State of the Art, New Preparation Methods and Challenges in Drug Delivery. *Expert Opinion on Drug Delivery*. 2012;9(5):497–508.
7. Dabholkar N, Waghule T, Krishna Rapalli V, et al. Lipid Shell Lipid Nanocapsules as Smart Generation Lipid Nanocarriers. *Journal of Molecular Liquids*. 2021;339:117145.
8. Urimi D, Hellsing M, Mahmoudi N, et al. Structural Characterization Study of a Lipid Nanocapsule Formulation Intended for Drug Delivery Applications Using Small-Angle Scattering Techniques. *Molecular Pharmaceutics*. 2022;19(4):1068–1077.
9. Xu L, Wang X, Liu Y, et al. Lipid Nanoparticles for Drug Delivery. *Advanced NanoBiomed Research*. 2022;2(2):2100109.
10. Batur E, Özdemir S, Durgun ME, et al. Vesicular Drug Delivery Systems: Promising Approaches in Ocular Drug Delivery. *Pharmaceuticals*. 2024;17(4):511.
11. Bulbake U, Doppalapudi S, Kommineni N, et al. Liposomal Formulations in Clinical Use: An Updated Review. *Pharmaceutics*. 2017;9(2):12.
12. Filipczak N, Pan J, Yalamarty SSK, et al. Recent Advancements in Liposome Technology. *Advanced Drug Delivery Reviews*. 2020;156:4–22.
13. Nemati M, Fathi-Azarbayjani A, Al-Salami H, et al. Bile Acid-Based Advanced Drug Delivery Systems, Bilosomes and Micelles as Novel Carriers for Therapeutics. *Cell Biochemistry and Function*. 2022;40(6):623–635.

14. Nwabuife JC, Hassan D, Madhaorao Pant A, et al. Novel Vancomycin Free Base – Sterosomes for Combating Diseases Caused by *Staphylococcus aureus* and Methicillin-Resistant *Staphylococcus aureus* Infections (S. Aureus and MRSA). *Journal of Drug Delivery Science and Technology.* 2023;79:104089.

15. Cieślak A, Wauthoz N, Nieto Orellana A, et al. Stealth Nanocarriers Based Sterosomes Using PEG Post-insertion Process. *European Journal of Pharmaceutics and Biopharmaceutics.* 2017;115:31–38.

16. Cui Z-K, Fan J, Kim S, et al. Delivery of siRNA via Cationic Sterosomes to Enhance Osteogenic Differentiation of Mesenchymal Stem Cells. *Journal of Controlled Release.* 2015;217:42–52.

17. Cui ZK, Kim S, Baljon JJ, et al. Design and Characterization of a Therapeutic Non-phospholipid Liposomal Nanocarrier with Osteoinductive Characteristics to Promote Bone Formation. *ACS Nano.* 2017;11(8):8055–8063.

18. Bukke SPN, Venkatesh C, Bandenahalli Rajanna S, et al. Solid Lipid Nanocarriers for Drug Delivery: Design Innovations and Characterization Strategies—A Comprehensive Review. *Discover Applied Sciences.* 2024;6(6):279.

19. Mehnert W, Mäder K. Solid Lipid Nanoparticles: Production, Characterization and Applications. *Advanced Drug Delivery Reviews.* 2012;64:83–101.

20. El Moukhtari SH, Garbayo E, Amundarain A, et al. Lipid Nanoparticles for siRNA Delivery in Cancer Treatment. *Journal of Controlled Release.* 2023;361:130–146.

21. Viegas C, Patrício AB, Prata JM, et al. Solid Lipid Nanoparticles vs. Nanostructured Lipid Carriers: A Comparative Review. *Pharmaceutics.* 2023;15(6):1593.

22. Müller RH, Radtke M, Wissing SA. Nanostructured Lipid Matrices for Improved Microencapsulation of Drugs. *International Journal of Pharmaceutics.* 2002;242(1):121–128.

23. Scioli Montoto S, Muraca G, Ruiz ME. Solid Lipid Nanoparticles for Drug Delivery: Pharmacological and Biopharmaceutical Aspects [Review]. *Frontiers in Molecular Biosciences.* 2020;7:587997.

24. Dave V, Tak K, Sohgaura A, et al. Lipid-Polymer Hybrid Nanoparticles: Synthesis Strategies and Biomedical Applications. *Journal of Microbiological Methods.* 2019;160:130–142.

25. Antimisiaris SG, Marazioti A, Kannavou M, et al. Overcoming Barriers by Local Drug Delivery with Liposomes. *Advanced Drug Delivery Reviews.* 2021;174:53–86.

26. Saffari M, Moghimi HR, Dass CR. Barriers to Liposomal Gene Delivery: From Application Site to the Target. *Iranian Journal of Pharmaceutical Research.* 2016;15(Suppl):3–17.

27. Liu C, Zhang L, Zhu W, et al. Barriers and Strategies of Cationic Liposomes for Cancer Gene Therapy. *Molecular Therapy Methods & Clinical Development.* 2020;18:751–764.

28. Agrawal SS, Baliga V, Londhe VY. Liposomal Formulations: A Recent Update. *Pharmaceutics.* 2025;17(1):36.

29. Nakamura N, Ohta S. Precise Control Methods of the Physicochemical Properties of Nanoparticles for Personalized Medicine. *Current Opinion in Biotechnology.* 2024;87:103108.

30. Colapicchioni V, Tilio M, Digiacomo L, et al. Personalized Liposome-Protein Corona in the Blood of Breast, Gastric and Pancreatic Cancer Patients. *International Journal of Biochemistry and Cell Biology.* 2016;75:180–187.

31. Cipolla D, Wu H, Gonda I, et al. Modifying the Release Properties of Liposomes Toward Personalized Medicine. *Journal of Pharmaceutical Sciences.* 2014;103(6):1851–1862.

32. Abri Aghdam M, Bagheri R, Mosafer J, et al. Recent Advances on Thermosensitive and pH-Sensitive Liposomes Employed in Controlled Release. *Journal of Controlled Release.* 2019;315:1–22.

33. Kim Y, Oh KT, Youn YS, et al. pH-Sensitive Twin Liposomes Containing Quercetin and Laccase for Tumor Therapy. *Biomacromolecules.* 2022;23(9):3688–3697.

34. Giordano A, Provenza AC, Reverchon G, et al. Lipid-Based Nanocarriers: Bridging Diagnosis and Cancer Therapy. *Pharmaceutics.* 2024;16(9):1158.

35. Kevadiya BD, Ottemann BM, Thomas MB, et al. Neurotheranostics as Personalized Medicines. *Advanced Drug Delivery Reviews.* 2019;148:252–289.

36. Mura S, Couvreur P. Nanotheranostics for Personalized Medicine. *Advanced Drug Delivery Reviews.* 2012;64(13):1394–1416.

37. Shah S, Famta P, Raghuvanshi RS, et al. Lipid Polymer Hybrid Nanocarriers: Insights into Synthesis Aspects, Characterization, Release Mechanisms, Surface Functionalization and Potential Implications. *Colloid and Interface Science Communications.* 2022;46:100570.

38. Date T, Nimbalkar V, Kamat J, et al. Lipid-Polymer Hybrid Nanocarriers for Delivering Cancer Therapeutics. *Journal of Controlled Release.* 2018;271:60–73.

39. Chang J, Chen X, Glass Z, et al. Integrating Combinatorial Lipid Nanoparticle and Chemically Modified Protein for Intracellular Delivery and Genome Editing. *Accounts of Chemical Research.* 2019;52(3):665–675.

40. Persano F, Gigli G, Leporatti S. Lipid-Polymer Hybrid Nanoparticles in Cancer Therapy: Current Overview and Future Directions. *Nano Express*. 2021;2(1):012006.
41. Bochicchio S, Lamberti G, Barba AA. Polymer–Lipid Pharmaceutical Nanocarriers: Innovations by New Formulations and Production Technologies. *Pharmaceutics*. 2021;13(2):198.
42. Kashapov R, Ibragimova A, Pavlov R, et al. Nanocarriers for Biomedicine: From Lipid Formulations to Inorganic and Hybrid Nanoparticles. *International Journal of Molecular Sciences*. 2021;22(13):7055.
43. Mukherjee A, Waters AK, Kalyan P, et al. Lipid–Polymer Hybrid Nanoparticles as a Next-Generation Drug Delivery Platform: State of the Art, Emerging Technologies, and Perspectives. *International Journal of Nanomedicine*. 2019;14(null):1937–1952.
44. Alberto AG, Shira G-P, Shadan M, et al. Thirty Years from FDA Approval of Pegylated Liposomal Doxorubicin (Doxil/Caelyx): An Updated Analysis and Future Perspective. *BMJ Oncology*. 2025;4(1):e000573.
45. Barenholz Y. Doxil® − The First FDA-Approved Nano-Drug: Lessons Learned. *Journal of Controlled Release*. 2012;160(2):117–134.
46. Thi TT, Suys EJA, Lee JS, et al. Lipid-Based Nanoparticles in the Clinic and Clinical Trials: From Cancer Nanomedicine to COVID-19 Vaccines. *Vaccines*. 2021. DOI:10.3390/vaccines9040359.
47. Yang P, Zhang L, Wang T, et al. Doxorubicin and Edelfosine Combo-Loaded Lipid–Polymer Hybrid Nanoparticles for Synergistic Anticancer Effect Against Drug-Resistant Osteosarcoma. *OncoTargets and Therapy*. 2020;13:8055–8067.
48. Waheed I, Ali A, Tabassum H, et al. Lipid-Based Nanoparticles as Drug Delivery Carriers for Cancer Therapy. *Frontiers in Oncology*. 2024;14:1296091.
49. Parvez S, Yadagiri G, Gedda MR, et al. Modified Solid Lipid Nanoparticles Encapsulated with Amphotericin B and Paromomycin: an Effective Oral Combination against Experimental Murine Visceral Leishmaniasis. *Scientific Reports*. 2020;10(1):12243.
50. Lv Q, Cheng L, Lu Y, et al. Thermosensitive Exosome–Liposome Hybrid Nanoparticle-Mediated Chemoimmunotherapy for Improved Treatment of Metastatic Peritoneal Cancer. *Advanced Science*. 2020;7(18):2000515.
51. Chen Y, Bathula SR, Li J, et al. Multifunctional Nanoparticles Delivering Small Interfering RNA and Doxorubicin Overcome Drug Resistance. *Cancer Journal of Biological Chemistry*. 2010;285(29):22639–22650.
52. Chehelgerdi M, Chehelgerdi M, Allela OQB, et al. Progressing Nanotechnology to Improve Targeted Cancer Treatment: Overcoming Hurdles in Its Clinical Implementation. *Molecular Cancer*. 2023;22(1):169.
53. Majumder J, Minko T. Multifunctional and Stimuli-Responsive Nanocarriers for Targeted Therapeutic Delivery. *Expert Opinion on Drug Delivery*. 2021;18(2):205–227.
54. Samaridou E, Heyes J, Lutwyche P. Lipid Nanoparticles for Nucleic Acid Delivery: Current Perspectives. *Advanced Drug Delivery Reviews*. 2020;154–155:37–63.
55. Swetha K, Kotla NG, Tunki L, et al. Recent Advances in the Lipid Nanoparticle-Mediated Delivery of mRNA Vaccines. *Vaccines*. 2023;11(3):658.
56. Tang Q, Yu B, Gao L, et al. Stimuli Responsive Nanoparticles for Controlled Anti-Cancer Drug Release. *Current Medicinal Chemistry*. 2018;25(16):1837–1866.
57. Hassanin I, Elzoghby A. Albumin-Based Nanoparticles: A Promising Strategy to Overcome Cancer Drug Resistance. *Cancer Drug Resistance*. 2020;3(4):930.
58. An X, Zhu A, Luo H, et al. Rational Design of Multi-Stimuli-Responsive Nanoparticles for Precise Cancer Therapy. *ACS Nano*. 2016;10(6):5947–5958.
59. Alameh MG, Tombácz I, Bettini E, et al. Lipid Nanoparticles Enhance the Efficacy of mRNA and Protein Subunit Vaccines by Inducing Robust T Follicular Helper Cell and Humoral Responses. *Immunity*. 2021;54(12):2877–2892.e7.
60. Lee Y, Jeong M, Park J, et al. Immunogenicity of Lipid Nanoparticles and Its Impact on the Efficacy of mRNA Vaccines and Therapeutics. *Experimental & Molecular Medicine*. 2023;55(10):2085–2096.
61. Tanaka H, Hagiwara S, Shirane D, et al. Ready-to-Use-Type Lyophilized Lipid Nanoparticle Formulation for the Postencapsulation of Messenger RNA. *ACS Nano*. 2023;17(3):2588–2601.
62. Blakney AK, McKay PF, Hu K, et al. Polymeric and Lipid Nanoparticles for Delivery of Self-Amplifying RNA Vaccines. *Journal of Controlled Release*. 2021;338:201–210.
63. Lam K, Leung A, Martin A, et al. Unsaturated, Trialkyl Ionizable Lipids are Versatile Lipid-Nanoparticle Components for Therapeutic and Vaccine Applications. *Advanced Materials*. 2023;35(15):2209624.
64. Fan C-Y, Wang S-W, Chen J-Y, et al. Synthesis and Assembly of mRNA-Bifunctional Lipid Nanoparticle (BLNP) for Selective Delivery of mRNA Vaccines to Dendritic Cells. *bioRxiv*. 2023:2023.12.26.572282.

65. Zhuang X, Qi Y, Wang M, et al. mRNA Vaccines Encoding the HA Protein of Influenza A H1N1 Virus Delivered by Cationic Lipid Nanoparticles Induce Protective Immune Responses in Mice. *Vaccines.* 2020;8(1):123.

66. Long J, Yu C, Zhang H, et al. Novel Ionizable Lipid Nanoparticles for SARS-CoV-2 Omicron mRNA Delivery. *Advanced Healthcare Materials.* 2023;12(13):2202590.

67. Broos K, Van der Jeught K, Puttemans J, et al. Particle-Mediated Intravenous Delivery of Antigen mRNA Results in Strong Antigen-specific T-cell Responses Despite the Induction of Type I Interferon. *Molecular Therapy Nucleic Acids.* 2016;5:e326.

68. Kauffman KJ, Mir FF, Jhunjhunwala S, et al. Efficacy and Immunogenicity of Unmodified and Pseudouridine-Modified mRNA Delivered Systemically with Lipid Nanoparticles In Vivo. *Biomaterials.* 2016;109:78–87.

69. Wang H-L, Wang Z-G, Liu S-L. Lipid Nanoparticles for mRNA Delivery to Enhance Cancer Immunotherapy. *Molecules.* 2022;27(17):5607.

70. Zhang H, You X, Wang X, et al. Delivery of mRNA Vaccine with a Lipid-Like Material Potentiates Antitumor Efficacy through Toll-Like Receptor 4 Signaling. *Proceedings of the National Academy of Sciences.* 2021;118(6):e2005191118.

71. Sato S. Understanding the Manufacturing Process of Lipid Nanoparticles for mRNA Delivery Using Machine Learning. *Chemical and Pharmaceutical Bulletin.* 2024;72(6):529–539.

72. Steiner D, Bunjes H. Influence of Process and Formulation Parameters on the Preparation of Solid Lipid Nanoparticles by Dual Centrifugation. *International Journal of Pharmaceutics X.* 2021;3:100085.

73. Roces CB, Lou G, Jain N, et al. Manufacturing Considerations for the Development of Lipid Nanoparticles Using Microfluidics. *Pharmaceutics.* 2020;12(11):1095.

74. Wilson DR, Mosenia A, Suprenant MP, et al. Continuous Microfluidic Assembly of Biodegradable Poly(Beta-Amino Ester)/Dna Nanoparticles for Enhanced Gene Delivery. *Journal of Biomedical Materials Research Part A.* 2017;105(6):1813–1825.

75. Roces CB, Port EC, Daskalakis NN, et al. Rapid Scale-Up and Production of Active-Loaded PEGylated Liposomes. *International Journal of Pharmaceutics.* 2020;586:119566.

76. Laramy MNO. Process Robustness in Lipid Nanoparticle Production: A Comparison of Microfluidic and Turbulent Jet Mixing. *Molecular Pharmaceutics.* 2023;20(8):4285–4296.

77. Amrani S, Tabrizian M. Characterization of Nanoscale Loaded Liposomes Produced by 2D Hydrodynamic Flow Focusing. *ACS Biomaterials Science & Engineering.* 2018;4(2):502–513.

78. Maeki M, Okada Y, Uno S, et al. Production of siRNA-Loaded Lipid Nanoparticles Using a Microfluidic Device. *Journal of Visualized Experiments.* 2022(181):e62999.

79. Nag K, Sarker MEH, Kumar S, et al. DoE-Derived Continuous and Robust Process for Manufacturing of Pharmaceutical-Grade Wide-Range LNPs for RNA-Vaccine/Drug Delivery. *Scientific Reports.* 2022;12(1):9394.

80. Matsuura-Sawada Y, Maeki M, Nishioka T, et al. Microfluidic Device-Enabled Mass Production of Lipid-Based Nanoparticles for Applications in Nanomedicine and Cosmetics. *ACS Applied Nano Materials.* 2022;5(6):7867–7876.

81. Sato Y. Development of Lipid Nanoparticles for the Delivery of Macromolecules Based on the Molecular Design of pH-Sensitive Cationic Lipids. *Chemical and Pharmaceutical Bulletin.* 2021;69(12):1141–1159.

82. Samaddar S, Mazur J, Sargent J, et al. Immunostimulatory Response of RWFV Peptide-Targeted Lipid Nanoparticles on Bladder Tumor Associated Cells. *ACS Applied Bio Materials.* 2021;4(4):3178–3188.

83. Hengelbrock A, Schmidt A, Strube J. Formulation of Nucleic Acids by Encapsulation in Lipid Nanoparticles for Continuous Production of mRNA. *Processes.* 2023;11(6):1718.

84. Webb C, Khadke S, Schmidt S, et al. The Impact of Solvent Selection: Strategies to Guide the Manufacturing of Liposomes Using Microfluidics. *Pharmaceutics.* 2019;11(12):653.

85. Aibani N, Patel P, Buchanan R, et al. Assessing the *In Vivo* Effectiveness of Cationic Lipid Nanoparticles with a Triple Adjuvant for Intranasal Vaccination Against the Respiratory Pathogen *Bordetella pertussis. Molecular Pharmaceutics.* 2022;19(6):1814–1824.

86. Das S, Devarajan PV. Enhancing Safety and Efficacy by Altering the Toxic Aggregated State of Amphotericin B in Lipidic Nanoformulations. *Molecular Pharmaceutics.* 2020;17(6):2186–2195.

87. Kent SJ, Li S, Amarasena TH, et al. Blood Distribution of SARS-CoV-2 Lipid Nanoparticle mRNA Vaccine in Humans. *ACS Nano.* 2024;18(39):27077–27089.

88. Pereira-Leite C. Exploring Stearic-Acid-Based Nanoparticles for Skin Applications—Focusing on Stability and Cosmetic Benefits. *Cosmetics.* 2023;10(4):99.

89. Sahin U, Oehm P, Derhovanessian E, et al. An RNA Vaccine Drives Immunity in Checkpoint-Inhibitor-Treated Melanoma. *Nature.* 2020;585(7823):107–112.

90. Li Z, Zhang XQ, Ho WW, et al. Enzyme-Catalyzed One-Step Synthesis of Ionizable Cationic Lipids for Lipid Nanoparticle-Based mRNA COVID-19 Vaccines. *ACS Nano*. 2022;16(11):18936–18950.

91. Skwarczynski M, Zhao G, Ozberk V, et al. Polyphenylalanine as a Self-Adjuvanting Delivery System for Peptide-Based Vaccines: The Role of Peptide Conformation. *Australian Journal of Chemistry*. 2022;76(8):429–436.

92. Pu X. Lipids Extracted from Mycobacterial Membrane and Enveloped PLGA Nanoparticles for Encapsulating Antibacterial Drugs Elicit Synergistic Antimicrobial Response Against Mycobacteria. *Molecular Pharmaceutics*. 2024;21(5):2238–2249.

93. Norling K, Bernasconi V, Hernández VA, et al. Gel Phase 1,2-Distearoyl-*sn*-Glycero-3-Phosphocholine-Based Liposomes Are Superior to Fluid Phase Liposomes at Augmenting Both Antigen Presentation on Major Histocompatibility Complex Class II and Costimulatory Molecule Display by Dendritic Cells *in Vitro*. *ACS Infectious Diseases*. 2019;5(11):1867–1878.

94. Debnath M. Protein Corona Formation on Lipid Nanoparticles Negatively Affects the NLRP3 Inflammasome Activation. *Bioconjugate Chemistry*. 2023;34(10):1766–1779.

95. Sun D, Schur R, Sears AE, et al. Non-Viral Gene Therapy for Stargardt Disease With ECO/pRHO-ABCA4 Self-Assembled Nanoparticles. *Molecular Therapy*. 2020;28(1):293–303.

96. Sun D, Sun Z, Jiang H, et al. Synthesis and Evaluation of pH-Sensitive Multifunctional Lipids for Efficient Delivery of CRISPR/Cas9 in Gene Editing. *Bioconjugate Chemistry*. 2018;30(3):667–678.

97. Kim D-Y, Wu Y, Shim G, et al. Genome-Editing-Mediated Restructuring of Tumor Immune Microenvironment for Prevention of Metastasis. *ACS Nano*. 2021;15(11):17635–17656.

98. Xu C, Lu Z-D, Luo Y-L, et al. Targeting of NLRP3 Inflammasome with Gene Editing for the Amelioration of Inflammatory Diseases. *Nature Communications*. 2018;9(1):4092.

99. Uddin MN, Roni MA. Challenges of Storage and Stability of mRNA-Based COVID-19 Vaccines. *Vaccines*. 2021;9(9):1033.

100. Dąbrowska M, Souto EB, Nowak I. Lipid Nanoparticles Loaded with Iridoid Glycosides: Development and Optimization Using Experimental Factorial Design. *Molecules*. 2021;26(11):3161.

101. Ribeiro LNDM, Couto VM, Fraceto LF, et al. Use of Nanoparticle Concentration as a Tool to Understand the Structural Properties of Colloids. *Scientific Reports*. 2018;8(1):982.

102. Gupta KM, Das S, Macbeath C. Encapsulation of Ferulic Acid in Lipid Nanoparticles as Antioxidant for Skin: Mechanistic Understanding through Experiment and Molecular Simulation. *ACS Applied Nano Materials*. 2020;3(6):5351–5361.

103. Dwiastuti R, Marchaban M, Istyastono EP, et al. Analytical Method Validation and Determination of Free Drug Content of 4-n-Butylresorcinol in Complex Lipid Nanoparticles Using RP-HPLC Method. *Indonesian Journal of Chemistry*. 2018;18(3):496.

104. Makoni PA, Kasongo KW, Walker RB. Short Term Stability Testing of Efavirenz-Loaded Solid Lipid Nanoparticle (SLN) and Nanostructured Lipid Carrier (NLC) Dispersions. *Pharmaceutics*. 2019;11(8):397.

105. Wang X, Wang X, Bai X, et al. Nanoparticle Ligand Exchange and Its Effects at the Nanoparticle–Cell Membrane Interface. *Nano Letters*. 2018;19(1):8–18.

106. Patel P, Santo KP, Burgess S, et al. Stability of Lipid Coatings on Nanoparticle-Decorated Surfaces. *ACS Nano*. 2020;14(12):17273–17284.

107. Kudsiova L, Lansley AB, Scutt G, et al. Stability Testing of the Pfizer-BioNTech BNT162b2 COVID-19 Vaccine: A Translational Study in UK Vaccination Centres. *BMJ Open Science*. 2021;5(1):e100203.

108. Shrestha SC, Ghebremeskel K, White K, et al. Formulation and Characterization of Phytostanol Ester Solid Lipid Nanoparticles for the Management of Hypercholesterolemia: An Ex Vivo Study. *International Journal of Nanomedicine*. 2021;16:1977–1992.

109. Gagliardi A, Giuliano E, Venkateswararao E, et al. Biodegradable Polymeric Nanoparticles for Drug Delivery to Solid Tumors. *Frontiers in Pharmacology*. 2021;12:601626.

110. Anselmo AC, Mitragotri S. Nanoparticles in the Clinic: An Update. *Bioengineering & Translational Medicine*. 2019;4(3):e10143.

111. Letourneur D, Joyce K, Chauvierre C, et al. Enabling MedTech Translation in Academia: Redefining Value Proposition with Updated Regulations. *Advanced Healthcare Materials*. 2020;10(1):e2001237.

112. Ilić T, Đoković J, Nikolić I, et al. Parenteral Lipid-Based Nanoparticles for CNS Disorders: Integrating Various Facets of Preclinical Evaluation Towards More Effective Clinical Translation. *Pharmaceutics*. 2023;15(2):443.

113. Ali M. Recent Developments in Nanoparticle Formulations for Resveratrol Encapsulation as an Anticancer Agent. *Pharmaceuticals*. 2024;17(1):126.

4 Self-Generating Nano-Emulsions for Enhanced Anticancer Delivery

Shakir U. Shakir, Saeed A.Khan, and Mutasem Rawas-Qalaji

4.1 INTRODUCTION

Cancer therapies are constrained by systemic toxicity, poor drug solubility, and multidrug resistance (MDR), leading to treatment failures in nearly 90% of metastatic cases [1]. Conventional treatment strategies such as chemotherapy, immunotherapy, radiation, and surgery indiscriminately target healthy and malignant cells, resulting in unbearable side effects including organ toxicity and immune suppression [2]. Moreover, most chemotherapeutic agents are kept in Biopharmaceutics Classification System (BCS) Class II and Class IV, characterized by poor water solubility, which often results in suboptimal therapeutic response and limited accumulation in tumor tissue. The emergence of MDR further complicates cancer treatment, as efflux transporters, such as P-glycoprotein (P-gp), actively expel drugs from cancer cells, rendering them ineffective.

Current nanotechnology-based drug delivery systems, such as polymeric micelles, liposomes, and lipid nanoparticles, have significantly improved the pharmacokinetics of anticancer agents. Current systems using these methods frequently experience stability problems while also failing to bypass MDR mechanisms effectively [3,4]. A new approach called SGNEs emerged to overcome existing challenges and allow for the effective oral delivery of anticancer agents.

SGNEs are lipid-based drug delivery systems that spontaneously form nano-sized droplets upon contact with aqueous media in the stomach. The ability of self-emulsification improves drug solubilization, absorption, and bioavailability, which are essential for the effective delivery of poorly soluble anticancer agents [5].

The self-emulsification process depends on the thermodynamic stability and the interactivity between oil, surfactants, and cosurfactants with aqueous media [6]. Surfactants reduce oil and water interfacial tensions that allow oil drops to naturally fragment into submicron sizes [7]. The decreased droplet size enhances gastrointestinal dispersion while accelerating biological membrane diffusion, which creates enhanced delivery advantages for lipophilic drugs [8]. Proper selection of surfactants that consider hydrophilic-lipophilic balance (HLB) values and surfactant-to-oil ratios brings about an effective self-emulsification process [5,9]. Furthermore, surfactants in SGNE formulations inhibit P-gp efflux pumps that improve the retention of chemotherapeutic agents inside cells. Cosurfactants help increase interface flexibility, which reduces droplet dispersion to prevent coalescence [10].

Bile salts along with lipases contribute vital functions in stabilizing SGNEs and the absorption of lipophilic drugs in the gastrointestinal tract (GIT). Emulsification processes under the influence of bile salts, such as sodium taurocholate, stabilize droplets with surfactant interactions while inhibiting phase separation. Lipases break down triglycerides from the oil phase, which influences drug release kinetics and absorption [11].

SGNEs provide several benefits by enabling targeted drug delivery, improving absorption and bioavailability, and reversing MDR, thereby maximizing therapeutic effects and minimizing drug-related side effects [12]. SGNEs enable therapeutic agents to bypass efflux pumps in resistant

DOI: 10.1201/9781003517870-4

cancer cells, thus enhancing their chemotherapeutic effect [13]. Selective targeting using SGNEs becomes possible through their functionalization with tumor-targeting ligands along with monoclonal antibodies, cell-penetrating peptides, and small molecules [14]. The self-emulsifying nature of SGNEs improves drug absorption mechanics and cell absorption pathways, thus making oral medication delivery effective. SGNEs also enable controlled drug release, regulating the drug release through diffusion from lipid droplets [15], which sustains drug availability at tumor sites with reduced off-target systemic effects [15–18].

Despite promising preclinical results, SGNEs encounter multiple hurdles that prevent their clinical applications. The main challenges for SGNEs include the lack of long-term stability, large-scale manufacturing feasibility, and *in vivo* biodistribution. While various studies have demonstrated improved drug solubility and bioavailability, limited clinical data and regulatory uncertainties hinder widespread application.

This chapter highlights the therapeutic potential of SGNEs, focusing on strategies to overcome barriers in cancer treatment. Table 4.1 shows a comparison between conventional and SGNEs-based cancer therapies.

4.2 SGNEs FOR CANCER THERAPY

The SGNEs, as an emerging drug delivery system, represent a unique approach to enhancing the solubility and absorption of hydrophobic chemotherapeutic agents [20–22]. SGNE platforms can be engineered for oral delivery (e.g., encapsulated tablets) or parenteral administration (e.g., injectable suspensions), making them adaptable for diverse cancer therapy protocols. Oral SGNEs bypass enzymatic degradation in the gut [23], improve solubility and absorption with subsequent

TABLE 4.1

Comparison between Conventional and Self-Generating Nanoemulsion (SGNE)-Based Cancer Therapies

Treatment Method	Conventional Chemotherapy, Radiation, Surgery	Self-Generating Nanoemulsions
Properties	• Systemic delivery • Generalized distribution • Drug circulates throughout the body, affecting both cancerous and healthy cells • High risk of side effects (e.g., nausea, immune suppression, organ toxicity) [15].	• Targeted delivery with localized drug release at tumor sites, • Minimized side effects on healthy cells, • Controlled-release for prolonged impact [16].
Challenges	• Generalized action, lacks targeting of specific cells. • Increased toxicity due to systemic exposure. • Poor solubility for some drugs, reducing effectiveness and bioavailability.	• Scalability challenges for large-scale production. • Requires stabilization to maintain nanoemulsion's particle stability [17].
Advantages	• Established manufacturing processes.	• Improved solubility of hydrophobic drugs, enhancing bioavailability [17]. • Reduced systemic toxicity due to specific targeting. • Enhanced tumor penetration via enhanced permeability and retention (EPR) effect [18]. • Overcomes multidrug resistance (MDR) by bypassing efflux pumps.
Particle Size	• Micro to macro-sized [15]	• Nano-sized (typically <100 nm) [19]

bioavailability of BCS class II or class IV anticancer drugs. For example, diindolylmethane-loaded SGNE demonstrated a superior bioavailability rate, which exceeded its conventional dosage form by 8.27 times [24]. Also, SGNEs target tumor microenvironments and the drug is released in response to specific tumor conditions, thus minimizing off-target effects [25]. SGNEs can optimize drug release kinetics while improving siRNA delivery efficiency in cancer models. For example, SGNE-based formulations increase transfection rates by approximately 40%–60%, as compared to conventional carriers [26].

SGNEs improve therapeutic efficacy in cancer by improving drug solubility, increasing bioavailability, enabling tumor-specific delivery, and reducing systemic toxicity. The following section discusses these mechanisms in detail.

4.2.1 Enhanced Drug Solubility and Bioavailability

Conventional drug formulations often suffer from erratic absorption, which in turn decreases the ultimate efficacy of the drug. In contrast, SGNEs increase drug solubilization and absorption by forming nano-sized emulsions in the GIT, as illustrated in Figure 4.1.

More than 40% of the newly developed drugs exhibit poor solubility, which ultimately leads to low bioavailability [27]. To overcome this challenge, various drug delivery systems, including lipid and surfactant-based formulations, were used to improve the solubility of hydrophobic drugs. SGNEs are one of these promising approaches that have garnered significant attention. This innovative technique improves solubility, facilitating better absorption in the GIT and ultimately improving the bioavailability of hydrophobic anticancer agents [28–30]. For instance, Ding D. et al. designed oral paclitaxel (PTX)-loaded SGNE, which demonstrated a 3.42-fold improvement in its absorption compared to the conventional formulation. Furthermore, the bioavailability of PTX increased by 2.13 times compared to PTX solution [31].

Similarly, Seo et al. examined the bioavailability and chemotherapeutic efficacy of docetaxel (DCT)-loaded SGNE. Their study revealed a notable enhancement in the bioavailability of DCT, that is, 17% bioavailability was observed for SGNE, compared to merely 2.6% for DCT solution [32,33]. SGNEs have also been investigated for oral administration of plant extracts. Nazari-vanani et al. formulated SGNE containing oil extracts of *Rosa damascena* and *Citrus aurantium L.* for pancreatic cancer treatment. Cytotoxicity assays on PANC1 and MCF7 cell lines confirmed enhanced toxicity, demonstrating the potential of SGNEs in delivering natural bioactive compounds [34].

Beyond the plant extract, SGNEs have been applied to various anticancer drugs. Ansari et al. designed brigatinib-loaded SGNE to enhance the solubility profile of this poorly soluble anticancer agent.

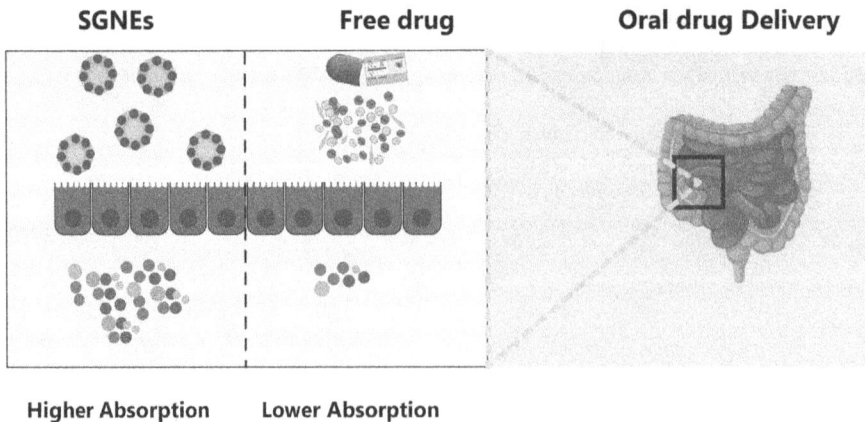

FIGURE 4.1 Schematic representation of free drug and SGNE formulation for oral drug delivery.

Their finding showed a remarkable 205-fold increase in solubility, with nearly double the therapeutic efficacy compared to control [35]. In another study, Kamel and Mahmoud developed mannitol-based, spray-dried SGNE loaded with rosuvastatin, resulting in an improved dissolution rate relative to the marketed product [36]. Similarly, Troung et al. were able to improve the solubility of erlotinib by adapting SGNE as a delivery system [37].

Qian et al. investigated myricetin-loaded SGNE, assessing its solubility across different excipients. Their study found a 6.33-fold improvement in bioavailability compared to a pure drug [38]. Similarly, Akhtar et al. demonstrated that etoposide-loaded SGNE formulations significantly enhanced permeability across Caco-2 monolayers, achieving a 2.6- and 11-fold increase in the apical-to-basolateral permeability coefficient compared to Etosid® (commercial etoposide injection) and plain drug solution, respectively. An *in vivo* pharmacokinetic study further confirmed a 3.2- and 7.9-fold increase in relative oral bioavailability compared to Etosid® and a drug suspension, respectively [39].

In another approach, Beg et al. incorporated a microbial surfactant into an SGNE formulation of PTX to mitigate the severe toxicity associated with conventional delivery systems of PTX. Utilizing surfactin as a biosurfactant, the optimized formulation achieved more than 85% *in vitro* drug release in 60 minutes, while maintaining a globule size of less than 250 nm. Biocompatibility studies indicated lower reactive oxygen species (ROS) generation compared to synthetic surfactant-based formulations, while *in vitro* cytotoxicity tests on MCF-7 cells revealed a significantly lower IC_{50} (2.65 μM) for PTX-surfactin SGNE compared to that of PTX suspension (18.5 μM), indicating improved anticancer efficacy [40].

Further studies have explored the co-delivery of multiple drugs. Nazari-Vanani et al. formulated SGNE for capecitabine and vorinostat using castor oil, polysorbate 80, and PEG 600. The surfactant type greatly affected droplet size (117±26 nm for Tween® 80 and 37±8 nm for PEG 600). Cytotoxicity assays demonstrated significant cancer cell apoptosis, highlighting the ability of SGNEs to enhance solubility, bioavailability, and therapeutic efficacy [41]. Similarly, a study by Usmani et al. investigated an SGNE formulation for the co-delivery of doxorubicin (DOX) and *Nigella sativa* oil to improve solubility and anticancer efficacy against HepG2 liver cancer cells. The optimized formulation demonstrated a small particle size (79.7 nm), enhanced drug release (96.97% in 32 hours), and resulted in significant cytotoxicity improvement (IC_{50}=2.5 μg/mL, p<0.05) compared to free DOX [42].

Karimi et al. developed erlotinib-loaded SGNE (Ert-SGNE) for oral treatment of pancreatic cancer. The optimized formulation, composed of olive oil, polysorbate 80, and PEG 600, formed NE with an average droplet size of 83.9 nm when mixed with water. In pharmacokinetic studies in rats, Ert-SGNE demonstrated significant bioavailability enhancement compared to free Ert [43]. Chakradhar JVUS et al. developed an SGNE formulation for Vemurafenib (VMF), a poorly soluble BRAF inhibitor with <1% oral bioavailability. Using a Quality by Design (QbD) approach, the ratios of Capryol® 90, polysorbate 80, and Transcutol® HP were optimized to enhance drug solubilization and emulsification efficiency. The optimized SGNE exhibited improved dissolution in both pH 1.2 and 6.8 media, achieving a 2.13-fold increase in bioavailability compared to the free drug [44].

Shin et al. developed a solid SGNE (S-SGNE) formulation for olaparib, a BCS Class IV anticancer drug, to enhance its solubility and oral absorption. Using Capmul® MCM, polysorbate 80, and PEG 400, the optimized S-SGNE formulation exhibited a mean droplet size of 87.0 nm and superior drug loading efficiency. The S-SGNE formulation demonstrated improved dissolution, stability, and a fourfold increase in intestinal permeability (using Caco-2 assay) compared to the raw drug, confirming its potential for the oral delivery of poorly soluble oncologic agents [45]. Wu et al. developed a cannabidiol (CBD)-loaded S-SGNE formulation to improve its solubility, bioavailability, and therapeutic potential for anticancer applications. CBD has demonstrated anticancer activity by inducing apoptosis, modulating oxidative stress, and inhibiting tumor progression. However, its poor solubility and bioavailability limit its clinical efficacy. The optimized S-SGNE formulation, using MCTs (medium-chain triglycerides) oil, Labrasol®, Tween® 80, and Transcutol®,

TABLE 4.2

Self-Generating Nanoemulsion (SGNE) Formulations for Enhancing the Solubility and Bioavailability of Anticancer Drugs

Drug/Formulation	Bioavailability Enhancement	Study Outcomes	Reference
Paclitaxel-SGNE	4.1-fold increase in absorption, 5.2-fold increase in bioavailability	3.2 times more drug accumulation in cancer cells, faster drug release (~4 h)	[47]
Docetaxel-SGNE	17% (compared to 2.6% for plain DCT)	Improved chemotherapeutic effect	[32,33]
Rose damascene and Citrus aurantium SGNE	Not specified	Enhanced cytotoxicity in PANC1 and MCF7 cancer cells	[34]
Brigatinib-SGNE	205-fold solubility enhancement	Almost double the efficacy compared to the control formulation	[35]
Rosuvastatin-SGNE (spray-dried)	Not specified	Improved dissolution rate compared to the marketed product	[36]
Erlotinib-SGNE	Not specified	Enhanced dissolution compared to pure drug powder	[37]
Myricetin-SGNE	6.33-fold bioavailability increase	Improved solubility and bioavailability	[38]
Etoposide-SGNE	2.6-fold (Caco-2 permeability), 7.9-fold (oral bioavailability)	Improved permeability and bioavailability compared to Etosid®	[39]
Paclitaxel-SGNE (Microbial Surfactant-based)	>85% drug release in 60 min	Improved solubility, reduced ROS generation, enhanced anticancer efficacy in MCF-7 cells	[40]
Capecitabine and Vorinostat-SGNE	Not specified	Significant improvement in cancer cell apoptosis	[41]
Erlotinib-SGNE (Olive oil, Tween® 80, PEG 600)	Significant increase	Enhanced bioavailability in rats	[43]
Vemurafenib-SGNE	2.13-fold increase	Improved dissolution in pH 1.2 and 6.8 media	[44]
Olaparib-SGNE (Solid-state)	Fourfold increase	Improved intestinal permeability (Caco-2 assay)	[45]
CBD-SGNE (Solid-state)	Increased C_{max}	Faster absorption, higher systemic exposure	[46]

enhanced drug absorption, increased C_{max}, and facilitated tumor targeting. Pharmacokinetic studies in Sprague Dawley rats confirmed faster drug absorption and higher systemic exposure compared to oil-based formulations. Collectively, these findings demonstrate SGNEs as promising nanocarriers for improving the tumor-targeting efficiency of lipophilic anticancer agents like CBD [46]. The various investigated SGNE formulations to enhance drug solubility and bioavailability are summarized in Table 4.2.

4.2.2 ENHANCED TUMOR TARGETING

Solid tumors have irregular vascular architecture characterized by hyperpermeable blood vessels and impaired lymphatic drainage. This unique pathological feature was exploited by SGNEs via the passively facilitated enhanced permeability and retention (EPR) effect and the accumulation of drugs in tumor tissues. Hence, tumor targeting is improved with reduced off-target effects.

Targeted drug delivery can improve the efficacy of anticancer drugs with reduced toxicity [48]. Nanodroplets persist in the body for an extended period, while evading the mononuclear phagocyte system. Negatively charged droplets are attracted toward positively charged membrane barriers [49]. SGNEs are taken up by the liver and spleen and can passively target these organs [50].

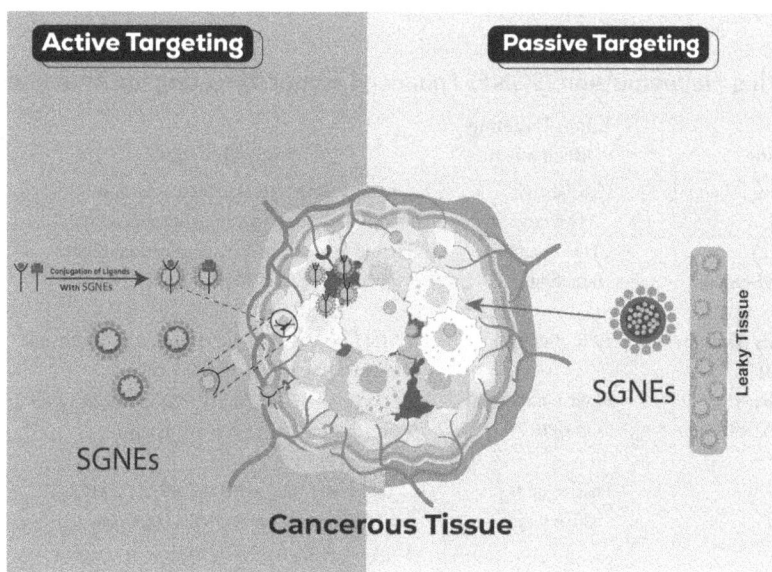

FIGURE 4.2 Illustration of active and passive targeting mechanisms of SGNEs for drug delivery to cancerous tissue.

Hydrophilic polymers are linked on the surface of the droplets to tailor them to provide stealth properties [51]. PEGylation, for instance, links polyethylene glycol (PEG) to the surface of the nanodroplets through their interaction with surfactant molecules. PEG provides a hydrophilic surface that attracts water, preventing opsonin binding. This, in turn, prevents opsonin from binding to the surface of nanocarriers, providing them with stealth properties [52]. In addition, it forms a barrier at the surface and inhibits enzyme degradation. Thus, it also provides steric hindrance [53]. Due to these properties, the administered drug reaches the intended site and minimizes off-target effects [54]. Furthermore, to enhance droplet retention in the GIT, mucoadhesive polymers like HPMC and thiolated chitosan can be utilized [55]. Thiolated mucoadhesive polymers interact with mucus glycoproteins and encapsulate the nanoemulsion droplets [56,57]. Active and passive targeting can be achieved by conjugating suitable ligands, such as peptides, antibodies, and nucleic acids, to bind to the target site's receptors and via EPR, respectively (Figure 4.2) [48,58].

Several studies have demonstrated that anticancer agents formulated into SGNEs can significantly improve tumor targeting. Kanwal et al. developed a curcumin-loaded SGNE formulation incorporating an absorption enhancer, which resulted in improved cellular uptake, enhanced anticancer activity, and significantly increased oral bioavailability [59]. Mucoadhesive SGNE formulations were explored to enhance drug permeability and site-specific delivery for cancer therapy. Batool et al. developed a hyaluronic acid-based mucoadhesive SGNE of tamoxifen, specifically designed for targeting CD44-overexpressed breast cancer cells. The formulation showed a 7.11-fold enhancement in drug permeation compared to raw tamoxifen and displayed significantly greater cytotoxicity against MCF-7 cells [60].

In a similar approach, Chaudhuri et al. designed omega-3 fatty acid-based SGNE of docetaxel (DCT) to improve tumor-targeting efficiency and therapeutic potential for breast cancer treatment. The SGNE formulation exhibited a 2.12-fold higher drug permeation across the intestinal membrane, which facilitated enhanced oral bioavailability. *In vitro* cytotoxicity studies on MDA-MB-231 breast cancer cells confirmed that the DCT-loaded-SGNE improved DCT-mediated cell death compared to the drug suspension [61]. Pandanus conoideus Lamk-loaded SGNE showed an improved effect against MCF-7 breast cancer cells. The optimized SGNE formulation contained red fruit oil, sugar monoester palmitate, propylene glycol, and Tween® 20. A particle size of $193.1 \pm 1.68\,\text{nm}$,

TABLE 4.3

Self-Generating Nanoemulsion (SGNE)-Enhanced Tumor Targeting for Anticancer Therapy

Drug/Formulation	Tumor Targeting Mechanism	Study Outcomes	Reference
Curcumin-SGNEs (Absorption enhancer-based)	Passive and active targeting	Improved cellular uptake, increased oral bioavailability, enhanced anticancer activity	[59]
Tamoxifen-SGNEs (Hyaluronic acid-based, mucoadhesive)	CD44 receptor targeting	7.11-fold increase in drug permeation, higher cytotoxicity against MCF-7 cells	[60]
Docetaxel-SGNEs (Omega-3 fatty acid-based)	EPR effect	2.12-fold increase in drug permeation, improved cytotoxicity in MDA-MB-231 cells	[61]
Red Fruit (*Pandanus conoideus* Lamk.)-SGNEs	Passive targeting via EPR	Zeta potential of $-43.26\,mV$, IC_{50} of $85.20\,\mu g/mL$, promising anticancer activity against MCF-7 cells	[62]
Venetoclax-SGNEs (VEN-SGNES)	Passive and active targeting	3–7 fold lower IC_{50} in MDA-MB-231, MCF-7, and T47D breast cancer cells, increased apoptosis & ROS generation	[63]
Pandanus tectorius leaf (PTL)-SGNEs	Passive targeting via EPR	Enhanced apoptosis in HeLa cells, improved tumor-targeted therapy in cervical cancer	[64]

a zeta potential of $-43.26\pm0.11\,mV$, and a PDI of 0.50 ± 0.01 were exhibited and resulted in an IC_{50} concentration of $85.20\,\mu g/mL$, which indicated its promising anticancer potential [62]. Rajana et al. developed a venetoclax-loaded SGNE (VEN-SGNE) to improve its tumor-targeting efficiency and therapeutic efficacy in breast cancer. The optimized formulation, with a particle size of $71.3\pm2.8\,nm$ and a polydispersity index of 0.113 ± 0.01, enhanced drug permeability and bioavailability. VEN-SGNE exhibited 3–4-fold, 6–7-fold, and 5–6-fold lower IC_{50} values in MDA-MB-231, MCF-7, and T47D breast cancer cells, respectively, compared to free VEN. VEN-SGNE resulted in increased cellular uptake, apoptosis induction, and ROS generation [63].

Natural bioactive compounds have also been incorporated into SGNES to enhance anticancer efficacy and tumor targeting. Kholieqoh et al. formulated Pandanus tectorius leaf (PTL)-loaded SGNE and demonstrated enhanced antioxidant and anticancer effects compared to crude PTL extracts. The optimized SGNE formulation induced early and late apoptosis in HeLa cells, confirming its potential for tumor-targeted therapy in cervical cancer. The findings further support the role of SGNES in delivering bioactive phytochemicals to improve cancer treatment outcomes [64]. SGNE-based formulations designed for enhanced tumor targeting are listed in Table 4.3.

4.3 COUNTERACTING MDR

MDR poses a significant challenge in cancer treatment. Overexpression of drug efflux pumps, decreased drug uptake, and altered drug metabolism result in MDR. As illustrated in Figure 4.3, cancer cells employ several defense mechanisms to limit drug efficacy, including the upregulation of adenosine triphosphate (ATP)-binding cassette transporters and the metabolic inactivation of drugs via enzymes such as glutathione S-transferase and CYP450. Additionally, mutations leading to receptor loss enhanced DNA repair mechanisms, and alterations in apoptotic pathways further contribute to MDR.

The persistent challenge of drug resistance necessitates innovative delivery systems that enhance bioavailability, enable site-specific targeting, and overcome formulation limitations such as poor solubility and short half-life. Therefore, SGNEs consisting of nanometer-sized droplets stabilized by emulsifiers have emerged as promising carriers for addressing MDR. Their small droplet size,

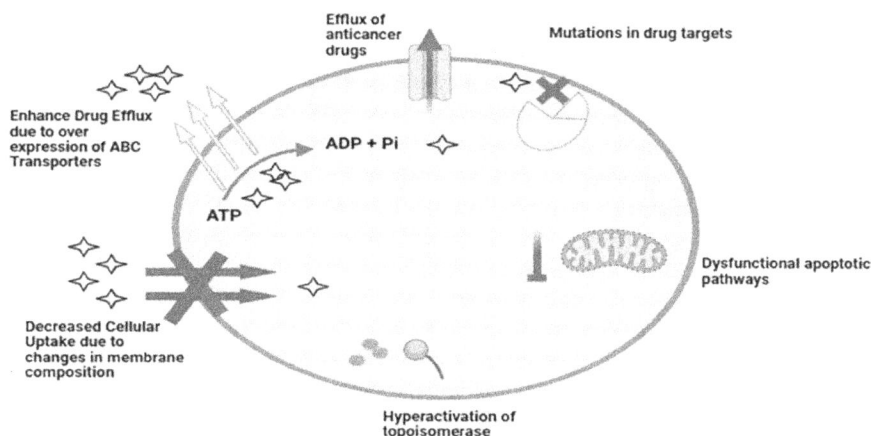

FIGURE 4.3 Mechanisms of cancer drug resistance, including enhanced drug efflux, decreased cellular uptake, mutations in drug targets, dysfunctional apoptotic pathways, and hyperactivation of topoisomerase.

high surface area, and stability enhance drug solubility, facilitate deep tumor penetration, and improve drug retention at target sites. SGNEs investigated for the delivery of antimicrobials demonstrated the ability to overcome drug resistance by increasing cellular uptake, disrupting biofilms, and modulating intracellular drug delivery [65]. These advantages can be translated into cancer therapy. SGNEs can bypass efflux pumps and facilitate intracellular drug accumulation. The nano-size of the droplets allows for better penetration into cancerous tissues, increasing drug bioavailability and therapeutic efficacy.

SGNEs employ several strategies to bypass cancer cell resistance mechanisms.

4.3.1 Efflux Pump Inhibition

P-glycoprotein (P-gp) is an ATP-dependent efflux transporter that reduces intracellular drug concentrations [66]. Overexpression of P-gp in cancer cells is a primary mechanism of MDR, leading to treatment failure [67]. Surfactants such as Tween® 80 affect P-gp ATPase activity [68]. Recent studies have identified new P-gp inhibitors that target its nucleotide-binding domains, showing promise in reversing MDR [69].

Chaturvedi et al. evaluated SGNEs as anti-MDR agents because they bypass P-gp and CYP3A4-mediated drug metabolism for better absorption and permeability. According to this research, long-chain triglycerides (LCTs) demonstrated lymphatic targeting ability so drugs can reach proliferating cancer cells without passing through the initial metabolic process. Many studies support that lipid-based carriers including SGNEs work as an effective approach to overcoming MDR in cancer therapy [70]. Shadab et al. developed a resveratrol-loaded SGNE for pancreatic cancer, which showed enhanced cytotoxicity by inhibiting NF-kB and COX-2 (key regulators of P-gp expression). The study showed that SGNE-mediated suppression of NF-kB could reduce efflux pump activity, leading to improved intracellular drug accumulation and therapeutic efficacy in pancreatic cancer [71]. Akhtar et al. developed an etoposide-loaded SGNE to bypass P-gp-mediated efflux and enhance drug transport across the intestinal membrane. An increase of 61.3% in intracellular uptake compared to plain drug solution was observed [72].

Reyna-Lázaro et al. developed multifunctional SGNEs for the co-delivery of small interfering RNAs (siRNAs) and anticancer drugs. Enhanced drug permeability, improved transfection efficiency, and a twofold increase in cytotoxicity compared to free APIs were recorded. Its role in circumventing efflux pump-mediated resistance was reinforced [26]. Dasatinib (DS)-loaded SGNE formulations were evaluated for their effects on HT29, SW420, and MCF7 cancer cells.

Dasatinib, a tyrosine kinase inhibitor, is subject to P-gp-mediated efflux. The results demonstrated enhanced cytotoxicity, with IC_{50} values in HT29 cells decreasing from 1.46 ± 0.05 mM (pure DS) to 0.56 ± 0.01 mM (DS-SGNE 3) and 0.81 ± 0.21 mM (DS-SGNE 4). Similarly, in SW820 cells, IC_{50} was reduced from 12.38 ± 1.40 mM (pure DS) to 10.60 ± 2.16 mM (DS-SGNE 3) and 10.97 ± 1.09 mM (DS-SGNE 4). Notably, in MCF7 cells, DS-SGNE 3 exhibited an IC_{50} value 23 times lower than pure DS, indicating significantly greater cytotoxicity. The high encapsulation efficiency (91.7%–97.5%) and improved intracellular drug retention ($61.3\pm2.6\%$ uptake compared to $15.3\pm1.62\%$ for pure DS) further suggest that SGNEs facilitated intracellular drug accumulation by inhibiting P-gp activity and improving drug permeability. These findings highlight the role of SGNEs in overcoming efflux-mediated MDR in cancer therapy [73].

4.3.2 Combinatorial Drug Delivery

Co-encapsulating chemotherapeutics with resistance-modulating agents amplifies efficacy by simultaneously targeting multiple mechanisms of drug resistance. This strategy enhances intracellular drug retention, prevents efflux-mediated clearance, and promotes synergistic cytotoxic effects. A notable example is the SGNE formulation of glycyrrhetinic acid derivative (CDODA-Me) and erlotinib (Ert), which was successfully used to overcome Ert resistance in non-small cell lung cancer (NSCLC). This formulation significantly improved oral bioavailability (from 22.13 to 151.76 µg/mL) and exhibited synergistic effects in wild-type and resistant H1975 and HCC827 cell lines (combination index <1). Mechanistically, the combination therapy inhibited phosphorylation of epidermal growth factor receptor (EGFR) and MET receptor tyrosine kinase, leading to enhanced apoptosis, intracellular ROS accumulation, and mitochondrial membrane potential depletion [74]. Patil et al. developed a Tadalafil (TDF) and Ketoconazole (KTZ) co-loaded SGNE to repurpose non-oncology drugs for hepatic cancer therapy. The optimized nanoemulsion had a particle size of 41 ± 5 nm, a PDI of 0.189 ± 0.064, and a zeta potential of -27.1 mV, ensuring stable and effective drug delivery. This combinatorial SGNE exhibited controlled drug release (52.37% for TDF and 52.31% for KTZ over 24 hours) and significantly enhanced cellular uptake compared to free drugs. The formulation demonstrated greater cytotoxicity against HepG2 cells than either drug alone, suggesting a synergistic anticancer effect. These findings highlight the potential of SGNEs-based combinatorial drug delivery systems in enhancing treatment outcomes for hepatic cancer [75].

Kazi et al. developed a neusilin-based bioactive SGNE (Bio-SGNE) co-loaded with curcumin (CUR) and piperine (PP) to enhance oral bioavailability and therapeutic efficacy. The optimized formulation, containing black seed oil, IMWITOR® 988, Transcutol® P, and Cremophor® RH40, exhibited excellent self-emulsification, a small droplet size, and a transparent appearance, ensuring efficient drug dispersion. The solidified Bio-SGNE formulation demonstrated enhanced dissolution of CUR (up to 60%) and PP (up to 77%), significantly improving drug retention and bioactivity. Since CUR downregulates P-gp and piperine inhibits CYP3A4-mediated metabolism, this combinatorial SGNES strategy provides a synergistic approach to overcoming MDR, enhancing intracellular drug accumulation, and preventing efflux-mediated clearance in cancer therapy [76]. Pangeni et al. developed a multiple nanoemulsion system for the oral co-delivery of oxaliplatin (OXA) and 5-fluorouracil (5-FU), addressing low drug bioavailability and MDR in colorectal cancer. The optimized formulation, with a droplet size of 20.3 ± 0.22 nm and a zeta potential of -4.65 ± 1.68 mV, significantly enhanced permeability across Caco-2 cells (4.80-fold for OXA and 4.30-fold for 5-FU) and improved oral bioavailability (9.19-fold and 1.39-fold, respectively). In a CT26 tumor-bearing mouse model, the nanoemulsion-based combination therapy resulted in 73.9% tumor growth inhibition, outperforming monotherapy with either drug. These findings confirm that SGNES-based combinatorial drug delivery systems can effectively enhance therapeutic outcomes in drug-resistant colorectal cancer [77].

Tripathi et al. developed a triple antioxidant SGNE co-loaded with quercetin, resveratrol, and genistein to improve oral bioavailability and synergistic anticancer effects. The optimized formulation had a particle size < 200 nm and PDI < 0.3, ensuring efficient self-emulsification and cellular uptake (internalization within 1 hour in Caco-2 cells). Pharmacokinetic studies indicated that the oral bioavailability of quercetin, resveratrol, and genistein increased by 4.27-fold, 1.5-fold, and 2.8-fold, respectively, in comparison to free antioxidants. In a dimethylbenzanthracene (DMBA)-induced breast cancer model, the SGNE formulation exhibited superior tumor suppression, confirming its potential as a combinatorial strategy for chemoprevention and treatment of breast cancer [78].

Ateeq et al. developed a solid SGNE (S-SGNE) co-loaded with docetaxel (DCT) and carvacrol (CV) to improve oral bioavailability and reduce systemic toxicity. The optimized formulation prepared using Nigella sativa oil, Cremophor® RH 40, and ethanol exhibited a globule size <200 nm and a 2.3-fold higher dissolution rate than free DCT. *Ex vivo* permeation studies showed significantly increased drug permeation (1077.02 ± 12.72 μg/cm²) over 12 hours. *In vitro* cytotoxicity studies revealed a 5.2-fold reduction in IC_{50} for MDA-MB-231 cells, confirming enhanced anticancer activity. Furthermore, *in vivo* pharmacokinetic studies demonstrated a 3.4-fold increase in C_{max} and improved bioavailability. The study also highlighted carvacrol's role in reducing docetaxel-induced toxicity, further enhancing its therapeutic potential [79]. Jain et al. developed a solid SGNE co-loaded with tamoxifen and quercetin (s-Tmx-QT-SGNE) that significantly enhanced oral bioavailability (~8-fold for tamoxifen and ~4-fold for quercetin). The optimized formulation exhibited 32-fold and 22-fold dose reduction indices, demonstrating superior synergistic cytotoxicity. In DMBA-induced breast cancer models, the SGNE formulation achieved ~80% tumor suppression, compared to only ~35% with tamoxifen citrate alone. Furthermore, hepatotoxicity—commonly associated with tamoxifen—was completely mitigated, as evidenced by normalized MMP-2 and MMP-9 levels and improved histopathological outcomes [80].

Grill et al. observed that Silibinin, when co-delivered with curcumin in SGNEs, not only significantly increased the plasma concentration of curcumin but also improved the overall bioavailability by 3.5 times compared to CUR SGNEs [81]. Jain et al. formulated an SGNE co-loaded with tamoxifen and quercetin to improve bioavailability and overcome MDR in breast cancer therapy. The optimized formulation exhibited a 9.63-fold and 8.44-times increase in cellular uptake of tamoxifen and quercetin, respectively, in comparison to free drugs. Quercetin, a known MDR modulator, enhances intracellular drug retention and inhibits P-gp activity, thereby improving tamoxifen bioavailability and anticancer efficacy. The SGNEs achieved rapid emulsification (<2 minutes) and maintained stability for 6 months, supporting its potential for oral delivery in combinatorial cancer therapy [82]. Similarly, Ceramella et al. developed a nanoemulsion for the co-encapsulation of cisplatin and quercetin to improve the anticancer properties of cisplatin, while mitigating its nephrotoxicity. The study demonstrated that quercetin, a potent antioxidant, reduced oxidative stress associated with cisplatin-induced toxicity, leading to enhanced safety and efficacy. SGNE facilitated synergistic anticancer effects against MDA-MB-231 breast cancer cells and reduced toxicity to normal renal HEK-293 cells. The co-delivery improved the intracellular uptake of both drugs and increased their therapeutic effect [83]. Kamal et al. developed a mesalamine-quercetin co-loaded SGNE formulation to enhance the therapeutic potential of both drugs in colorectal cancer. The optimized SGNE formulation contained Capmul® MCM, Tween® 80, and Transcutol® P. A particle size of 154.4 nm, a PDI of 0.329, and a zeta potential of −17.25 mV were achieved. Improved drug release, bioavailability, and anti-inflammatory effects resulted in a significant decrease in TNF-α, NO, and IL-6 levels by 2.8, 1.7, and 3.4 times, respectively, compared to the diseased group. These findings showed the role of SGNES-based combinatorial drug delivery in enhancing therapeutic efficacy for cancer [84].

4.3.3 ENDOCYTOSIS

The SGNEs enter cells via endocytosis to bypass membrane efflux pumps. This route enhances intracellular drug accumulation, as evidenced by a 3.5-fold increase in doxorubicin delivery to resistant

TABLE 4.4

Self-Generating Nanoemulsion (SGNE) Strategies for Overcoming Multidrug Resistance (MDR) in Cancer Therapy

Drug/Formulation	MDR Mechanism Targeted	Study Outcomes	Reference
Resveratrol-SGNEs	NF-kB and COX-2 inhibition (reduces P-gp expression)	Enhanced cytotoxicity in pancreatic cancer, improved drug retention	[71]
Etoposide-SGNEs	P-gp inhibition and ATPase suppression	61.3% increase in intracellular uptake, improved GIT permeability	[72]
siRNA and Anticancer Drug-SGNEs	Efflux pump inhibition, increased transfection efficiency	Twofold increase in cytotoxicity, improved intestinal absorption	[26]
Dasatinib-SGNEs	P-gp inhibition, increased intracellular retention	23.35× lower IC_{50} in MCF7 cells, higher drug accumulation	[73]
CDODA-Me and Erlotinib-SGNEs	EGFR & MET inhibition, ROS accumulation	Synergistic effects in NSCLC, improved oral bioavailability	[74]
Tadalafil and Ketoconazole-SGNEs	MDR modulation, enhanced cellular uptake	Increased cytotoxicity against HepG2 cells	[75]
Curcumin and Piperine-SGNEs	P-gp inhibition (curcumin), CYP3A4 suppression (piperine)	Enhanced solubility, bioavailability, and MDR reversal	[76]
Oxaliplatin and 5-FU-SGNEs	Efflux pump inhibition, enhanced permeability	73.9% tumor growth inhibition, 9.19-fold bioavailability improvement	[77]
Tamoxifen and Quercetin-SGNEs	P-gp inhibition, reduced hepatic metabolism	1.69 fold increased oral bioavailability, 32× enhanced cytotoxicity	[82]
Cisplatin and Quercetin-SGNEs	Oxidative stress reduction, enhanced uptake	Increased therapeutic efficacy, reduced nephrotoxicity	[83]

leukaemia cells using albumin-functionalized SGNEs [85]. Similarly, amidated pluronic-decorated SGNE showed enhanced ciprofloxacin uptake into intracellular Salmonella typhi reservoirs. Arshad et al. demonstrated that NH2-F127 SGNE significantly improved macrophage uptake, reducing bacterial survival to just 2% and effectively disrupting resistant biofilms in the gall bladder [86]. Examples of SGNE formulations counteracting MDR are shown in Table 4.4.

4.4 CONSIDERATIONS FOR THE FORMULATION OF SGNEs

SGNEs enhance the solubility, bioavailability, and therapeutic effect of hydrophobic drugs. Key ingredients used in the development of SGNEs-based-formulations include: surfactants, cosurfactants, and oils that play a crucial role in drug encapsulation efficiency and absorption [87]. The nature of the oil used in SGNE formulations directly affects systemic absorption and drug loading, as demonstrated by Zhao et al., who developed a Zedoary turmeric oil-based SNEDDS where the inclusion of a secondary oil phase (ethyl oleate) significantly enhanced drug loading and improved the oral bioavailability of germacrone, a key bioactive marker, by 1.7-fold in AUC and 2.5-fold in Cmax compared to unformulated oil [88]. The emulsification capacity and stability of SGNE-based formulations are greatly influenced by the choice of surfactants and cosurfactants, as demonstrated by Hayat et al., who developed a dual-drug-loaded SGNEs containing curcumin and naringin using Tween® 80, Labrasol®, and Transcutol®. Their selection not only provided thermodynamically stable emulsions with high transmittance and appropriate droplet size but also enhanced skin permeation and significantly improved wound healing activity compared to conventional formulations [89]. The encapsulated drug must be soluble in the oils used in the SGNEs to hinder medication crystallization during physiological dilution, as highlighted by Rehman et al., who emphasized that the miscibility of the drug within the lipidic phase is a critical factor for ensuring system stability and maintaining the drug in a solubilized state throughout the emulsification process [90]. Suitable oils, surfactants, and cosurfactants are first identified to formulate a stable SGNE-based formulation. Ternary phase diagrams are commonly employed to determine the SGNE zone for the chosen compounds [91]. In ternary mixtures, the proportion of the selected oil, surfactant, or cosurfactant is adjusted, while keeping the other two components at fixed concentrations. This approach allows for the determination of the optimal ingredients' ratio for the formulation of a stable SGNE formulation.

4.4.1 OIL PHASE SELECTION

Various natural oils and lipids are used in the formulation of SGNEs. The oil selection is primarily dependent on the drug's solubility profile along with its physical and chemical attributes, such as charge distribution, surface tension with aqueous media, thickness, mass per unit volume, phase transitions, and molecular stability [92]. The choice of oil significantly affects the self-emulsifying ability of SGNEs and drug precipitation in the GIT, thereby enhancing lymphatic transport through the intestinal walls [93]. Natural plant-derived oils, such as soybean, castor, coconut, and palm oils, have been investigated for SGNE-based formulations [94,95]. Holm et al. developed a halofantrine-loaded SGNE formulation using soybean oil, Maisine 35-1®, polysorbate 80, and Cremophor® EL, which spontaneously formed a transparent, single-phase liquid [96]. However, unmodified oils generally exhibit low drug-loading capacity, limiting their suitability for certain formulations [88].

To improve the dissolution of poorly water-soluble drugs, both medium and long-chain triglycerides (MCTs and LCTs) are utilized. MCTs primarily consist of mono- and diglycerides with lipid chain lengths of C8–C10, while LCTs contain triglycerides with chains longer than C10 [97]. Commonly used MCTs derivatives include Capmul® PG-8 (propylene glycol monocaprylate), Capmul® MCM (mono- and diglycerides of caprylic/capric acid), Captex® 300, Captex® 200, and Captex® 355, whereas LCTs include oleic acid, ethyl oleate, olive oil, and sesame oil. MCTs facilitate drug absorption via the hepatic portal system, leading to rapid systemic circulation, while LCTs enhance drug absorption through intestinal lymphatic transport [98]. Due to their superior solubility profile, self-emulsification capacity, and chemical stability, MCTs are often preferred for SGNE formulations over LCTs [94].

The lipid chain length plays a crucial role in optimizing SGNE design. In a study evaluating the impact of glycerides on SGNE formation, Prajapati et al. demonstrated that monoglycerides produced clear or translucent microemulsions, whereas di- and triglycerides resulted in an additional gel phase [98]. Furthermore, medium-chain mixed-glyceride polar oils have shown lower oxidation potential and superior drug solubilization properties, making them ideal candidates for SGNE formulations [30]. Additionally, phospholipids have been successfully employed as oil components in stable SGNE formulations. Studies indicate that phospholipids can enhance both aqueous and lipophilic drug solubility, thereby improving therapeutic efficacy [99,100].

Studies indicate that MCT-based formulations exhibit faster emulsification and enhanced lymphatic transport compared to LCTs, leading to improved bioavailability of lipophilic drugs [101]. Despite that, lipid oxidation remains a big problem in SGNEs stability. Previous research has shown that unsaturated fatty acids are prone to oxidation and will degrade the occurring lipid as well as the encapsulated drug, thus reducing efficacy [102]. To overcome this problem, antioxidants like tocopherols, butylated hydroxytoluene, and ascorbyl palmitate can be added to SGNEs in order to inhibit lipid peroxidation and to prolong shelf life [103,104].

4.4.2 ROLE OF SURFACTANTS AND COSURFACTANTS

The formation and stabilization of SGNEs depend on the use of surfactants. These amphiphilic molecules have a hydrophilic polar head and hydrophobic hydrocarbon tail, which allows them to minimize interfacial free energy and surface tension. Surfactants are generally known as cationic, anionic, zwitterionic, or nonionic and are further classified by their HLB value. Surfactants with HLB > 10 are classified as hydrophilic surfactants and HLB < 10 as lipophilic surfactants.

Normally, one prefers nonionic surfactants for their lower toxicity and propensity to form more stable emulsions over varying ranges of pH and ionic strengths. Research suggests that surfactant choice is predominantly dependent on its HLB value and those with HLB values less than 12 can induce the formation of SGNEs with droplet size less than 100 nm when diluted with natural digestive fluids found in the GIT. Regularly employed surface-active agents consist of ethoxylated sorbitan esters (Tween® 20, Tween® 40, Tween® 80), PEGylated castor oil derivatives (Cremophor® EL, Cremophor® RH40), and caprylocaproyl macrogol glycerides (Labrasol®) [105].

The emulsion droplet size is determined by the incorporated proportion of the surface-active agent. In general, the droplet size decreases as surfactant concentration increases through the reduction of the interfacial tension at the oil–water interface. However, excessive surfactant proportion can lead to an increase in droplet size due to interfacial disruption, which results in greater water penetration into oil droplets [30].

A combination of surfactants often produces a synergistic effect, enhancing emulsion stability and performance. This synergy arises from non-ideal mixing within micelles, leading to a reduced critical micelle concentration and surface tension at the interface [105,106]. Weerapol et al. investigated the influence of HLB and the molecular structure of surfactants on SGNE formation. Through the evaluation of different lipids and emulsifiers, they formulated a liquid SGNE loaded with nifedipine using mixed surfactants, such as Tween®/Span® or Cremophor®/Span®. Their findings highlighted that droplet size is largely affected by the surfactant's molecular structure. When the HLB value was approximately 10, formulations containing Cremophor® and lipophilic surfactants with longer hydrocarbon chains (C18, Span® 80) formed nanosized emulsions, whereas those with shorter hydrocarbon chains (C12, Span® 20) resulted in micrometer-sized emulsions [106,107]. Additionally, the droplet size of formed SGNE decreased when the HLB of mixed surfactants (Cremophor®/Span®) exceeded 9 in formulations with Span® 80 and 11.5 in formulations with Span® 20, likely due to increased hydrophilicity.

Mahdi et al. investigated the impact of nonionic surfactant blends on the pseudo-ternary phase diagram of palm kernel oil esters. Their findings revealed that the improved solubilization capacity and increased surface area of water-in-oil microemulsions were due to the structural resemblance between the lipophilic tail of Tween® 80 and the oleyl group in palm kernel oil esters. They concluded that the phase behavior of formulations based on palm kernel oil esters is governed not only by the HLB value but also by the molecular similarity between the surfactant and the oil phase [105]. Similarly, Wang et al. reported that surfactant pairs with similar molecular structures (e.g., Tween® and Span®) at optimal HLB values, combined with suitable oil phases, produce varying particle dimensions based on the surfactant molecular structure [108]. Dai et al. investigated that the molecular structure of surfactants has an important role in the determination of droplet size of the final emulsion [109].

Cosurfactants work with surfactants to enhance drug solubility and the distribution of hydrophilic emulsifiers within the lipid phase, thereby fostering uniformity and formulation stability [94]. Cosurfactants can increase interfacial fluidity by integrating into the surfactant film. Surfactants and cosurfactants primarily accumulate at interfaces, lowering interfacial energy, creating a protective mechanical layer to prevent coalescence, and improving the thermodynamic stability of NEs [87]. Secondary emulsifiers significantly improve the solubility of lipophilic drugs. These emulsifiers help maintain the stability of emulsions, preventing phase separation and ensuring a uniform distribution of the drug within the formulation [110].

4.4.3 SOLIDIFICATION TECHNIQUES

Several solidification techniques have been explored to transform liquid SGNEs into solid SGNEs, while retaining their self-emulsification properties. One of the most widely used techniques is adsorption onto solid carriers, where porous excipients such as hydrophilic fumed silica (Aerosil® 200), amorphous magnesium alumino-metasilicate (Neusilin®), and magnesium alumino-metasilicate absorb the liquid formulation, forming a free-flowing powder. This method has been extensively studied due to its ease of manufacturing and scalability, although concerns regarding carrier–drug interactions and reduced emulsification efficiency have been noted [111]. Spray-drying is another effective solidification approach that involves atomization of the liquid formulation into fine droplets, followed by rapid drying using excipients such as mannitol, lactose, or polyvinylpyrrolidone (PVP) [112]. Research indicates that spray-dried SGNEs demonstrate improved dissolution rates and enhanced bioavailability due to the formation of amorphous solid dispersions. For instance, Patel et al. developed optimized liquid and spray-dried solid SGNEs for ketoprofen using a QbD-DM3 rational design approach. The spray-dried SGNEs exhibited an amorphous nature and showed a 2.37-fold increase in drug release compared to pure ketoprofen in 0.1 N HCl, thereby supporting the potential of solid SGNEs to enhance drug dissolution and meet critical quality parameters, including globule size <100 nm and emulsification time <30 seconds [113]. Sharma et al. demonstrated that Aerosil® 200-based solid SGNEs maintained nanoemulsion characteristics under stress conditions and significantly improved both dissolution rate and oral bioavailability—achieving a 3.28-fold increase in bioavailability compared to the marketed formulation [114]. However, for certain formulations, high processing temperatures may degrade thermosensitive drugs and surfactants [113]. Besides, freeze-drying (lyophilization) has also been explored for solidifying SGNEs, especially for moisture-sensitive drugs. Sublimation under frozen conditions offers very high stability and reconstitution properties, but large-scale production is one of its major limitations [115].

4.4.4 STABILITY CONSIDERATIONS

To prevent the degradation of NE formulation, several factors must be considered. Lipid oxidation, surfactant degradation, and phase separation due to temperature fluctuations will all detrimentally impact the drug release profile. Li et al. showed that SGNE stability is improved by storage at controlled temperatures (15°C–25°C) and the use of moisture-resistant packaging. Moreover, hygroscopic surfactants may absorb moisture that changes emulsification behavior and cause instability [116]. Other factors, such as pH and ionic strength of the gastrointestinal environment, can affect the self-emulsification of SGNEs. In physiological fluid of high ionic strength, the electrical double layer around the emulsion droplets is compressed, and repulsive forces are reduced, which can promote aggregation [117]. Surfactants can be degraded by extreme pH conditions or alter droplet charge, and such changes can impact the integrity of the formulation [118]. The incorporation of pH-stable surfactants and precipitation inhibitors such as HPMC and PVP has been shown to mitigate these effects and enhance the robustness of SGNE formulations [119].

4.4.5 *IN VIVO* PERFORMANCE

However, droplet size, instability due to enzymatic degradation, and risks of drug precipitation during its emulsification in GIT fluids govern the *in vivo* performance of SGNEs. Smaller droplet sizes (<200 nm) improve drug absorption because of their increased surface area, but can also be more prone to coalescence [120]. Murshed et al demonstrated that the use of lipase inhibitors such as orlistat to inhibit gastric and pancreatic lipases can reduce lipid digestion and therefore alter drug release kinetics from SDNEs [121]. Supersaturation of the drug in the GIT environment can cause drug precipitation upon dilution and result in reduced bioavailability, which is one of the most significant concerns with the oral administration of SGNEs [122]. Formulations with crystallization inhibitors, such as polymers (HPMC and PVP), were investigated by Dong et al to stabilize nanocarriers and overcome the precipitation issue [123].

4.4.6 Regulatory and Quality Control

The safety, stability, and efficacy of SGNE drug delivery systems need to be followed by stringent regulatory approvals. Agencies such as the Food and Drug Administration (FDA) and European Medicines Agency (EMA) require extensive characterization data (droplet size distribution, zeta potential, and long-term stability). Furthermore, the excipients used in the formulation must be within allowable or acceptable limits for pharmaceutical applications. These are some of the main reasons for the limited number of approved SNGE-based products, despite their great potential in cancer therapy. The challenge is primarily due to the formulation complexity, potential toxicity concerns, and extensive *in vivo* studies required to demonstrate their safety and efficacy profiles. Tumors are heterogenic, and there are biological barriers NE must cross to reach the site of action, making the approval of NE more complex and demanding. Furthermore, as you scale up, quality control requirements become more difficult to achieve, where small differences in process parameters can result in a lack of consistency in the final product. Proper, robust, and reproducible characterization techniques of produced SGNEs are crucial for meeting regulatory specifications. However, the accuracy and feasibility of executing these characterization requirements can impede the capability of quality monitoring during scale-up.

4.5 CLINICAL TRANSLATION AND COMMERCIAL STANDPOINT

SGNEs have not yet achieved widespread clinical application in anticancer therapy, largely due to challenges with stability, toxicity, regulatory approval, and scalability. However, for indications other than cancer, SGNE-based products are already available and successfully marketed, which confirms their feasibility in enhancing solubility, absorption, and bioavailability. Several SGNE-based formulations have been commercialized for various therapeutic applications, as listed in Table 4.5 [124]. These formulations have significantly improved drug dissolution and systemic absorption, paving the way for broader clinical applications. While no SGNE-based anticancer therapy has been approved by the FDA yet, ongoing research is targeted to optimize stability, tumor-targeting capabilities, and drug release kinetics to overcome existing barriers [125]. Advancements in preclinical research, clinical trials, and industrial collaborations suggest that anticancer SGNEs could soon transition from experimental research to clinical reality, offering a promising approach for enhancing chemotherapy efficacy, while minimizing off-target effects.

TABLE 4.5

Marketed Self-Generating Nanoemulsion (SGNE)-Based Formulations

Brand Name	Active Ingredient	Company	Indication
Gengraf®	Cyclosporine	AbbVie Inc.	Immunosuppressant (Transplantation)
Lipirex®	Atorvastatin	Highnoon Laboratories	Hyperlipidemia
Convulex®	Sodium Valproate	Gerot Lannach UK	Epilepsy, Bipolar Disorder
Norvir®	Ritonavir	AbbVie Inc.	HIV Treatment
Rocaltrol®	Calcitriol	Aphena Pharma Solutions	Osteoporosis, Hypocalcemia
Sandimmune®	Cyclosporine	Novartis	Immunosuppressant
Neoral®	Cyclosporine	Novartis	Improved formulation of Sandimmune®
Vesanoid®	Tretinoin	Roche Laboratories	Acute Promyelocytic Leukemia
Accutane®	Isotretinoin	Roche Laboratories	Severe Acne
Agenerase®	Amprenavir	GlaxoSmithKline	HIV Treatment

REFERENCES

1. Bukowski, K., Kciuk, M., and Kontek, R, Mechanisms of multidrug resistance in cancer chemotherapy. *International Journal of Molecular Sciences*, 2020. **21**(9): p. 3233.
2. Kim, J.H., et al., Application of nanotechnology and phytochemicals in anticancer therapy. *Pharmaceutics*, 2024. **16**(9): p. 1169.
3. Agrawal, S.S., Baliga, V., and Londhe, V.Y., Liposomal formulations: A recent update. *Pharmaceutics*, 2024. **17**(1): p. 36.
4. Majidinia, M., et al., Overcoming multidrug resistance in cancer: Recent progress in nanotechnology and new horizons. *International Union of Biochemistry and Molecular Biology Life*, 2020. **72**(5): pp. 855–871.
5. Sailor, G.U., Self-nanoemulsifying drug delivery systems (snedds): An innovative approach to improve oral bioavailability. In Shah N (ed.), *Nanocarriers: Drug Delivery System: An Evidence Based Approach*, 2021: pp. 255–280.
6. Kadian, R. and Nanda, R., A comprehensive insight on self emulsifying drug delivery systems. *Recent Advances in Drug Delivery and Formulation*, 2022. **16**(1): pp. 16–44.
7. Chu, Y., et al., Experimental study on self–emulsification of shale crude oil by natural emulsifiers. *Journal of Dispersion Science and Technology*, 2024. **45**(5): pp. 859–869.
8. Dangre, P.V. and Chalikwar, S.S., Self-nanoemulsifying drug delivery system: Formulation development and quality attributes. In Vakhrushev, A. V., Kodolov, V. I., Haghi, A. K., Ameta, S. C. (eds.), *Carbon Nanotubes and Nanoparticles*, 2019. Apple Academic Press. pp. 241–258.
9. Suyal, J., Kumar, B, and Jakhmola, V., Novel approach self-nanoemulsifying drug delivery system: A review. *Advances in Pharmacology and Pharmacy*, 2023. **11**(2): pp. 131–139.
10. Salawi, A., Self-emulsifying drug delivery systems: A novel approach to deliver drugs. *Drug Delivery*, 2022. **29**(1): pp. 1811–1823.
11. Taira, M., et al., Stability of liposomal formulations in physiological conditions for oral drug delivery. *Drug Delivery*, 2004. **11**(2): pp. 123–128.
12. Chen, L.Z., et al., Valsartan loaded solid self-nanoemulsifying delivery system to enhance oral absorption and bioavailability. *AAPS PharmSciTech*, 2025. **26**(1): p. 45.
13. Maimaitijiang, A., et al., Progress in research of nanotherapeutics for overcoming multidrug resistance in cancer. *International Journal of Molecular Sciences*, 2024. **25**(18): pp. 9973–10002.
14. Spleis, H., et al., Hydrophobic ion pairing of small molecules: How to minimize premature drug release from SEDDS and reach the absorption membrane in intact form. *ACS Biomaterials Science and Engineering*, 2023. **9**(3): pp. 1450–1459.
15. Soni, A., et al., Nano-biotechnology in tumour and cancerous disease: A perspective review. *Journal of Cellular and Molecular Medicine*, 2023. **27**(6): pp. 737–762.
16. Tran, P., et al., Recent advances of nanotechnology for the delivery of anticancer drugs for breast cancer treatment. *Journal of Pharmaceutical Investigation*, 2020. **50**: pp. 261–270.
17. Malik, S., et al., Application of nanoemulsion in cancer treatment. In Ramalingam, K. (ed.), *Handbook of Research on Nanoemulsion Applications in Agriculture, Food, Health, and Biomedical Sciences*, 2022. IGI Global. pp. 237–259.
18. Al-Thani, A.N., et al., Nanoparticles in cancer theragnostic and drug delivery: A comprehensive review. *Life Sciences*, 2024;352: 122899.
19. Raj, S., et al. Specific targeting cancer cells with nanoparticles and drug delivery in cancer therapy. In *Seminars in Cancer Biology*, 2021. Elsevier, pp. 166–177. Amsterdam, Netherlands.
20. Liu, Z., et al., Unravelling the enigma of sirna and aptamer mediated therapies against pancreatic cancer. *Molecular Cancer*, 2023. **22**(1): p. 8.
21. Kohli, K., et al., Self-emulsifying drug delivery systems: An approach to enhance oral bioavailability. *Drug Discovery Today*, 2010. **15**(21–22): pp. 958–965.
22. Zeng, L., et al., Advancements in nanoparticle-based treatment approaches for skin cancer therapy. *Molecular Cancer*, 2023. **22**(1): p. 10.
23. Rathore C., et al., Self-nanoemulsifying drug delivery system (SNEDDS) mediated improved oral bioavailability of thymoquinone: Optimization, characterization, pharmacokinetic, and hepatotoxicity studies. *Drug Delivery and Translational Research*, 2023. **13**(1): pp. 292–307.
24. Natesh, J., et al., Development of a self-nanoemulsifying drug delivery system of diindolylmethane for enhanced bioaccessibility, bioavailability and anti-breast cancer efficacy. *Journal of Drug Delivery Science and Technology*, 2024. **93**: p. 105435.

25. Guo, Y., et al., Self-assembled nanoparticle-mediated cascade chemotherapy and chemodynamic therapy for enhanced tumor therapy. *Colloids and Surfaces A: Physicochemical and Engineering Aspects*, 2024. **692**: p. 133967.

26. Reyna-Lázaro, L., et al., Pharmaceutical nanoplatforms based on self-nanoemulsifying drug delivery systems for optimal transport and co-delivery of sirnas and anticancer drugs. *Journal of Pharmaceutical Sciences*, 2024. **113**(7): pp. 1907–1918.

27. Xie, B., et al., Solubilization techniques used for poorly water-soluble drugs. *Acta Pharmaceutica Sinica B*, 2024. **14**(11): pp. 4683–4716.

28. Chakraborty, S., et al., Lipid–an emerging platform for oral delivery of drugs with poor bioavailability. *European Journal of Pharmaceutics and Biopharmaceutics*, 2009. **73**(1): pp. 1–15.

29. Fahr, A. and Liu, X., Drug delivery strategies for poorly water-soluble drugs. *Expert Opinion on Drug Delivery*, 2007. **4**(4): pp. 403–416.

30. Pouton, C.W. and Porter, C.J., Formulation of lipid-based delivery systems for oral administration: Materials, methods and strategies. *Advanced Drug Delivery Reviews*, 2008. **60**(6): pp. 625–637.

31. Ding, D., et al., Integration of phospholipid-drug complex into self-nanoemulsifying drug delivery system to facilitate oral delivery of paclitaxel. *Asian Journal of Pharmaceutical Sciences*, 2019. **14**(5): pp. 552–558.

32. Seo, Y.G., et al., Development of docetaxel-loaded solid self-nanoemulsifying drug delivery system (snedds) for enhanced chemotherapeutic effect. *International Journal of Pharmaceutics*, 2013. **452**(1–2): pp. 412–420.

33. Lopes, G. and Lima, C.M.S.R., Docetaxel in the management of advanced pancreatic cancer. *Seminars in Oncology*, 2005. **2**(Suppl 4): pp. S10–S23.

34. Nazari-Vanani, R., Azarpira, N., and Heli, H., Development of self-nanoemulsifying drug delivery systems for oil extracts of citrus aurantium l. Blossoms and rose damascena and evaluation of anticancer properties. *Journal of Drug Delivery Science and Technology*, 2018. **47**: pp. 330–336.

35. Ansari, M.J., et al., Formulation and evaluation of self-nanoemulsifying drug delivery system of brigatinib: Improvement of solubility, in vitro release, ex-vivo permeation and anticancer activity. *Journal of Drug Delivery Science and Technology*, 2021. **61**: p. 102204.

36. Kamel, A.O. and Mahmoud, A.A., Enhancement of human oral bioavailability and in vitro antitumor activity of rosuvastatin via spray dried self-nanoemulsifying drug delivery system. *Journal of Biomedical Nanotechnology*, 2013. **9**(1): pp. 26–39.

37. Truong, D.H., et al., Development of solid self-emulsifying formulation for improving the oral bioavailability of erlotinib. *AAPS Pharmscitech*, 2016. **17**: pp. 466–473.

38. Qian, J., et al., Self-nanoemulsifying drug delivery systems of myricetin: Formulation development, characterization, and in vitro and in vivo evaluation. *Colloids and Surfaces B: Biointerfaces*, 2017. **160**: pp. 101–109.

39. Akhtar, N., et al., Self-nanoemulsifying lipid carrier system for enhancement of oral bioavailability of etoposide by p-glycoprotein modulation: In vitro cell line and in vivo pharmacokinetic investigation. *Journal of Biomedical Nanotechnology*, 2013. **9**(7): pp. 1216–1229.

40. Beg, S., et al., Natural microbial surfactant containing self-nanoemulsifying formulation with improved performance of paclitaxel therapy: A newer avenue in breast cancer treatment. *Journal of Drug Delivery Science and Technology*, 2023. **90**: p. 105105.

41. Nazari-Vanani, R., et al., An in vitro study on anticancer efficacy of capecitabine- and vorinostat-incorporated self-nanoemulsions. *Journal of Biomedical Physics and Engineering*, 2024. 15(4): pp. 353–368.

42. Usmani, A., et al., Development and evaluation of doxorubicin self nanoemulsifying drug delivery system with nigella sativa oil against human hepatocellular carcinoma. *Artificial Cells, Nanomedicine, and Biotechnology*, 2019. **47**(1): pp. 933–944.

43. Karimi, M., et al., Investigation of bioavailability and anti-pancreatic cancer efficacy of a self-nanoemulsifying erlotinib delivery system. *Therapeutic Delivery*, 2025. **16**(3): pp. 1–10.

44. Jvus, C., et al., A quality by design approach for developing SNEDDS loaded with vemurafenib for enhanced oral bioavailability. *AAPS PharmSciTech*, 2024. **25**(1): p. 14.

45. Shin, Y., et al., Development and characterization of Olaparib-loaded solid self-nanoemulsifying drug delivery system (s-snedds) for pharmaceutical applications. *AAPS PharmSciTech*, 2024. **25**(7): p. 221.

46. Wu, F., et al., Formulation and evaluation of solid self-nanoemulsifying drug delivery system of cannabidiol for enhanced solubility and bioavailability. *Pharmaceutics*, 2025. **17**(3): p. 340.

47. Patel, K., et al., Medium chain triglyceride (MCT) rich, paclitaxel loaded self nanoemulsifying preconcentrate (PSNP): A safe and efficacious alternative to taxol®. *Journal of Biomedical Nanotechnology*, 2013. **9**(12): pp. 1996–2006.

48. Park, J.H., et al., Targeted delivery of low molecular drugs using chitosan and its derivatives. *Advanced Drug Delivery Reviews*, 2010. **62**(1): pp. 28–41.

49. Hörmann, K. and Zimmer, A., Drug delivery and drug targeting with parenteral lipid nanoemulsions—A review. *Journal of Controlled Release*, 2016. **223**: pp. 85–98.

50. Nikonenko, B., et al., Therapeutic efficacy of sq641-ne against mycobacterium tuberculosis. *Antimicrobial Agents and Chemotherapy*, 2014. **58**(1): pp. 587–589.

51. Hua, S., et al., Advances in oral nano-delivery systems for colon targeted drug delivery in inflammatory bowel disease: Selective targeting to diseased versus healthy tissue. *Nanomedicine: Nanotechnology, Biology and Medicine*, 2015. **11**(5): pp. 1117–1132.

52. Lainé, A.-L., et al., Conventional versus stealth lipid nanoparticles: Formulation and in vivo fate prediction through fret monitoring. *Journal of Controlled Release*, 2014. **188**: pp. 1–8.

53. Mills, J.A., et al., Nanoparticle based medicines: Approaches for evading and manipulating the mononuclear phagocyte system and potential for clinical translation. *Biomaterials Science*, 2022. **10**(12): pp. 3029–3053.

54. Feeney, O.M., et al., 'Stealth'lipid-based formulations: Poly (ethylene glycol)-mediated digestion inhibition improves oral bioavailability of a model poorly water soluble drug. *Journal of Controlled Release*, 2014. **192**: pp. 219–227.

55. Khan, S.A. and Schneider, M., Nanoprecipitation versus two step desolvation technique for the preparation of gelatin nanoparticles. In *Colloidal Nanocrystals for Biomedical Applications VIII*, 2013. SPIE. Colloidal Nanocrystals for Biomedical Applications VIII, 85950H.

56. Dünnhaupt, S., et al., Nano-carrier systems: Strategies to overcome the mucus gel barrier. *European Journal of Pharmaceutics and Biopharmaceutics*, 2015. **96**: pp. 447–453.

57. Barthelmes, J., et al., Thiomer nanoparticles: Stabilization via covalent cross-linking. *Drug Delivery*, 2011. **18**(8): pp. 613–619.

58. Anton, N. and Vandamme, T.F., The universality of low-energy nano-emulsification. *International Journal of Pharmaceutics*, 2009. **377**(1–2): pp. 142–147.

59. Kanwal, T., et al., Design of absorption enhancer containing self-nanoemulsifying drug delivery system (snedds) for curcumin improved anti-cancer activity and oral bioavailability. *Journal of Molecular Liquids*, 2021. **324**: p. 114774.

60. Batool, A., et al., Formulation and evaluation of hyaluronic acid-based mucoadhesive self nanoemulsifying drug delivery system (snedds) of tamoxifen for targeting breast cancer. *International Journal of Biological Macromolecules*, 2020. **152**: pp. 503–515.

61. Chaudhuri, A., et al., Designing and development of omega-3 fatty acid based self-nanoemulsifying drug delivery system (snedds) of docetaxel with enhanced biopharmaceutical attributes for management of breast cancer. *Journal of Drug Delivery Science and Technology*, 2022. **68**: p. 103117.

62. Setiawan, S.D., et al., Study of self nano-emulsifying drug delivery system (SNEDDS) loaded red fruit oil (*Pandanus conoideus* lamk.) as an eliminated cancer cell MCF-7. *International Journal of Drug Delivery Technology*, 2018. **8**(4): pp. 229–232.

63. Rajana, N., et al., Quality by design approach-based fabrication and evaluation of self-nanoemulsifying drug delivery system for improved delivery of venetoclax. *Drug Delivery and Translational Research*, 2024. **14**(5): pp. 1277–1300.

64. Kholieqoh, A.H., et al., SNEDDS to improve the bioactivities of pandanus tectorius leaves: Optimization, antioxidant, and anticancer activities via apoptosis induction in human cervical cancer cell line. *Journal of Applied Pharmaceutical Science*, 2024. **14**(10): pp. 175–189.

65. Garcia, C.R., et al., Nanoemulsion delivery systems for enhanced efficacy of antimicrobials and essential oils. *Biomaterials Science*, 2022. **10**(3): pp. 633–653.

66. Shah, D., Ajazuddin, and Bhattacharya, S., Role of natural p-gp inhibitor in the effective delivery for chemotherapeutic agents. *Journal of Cancer Research and Clinical Oncology*, 2023. **149**(1): pp. 367–391.

67. Chen, T., et al., Dasatinib reverses the multidrug resistance of breast cancer mcf-7 cells to doxorubicin by downregulating p-gp expression via inhibiting the activation of ERK signaling pathway. *Cancer Biology & Therapy*, 2015. **16**(1): pp. 106–114.

68. Rathod, S., et al., Non-ionic surfactants as a P-glycoprotein(P-gp) efflux inhibitor for optimal drug delivery-A concise outlook. *AAPS PharmSciTech*, 2022. **23**(1): p. 55.

69. Patel, D., et al., Exploring the potential of p-glycoprotein inhibitors in the targeted delivery of anticancer drugs: A comprehensive review. *European Journal of Pharmaceutics and Biopharmaceutics*, 2024. **198**: p. 114267.

70. Chaturvedi, S., Verma, A., and Saharan, V.A., Lipid drug carriers for cancer therapeutics: An insight into lymphatic targeting, P-gp, CYP3A4 modulation and bioavailability enhancement. *Advanced Pharmaceutical Bulletin*, 2020. **10**(4): p. 524.

71. Shadab Md, et al., Resveratrol loaded self-nanoemulsifying drug delivery system (snedds) for pancreatic cancer: Formulation design, optimization and in vitro evaluation. *Journal of Drug Delivery Science and Technology*, 2021. **64**: p. 102555.

72. Akhtar, N., et al., Potential of a novel self nanoemulsifying carrier system to overcome p-glycoprotein mediated efflux of etoposide: In vitro and ex vivo investigations. *Journal of Drug Delivery Science and Technology*, 2015. **28**: pp. 18–27.

73. Baghdadi, R.A., et al., Evaluation of the effects of a dasatinib-containing, self-emulsifying, drug delivery system on HT29 and SW420 human colorectal carcinoma cells, and MCF7 human breast adenocarcinoma cells. *Journal of Taibah University Medical Sciences*, 2024. **19**(4): pp. 806–815.

74. Nottingham, E., et al., The role of self-nanoemulsifying drug delivery systems of cdoda-me in sensitizing erlotinib-resistant non–small cell lung cancer. *Journal of Pharmaceutical Sciences*, 2020. **109**(6): pp. 1867–1882.

75. Patil, O.B., et al., Development of stable self-nanoemulsifying composition and its nanoemulsions for improved oral delivery of non-oncology drugs against hepatic cancer. *OpenNano*, 2022. **7**: p. 100044.

76. Kazi, M., et al., Bioactive self-nanoemulsifying drug delivery systems (bio-snedds) for combined oral delivery of curcumin and piperine. *Molecules*, 2020. **25**(7): p. 1703.

77. Pangeni, R., et al., Multiple nanoemulsion system for an oral combinational delivery of oxaliplatin and 5-fluorouracil: Preparation and in vivo evaluation. *International Journal of Nanomedicine*, 2016: pp. 6379–6399.

78. Tripathi, S., et al., Triple antioxidant snedds formulation with enhanced oral bioavailability: Implication of chemoprevention of breast cancer. *Nanomedicine: Nanotechnology, Biology and Medicine*, 2016. **12**(6): pp. 1431–1443.

79. Ateeq, M.A.M., et al., Self-nanoemulsifying drug delivery system (snedds) of docetaxel and carvacrol synergizes the anticancer activity and enables safer toxicity profile: Optimization, and in-vitro, ex-vivo and in-vivo pharmacokinetic evaluation. *Drug Delivery and Translational Research*, 2023. **13**(10): pp. 2614–2638.

80. Jain, A.K., Thanki, K., and Jain, S., Solidified self-nanoemulsifying formulation for oral delivery of combinatorial therapeutic regimen: Part II in vivo pharmacokinetics, antitumor efficacy and hepatotoxicity. *Pharmaceutical Research*, 2014. **31**: pp. 946–958.

81. Grill, A.E., Koniar, B, Panyam, J., Co-delivery of natural metabolic inhibitors in a self-microemulsifying drug delivery system for improved oral bioavailability of curcumin. *Drug Delivery and Translational Research*, 2015. **4**(4):344–352.

82. Jain, A.K., Thanki, K, and Jain, S., Solidified self-nanoemulsifying formulation for oral delivery of combinatorial therapeutic regimen: Part I. Formulation development, statistical optimization, and in vitro characterization. *Pharmaceutical Research*, 2014. **31**: pp. 923–945.

83. Ceramella, J., et al., A winning strategy to improve the anticancer properties of cisplatin and quercetin based on the nanoemulsions formulation. *Journal of Drug Delivery Science and Technology*, 2021. **66**: p. 102907.

84. Kamal, R., et al., Formulation and evaluation of mesalamine-quercetin loaded self-nano emulsifying drug delivery systems (snedds) for the management of colon cancer. *BioNanoScience*, 2025. **15**(1): pp. 1–15.

85. Koirala, M. and Dipaola, M., Overcoming cancer resistance: Strategies and modalities for effective treatment. *Biomedicines*, 2024. **12**(8): p. 1801.

86. Arshad, R., et al., Amidated pluronic decorated muco-penetrating self-nano emulsifying drug delivery system (SNEDDS) for improved anti-salmonella typhi potential. *Pharmaceutics*, 2022. **14**(11): p. 2433.

87. Singh, D., Self-nanoemulsifying drug delivery system: A versatile carrier for lipophilic drugs. *Pharmaceutical Nanotechnology*, 2021. **9**(3): pp. 166–176.

88. Zhao, Y., et al., Self-nanoemulsifying drug delivery system (SNEDDS) for oral delivery of zedoary essential oil: Formulation and bioavailability studies. *International Journal of Pharmaceutics*, 2010. **383**(1–2): pp. 170–177.

89. Hayat, A., et al., Design and development of a self-nanoemulsifying drug delivery system for co-delivery of curcumin and naringin for improved wound healing activity in an animal model. *Planta Medica*, 2024. **90**(12): pp. 959–970.

90. Rehman, F.U., et al., From nanoemulsions to self-nanoemulsions, with recent advances in self-nanoemulsifying drug delivery systems (snedds). *Expert Opinion on Drug Delivery*, 2017. **14**(11): pp. 1325–1340.

91. Elsheikh, M.A., et al., Nanoemulsion liquid preconcentrates for raloxifene hydrochloride: Optimization and in vivo appraisal. *International Journal of Nanomedicine*, 2012, **7**: pp. 3787–3802.

92. Rao, J. and Mcclements, D.J., Formation of flavor oil microemulsions, nanoemulsions and emulsions: Influence of composition and preparation method. *Journal of Agricultural and Food Chemistry*, 2011. **59**(9): pp. 5026–5035.

93. Čerpnjak, K., et al., Lipid-based systems as promising approach for enhancing the bioavailability of poorly water-soluble drugs. *Acta Pharmaceutica*, 2013. **63**(4): pp. 427–445.

94. Lu, L.-Y., et al., Pomegranate seed oil exerts synergistic effects with trans-resveratrol in a self-nanoemulsifying drug delivery system. *Biological and Pharmaceutical Bulletin*, 2015. **38**(10): pp. 1658–1662.

95. Holm, R., et al., Influence of bile on the absorption of halofantrine from lipid-based formulations. *European Journal of Pharmaceutics and Biopharmaceutics*, 2012. **81**(2): pp. 281–287.

96. Singh, B., et al., Self-emulsifying drug delivery systems (SEDDS): Formulation development, characterization, and applications. *Critical Reviews™ in Therapeutic Drug Carrier Systems*, 2009. 26(5):427–521.

97. Porter, C.J. and Charman, W.N., Intestinal lymphatic drug transport: An update. *Advanced Drug Delivery Reviews*, 2001. **50**(1–2): pp. 61–80.

98. Prajapati, H.N., Dalrymple, D.M., and Serajuddin, A.T., A comparative evaluation of mono-, di-and triglyceride of medium chain fatty acids by lipid/surfactant/water phase diagram, solubility determination and dispersion testing for application in pharmaceutical dosage form development. *Pharmaceutical Research*, 2012. **29**: pp. 285–305.

99. Zhang, J., et al., Biodistribution, hypouricemic efficacy and therapeutic mechanism of morin phospholipid complex loaded self-nanoemulsifying drug delivery systems in an experimental hyperuricemic model in rats. *Journal of Pharmacy and Pharmacology*, 2016. **68**(1): pp. 14–25.

100. Zhou, H., et al., A new strategy for enhancing the oral bioavailability of drugs with poor water-solubility and low liposolubility based on phospholipid complex and supersaturated sedds. *PLoS One*, 2013. **8**(12): p. e84530.

101. Tian, C.-T., et al., Long chain triglyceride-lipid formulation promotes the oral absorption of the lipidic prodrugs through coincident intestinal behaviors. *European Journal of Pharmaceutics and Biopharmaceutics*, 2022. **176**: pp. 122–132.

102. Islam, F., et al., Functional roles and novel tools for improving-oxidative stability of polyunsaturated fatty acids: A comprehensive review. *Food Science & Nutrition*, 2023. **11**(6): pp. 2471–2482.

103. Azeem, A., et al., Nanoemulsion components screening and selection: A technical note. *AAPS Pharmscitech*, 2009. **10**: pp. 69–76.

104. Martin, C., et al., Improved biodistribution and enhanced immune response of subunit vaccine using a nanostructure formed by self-assembly of ascorbyl palmitate. *Nanomedicine*, 2024. **58**: p. 102749.

105. Mahdi, E.S., et al., Effect of surfactant and surfactant blends on pseudoternary phase diagram behavior of newly synthesized palm kernel oil esters. *Drug Design, Development and Therapy*, 2011. **5**: pp. 311–323.

106. Weerapol, Y., et al., Self-nanoemulsifying drug delivery system of nifedipine: Impact of hydrophilic–lipophilic balance and molecular structure of mixed surfactants. *AAPS Pharmscitech*, 2014. **15**: pp. 456–464.

107. Weerapol, Y., et al., Spontaneous emulsification of nifedipine-loaded self-nanoemulsifying drug delivery system. *AAPS PharmSciTech*, 2015. **16**: pp. 435–443.

108. Wang, L., et al., Design and optimization of a new self-nanoemulsifying drug delivery system. *Journal of Colloid and Interface Science*, 2009. **330**(2): pp. 443–448.

109. Dai, L., W. Li, and Hou, X., Effect of the molecular structure of mixed nonionic surfactants on the temperature of miniemulsion formation. *Colloids and Surfaces A: Physicochemical and Engineering Aspects*, 1997. **125**(1): pp. 27–32.

110. Karwal, R., et al., Current trends in self-emulsifying drug delivery systems (SEDDS) to enhance the bioavailability of poorly water-soluble drugs. *Critical Reviews in Therapeutic Drug Carrier Systems*, 2016. **33**(1): 1–39.

111. Uttreja, P., et al., Self-emulsifying drug delivery systems (sedds): Transition from liquid to solid—A comprehensive review of formulation, characterization, applications, and future trends. *Pharmaceutics*, 2025. **17**(1): p. 63.

112. Li, L., Yi, T., and Lam, C.W.-K., Effects of spray-drying and choice of solid carriers on concentrations of Labrasol® and Transcutol® in solid self-microemulsifying drug delivery systems (SMEDDS). *Molecules*, 2013, 18(1):545–560.

113. Patel, V.D., et al., Optimized L-SNEDDS and spray-dried S-SNEDDS using a linked QbD-DM3 rational design for model compound ketoprofen. *International Journal of Pharmaceutics*, 2023. **631**: p. 122494.

114. Sharma, P., et al., Impact of solid carriers and spray drying on pre/post-compression properties, dissolution rate and bioavailability of solid self-nanoemulsifying drug delivery system loaded with simvastatin. *Powder Technology*, 2018. **338**: pp. 836–846.

115. Liu, Y., Zhang, Z., and Hu, L., High efficient freeze-drying technology in food industry. *Critical Reviews in Food Science and Nutrition*, 2022. **62**(12): pp. 3370–3388.

116. Li, X.-L., et al., Fortification of polysaccharide-based packaging films and coatings with essential oils: A review of their preparation and use in meat preservation. *International Journal of Biological Macromolecules*, 2023. **242**: p. 124767.

117. Fatouros, D.G., Bergenstah, B., and Mullertz, A., Morphological observations on a lipid-based drug delivery system during in vitro digestion. *European Journal of Pharmaceutical Sciences*, 2007. **31**(2): pp. 85–94.

118. Morales, A.M., et al., End-to-end approach to surfactant selection, risk mitigation, and control strategies for protein-based therapeutics. *The AAPS Journal*, 2022. **25**(1): p. 6.

119. Verma, R., et al., Exploring the role of self-nanoemulsifying systems in drug delivery: Challenges, issues, applications and recent advances. *Current Drug Delivery*, 2023. **20**(9): pp. 1241–1261.

120. Mcclements, D.J. and Xiao, H., Potential biological fate of ingested nanoemulsions: Influence of particle characteristics. *Food & Function*, 2012. **3**(3): pp. 202–220.

121. Murshed, M., et al., Controlling drug release by introducing lipase inhibitor within a lipid formulation. *International Journal of Pharmaceutics*, 2022. **623**: p. 121958.

122. Borhade, V., et al., Clotrimazole nanoemulsion for malaria chemotherapy. Part I: Preformulation studies, formulation design and physicochemical evaluation. *International Journal of Pharmaceutics*, 2012. **431**(1–2): pp. 138–148.

123. Dong, B., Lim, L.M., and Hadinoto, K., Enhancing the physical stability and supersaturation generation of amorphous drug-polyelectrolyte nanoparticle complex via incorporation of crystallization inhibitor at the nanoparticle formation step: A case of HPMC versus PVP. *European Journal of Pharmaceutical Sciences*, 2019. **138**: p. 105035.

124. Park, H., Ha E.S., and Kim, M.S., Current status of supersaturable self-emulsifying drug delivery systems. *Pharmaceutics*, 2020. **12**(4): p. 365.

125. Lenane, H.B., et al., A retrospective biopharmaceutical analysis of >800 approved oral drug products: Are drug properties of solid dispersions and lipid-based formulations distinctive? *Journal of Pharmaceutical Sciences*. 2020. **109**(11): pp. 3248–3261.

5 Active Targeting of Nanoparticles is a Smart Way for Treating Cancer

Ahmed A.H. Abdellatif, Mohamed S. Abdel-Bakky,
Shaaban K. Osman, and Abdellatif Bouazzaoui

5.1 INTRODUCTION

It is important that the biologically active molecule, delivered in any dosage form, is capable of crossing biological barriers to reach the site of action. Many components form the biological barriers, including the skin, mucosa, cell membrane, epithelium, endothelium, small and large intestine, and blood–brain barrier. Many biologically active molecules fail to reach their targets; therefore, they are classified as ineffective. The location-specific delivery of therapeutic agents is currently of great interest to many researchers, so that many diseases can be efficiently overcome with few side effects to treat many diseases such as cancer and genetic diseases [1]. Low bioavailability, side effects, and biological degradation are the common drawbacks of the active pharmaceutical ingredients; therefore, novel drug carrier systems have become imperative to overcome these downsides [2]. Therefore, drugs or genes have been introduced to targeted delivery systems to be delivered into specific cells or tissues. Two advanced methods have been developed to develop effective interventions, including active and passive drug targeting choices. Both drug targeting strategies are used in drug delivery systems to increase drugs specificity and efficacy of the medication of certain diseases like infections or tumors [3].

Passive targeting employs the natural physiological and pathological differences between healthy and diseased tissues to drive active medication into the target site [4]. This strategy depends on the enhanced permeability and retention (EPR) action occurring with inflammation or cancer. In this condition, blood vessels become permeable and reduce lymphatic drainage, permitting the accumulation of nanoparticles [5]. Passive drug targeting lacks selectivity, that is, drugs may reach healthy tissues and have limited applicability to tissues without EPR action. In contrast, modifying the drug delivery system in the case of active targeting is the cornerstone in this strategy to interact precisely with the target site via the molecular or biochemical recognition [6]. The commonly used legends in this strategy are peptides, antibodies, or other legends that have an affinity to specific receptors expressed on their target cell membrane. Legends reach intracellular locations by endocytosis mediated through high-affinity binding to their receptors [7]. The main advantages of this strategy are the higher selectivity for target sites that diminish off-target effects and the overwhelming many limitations of passive targeting. The most common limitations of active targeting are its complexity and its relatively expensive strategy to develop. Furthermore, its affinity and stability can be modified in response to the biological conditions (Table 5.1).

In the current work, we discuss innovative therapeutic approaches that enhance the accuracy of drug delivery systems, including passive and active targeting methods to aid these nanoparticles to be delivered to the required bindings, principally cancer, effectively. To understand active and passive approaches, we discussed in brief the nanoparticle examples used, advantages and drawbacks for both, and their role in treating chronic diseases, as well as the mechanism of nanoparticle internalization of active targeting.

DOI: 10.1201/9781003517870-5

TABLE 5.1
Key Differences between Passive and Active Targeting Strategies

Feature	Passive Targeting	Active Targeting
Specificity	Low (depends on physiological features)	High (ligand–receptor interactions)
Mechanism	EPR effect, size-based accumulation	Molecular recognition (e.g., ligands)
Complexity	Simple	Complex
Applications	Tumor targeting, inflammation	Precision therapies, personalized medicine

5.2 ACTIVE VERSUS PASSIVE TARGETING AND THEIR ROLE IN TREATING CHRONIC DISEASES

Passive targeting can be defined as the normal distribution of drugs or molecules through the body, depending on the physiological mechanisms such as the bloodstream and the permeability of vascular systems [8,9]. For example, chemotherapeutic agents usually circulate in the bloodstream and slowly accumulate in cancer areas because of the blood vessel leakage characteristic of cancer tissues, which is defined as the EPR effect [10,11]. This process permits drugs to spread through the diseased tissues without the need for complex modification; this kind of targeting is considered non-specific distribution to healthy tissues as well. Accordingly, although passive targeting can offer some benefits, it also faces challenges, including general toxicity and concentrated therapeutic efficacy [12].

Active targeting can be formulated through bioconjugation techniques to easily transport the therapeutics to the target site. This kind of bioconjugation helps them to easily find the target organs, cells, and tissues in the whole body [13]. These bioconjugations proceeded through the attachment of the active ligands, such as peptides or polypeptides, to the nanoparticles for targeting the specific receptors on the targeted cells. Drugs can also be carried out on the formed nanoparticle (NP)-ligands inside the formed NPs or in between the branched ligands [14]. The purpose of using targeting ligands is to determine and regulate the specificity of the healing. The active targeting holds great promise for treating cancer, autoimmune diseases, and diabetes mellitus. Targeting affected cells can improve therapeutic outcomes, while decreasing damage to healthy cells. For example, monoclonal antibodies were designed to target different specific tumor antigens, thus decreasing side effects typically associated with standard chemotherapy [15]. Both passive and active targeting offer distinctive advantages and challenges in the treatment of chronic disorders. Passive targeting is simple and easy to develop, making it attractive for prevalent initial treatment choices. Nevertheless, lacking specificity may lead to substantial side effects, which limit their continuing use for chronic conditions where patient quality of life is a serious concern [15,16].

Furthermore, active targeting has the power to achieve greater specificity and efficacy. They frequently imply higher development costs and complications in preparation [17,18]. The developing medication in the form of nanomedicine opens the gate for novel carriers for targeting sites. Liposomes, metal NPs, and lipid NPs are able to incorporate with ligands for both active and passive targeting functionalities, thus expanding therapeutic outcomes [19–21].

Recent advancements in drug delivery technology continue to bridge the gap between these two approaches. Hybrid systems that include both passive and active targeting approaches are being researched extensively, showing promising results in enhancing drug delivery precision and effectiveness [21,22]. For instance, integrating targeting ligands and leveraging the EPR effect can allow for a multi-faceted approach that responds to the biopharmaceutical's behavior in complex biological systems [3,12,23]. As researchers delve further into optimizing these targeting strategies, the future of chronic disease treatment looks promising. By balancing the strengths of passive and active targeting mechanisms, we may be on the verge of transitioning towards more personalized, effective, and less harmful therapeutic modalities. This holistic approach holds the potential not only to enhance treatment outcomes but also to significantly improve the overall quality of life for those facing chronic diseases [12,24].

5.3 PASSIVE TARGETING IN NP DRUG DELIVERY

NP technology has emerged as a revolutionary approach to drug delivery. One promising application lies in the mechanism of passive targeting, which harnesses the unique properties of NPs to improve the efficacy of therapeutic agents while minimizing side effects [2,12,25]. Passive targeting denotes the NPs accumulated in target tissues because physiological differences between healthy and diseased tissues [2]. A key typical of tumors and other unhealthy cells is the improved EPR effect [2,26,27]. There are many types of NPs that have been reported and formulated for passive targeting. Figure 5.1a shows the shape and structure of NPs.

5.3.1 LIPOSOMES (LPS)

LPS are spherical shapes of vesicles comprised of different types of lipids. They can be encapsulated with different drugs to promote their solubility and bioavailability [28]. LPS have a nanoscale size and can exert the EPR effect, which leads to high concentrations of drug in cancer tissues. Doxil is an example of a drug that was formulated in liposomal formulations of doxorubicin. It is a crucial example where the mechanism of targeting is passive, which resulted in significantly reduced side effects in comparison to the traditional dosage form of chemotherapy [29].

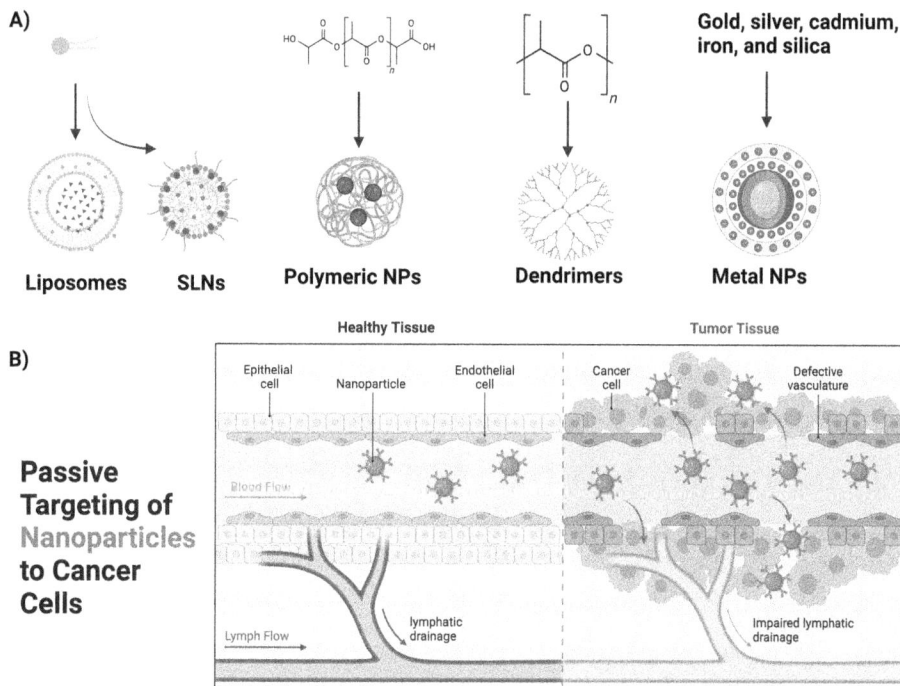

FIGURE 5.1 (a) Shapes of different types of nanoparticles that are reported to be selected for passive targeting. (b) The process of nanoparticles internalization through the passive targeting process in cancer cells. (Created in BioRender. Abdellatif, A. (2024) https://BioRender.com/w07d509).

5.3.2 Polymeric NPs (PNPs)

PNPs are biodegradable NPs prepared from polymers such as poly(lactic-co-glycolic acid) (PLGA), exploiting the passive directing effectually [30,31]. PLGA can encapsulate drugs with different solubilities (hydrophilic and hydrophobic drugs). This kind of formulation slowly releases drugs within tumor sites. This kind of property has been verified in the delivery of anti-tumor medications that showed improved therapeutic effects in both preclinical and clinical situations [32,33].

5.3.3 Dendrimers

Dendrimers are substantially branched polymers. They look like tree-like formations. They can be modified to carry numerous drugs [34]. They exert different nanoscale sizes and surface functionalities that enable them to be collected in tumor sites depending on the EPR action [35]. Several studies proved that dendritic NPs can successfully enhance the penetrability of drugs via the blood–brain barrier, making them fundamental in curing neurological disorders [34,35].

5.3.4 Inorganic NPs

Inorganic NPs **and** metal NPs like silver, gold, iron, cadmium, and silica NPs are frequently used for therapeutic as well as imaging purposes in treating cancer [15,25,36,37]. These types of NPs have special characteristics, such as small size, which facilitates their passive targeting to tumors. Furthermore, gold NPs (AuNPs) can be prepared to improve photothermal therapy, where they attract light and transform it to heat, a property of Au to selectively destroy cancer cells without affecting the surrounding healthy tissues [38,39].

5.3.5 Nanostructured Solid Lipid NPs (SLNs)

SLNs combine the benefits of liposomes and solid lipid transporters and can be encapsulated with drugs, while keeping a solid structure. They help in effective passive targeting due to their small nanometer-sized dimensions. They can also enhance the oral bioavailability of drugs with poor solubility that lead to improved therapeutic outcomes [22,32].

For internalization of NPs via passive targeting, the EPR effect plays the main role in their internalization. The NPs are significantly affected by the shape, size, and surface properties. Mostly, NPs ranging (10–200 nm) in diameter are predicted to be internalized by passive targeting. The smaller sizes are more expected to be cleared from the bloodstream through the kidneys, while the bigger NPs might not enter the tumor site efficiently. Hence, enhancing the NP size is critical for exploiting their accumulation in cancer cells passively (Figure 5.1b). Neutral or modestly positive surface charges can be penetrated through cellular absorption by lowering repulsive interactions between the NPs and the negatively charged membrane [12,40].

5.4 ADVANTAGES OF PASSIVE TARGETING

The advantages of utilizing NPs with passive targeting mechanisms are considerable. They offer improved drug solubility, extended circulation time, and lower toxicity to healthy tissues. By focusing on transporting the therapeutic agents into the tumor sites, NPs effectively attain higher local drug concentrations and enhance drug efficacy [41]. Moreover, passive targeting comprises fewer adjustments and inappropriate targeting ligands in comparison to active targeting mechanisms, making the process easier and inexpensive, which leads to cost-effective pharmaceutical development without affecting the safety and effectiveness [2,22]. This field of passive targeting nanomedicine can be used in diagnosis and treatment in numerous medical circumstances, specifically

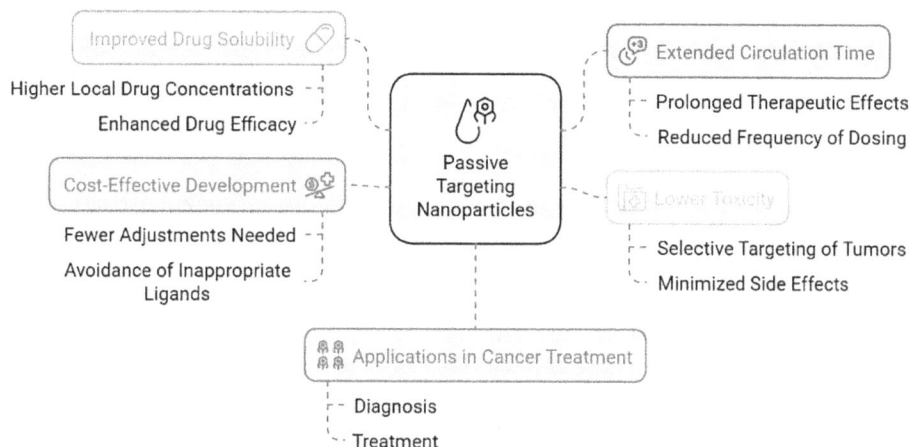

FIGURE 5.2 Advantages of passive targeting (Graph created using napkin.ai).

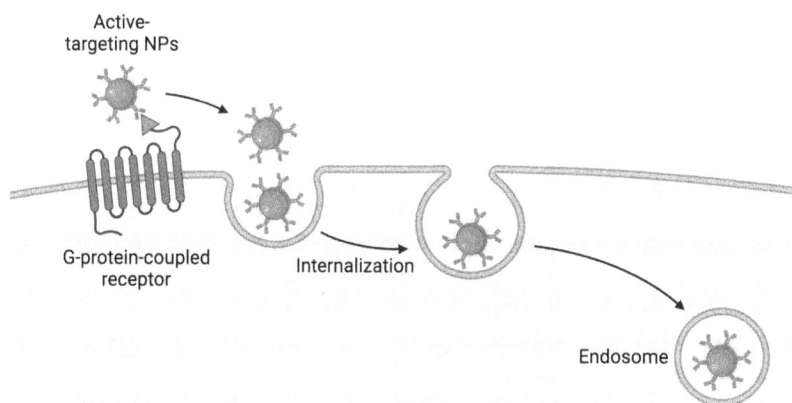

FIGURE 5.3 The process of internalization of NPs through G-protein coupled receptors as an example of active targeting (Created in BioRender. Abdellatif, A. (n.d.) https://BioRender.com/s73v942).

malignancy. One of the revolutionary developments in this area is the formulation of NPs that can target cells or tissues (Figure 5.2) [41].

5.5 ACTIVE TARGETING IN DRUG DELIVERY

Active targeting involves modification of NPs that can specifically be recognized and bound to definite biological indicators on target cells. This formulation is predictably accomplished by the conjugation of NPs with targeting ligands/molecules that can guide NPs selectively to receptors overexpressed on targeted diseased cells. By navigating therapeutic agents straightforwardly to the affected targeted sites, this kind of technique can develop treatment outcomes and reduce systemic toxicity [42] (Figure 5.3).

5.5.1 GOLD NPS (AUNPS)

Scientists developed AuNPs coated with antibodies for targeting cancer-specific antigens such as HER2 and HER3, which are overexpressed in certain breast cancers. This tailored technique has established potential in selectively delivering chemotherapeutic drugs to cancer cells, boosting cell death, while limiting injury to healthy organs [43,44]. AuNPs were coated with somatostatin hormone for targeting of somatostatin receptors in different cell lines [16].

5.5.2 LIPOSOMAL TARGETING

Liposomes can be conjugated with different ligands that can actively target the receptors on cell cancer types that are positively expressed by the receptor. For example, PEGylated liposomes capped with folic acid have been reported to target ovarian cancer cells, which habitually overexpress folate receptors [45]. Different studies showed that treatment with liposomes increases cellular absorption of the medication and develops therapeutic outcomes [45,46].

5.5.3 BIODEGRADABLE POLYMERIC NPs

Polymeric NPs can also be defined as biodegradable polymeric NPs (BPNs). They have been broadly exploited for different targeted drug delivery. PLGA is an example of BNPs that can be capped with different types of peptides, polypeptides, or antibodies to target specific membrane antigen (PSMA) such as prostate cancer cells [47], liver cancer [48], and colorectal cancer [49]. This technique improves the distribution of anticancer medicines directly to the tumor location, allowing for more efficient treatment.

5.5.4 MESOPOROUS SILICA NPs

Silica NPs can be defined as mesoporous silica NPs (MSNs). They can be tailored after decoration of their surfaces with different targeting ligands for targeted delivery. MSNs were decorated with transferrin for targeting different cells expressing transferrin receptors, such as brain tumors [50]. Also, MSNs were decorated with different targeting agents such as 1-methyl-D-tryptophan [51] and 17alpha-Ethynylestradiol [52] for targeting breast cancer, which indeed improved the drug accumulation inside tumors and reduced side effects.

5.5.5 CARBON NANOTUBES (CNTs)

CNTs are being studied for numerous types of active targeting purposes. It was reported that CNTs have been functionalized with various ligands to target specific tumor markers. CNTs have a distinctive structure for effective drug targeting and imaging applications [53]. Active targeting using CNTs has promising specific and effective treatment options for numerous diseases, especially in cancer [54]. The forthcoming generation of active targeting NPs seems promising, with the potential of completely altering the landscape of medical therapies.

5.6 TYPES OF CANCERS TARGETED BY ACTIVE TARGETING

One of the most widespread cancers among women is breast cancer. Breast cancer can be targeted using different ligands, such as trastuzumab (Herceptin) that binds specifically to the HER2 receptor [55]. Lung cancer, particularly non-small cell lung cancer, can also be targeted by active targeting strategies such as targeting the epidermal growth factor receptor [56]. Figure 5.4 shows the different types of cancers and shapes of NPs that can be targeted actively. Osimertinib is selected to inhibit tumor growth in patients who have specific genetic mutations [57]. Further, monoclonal antibodies can bind with vascular endothelial growth factor for healing of colorectal cancer. It was reported that bevacizumab inhibited the tumor growth through blocking the blood supply in cancer cells [58].

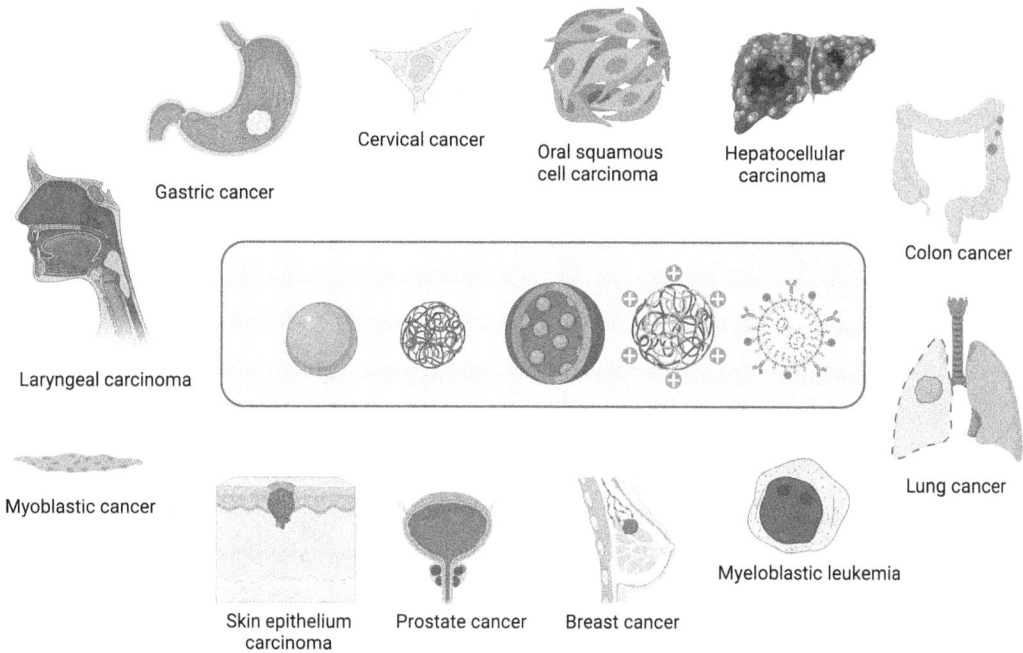

FIGURE 5.4 The different types of cancers that can be actively targeted, eventually leading to more modified and operative treatment options for patients (Created in BioRender. Abdellatif, A. (2025) https://BioRender. com/x40k791).

Moreover, prostate cancer can be targeted using PSMA as a targeting agent [59]. Certain leukemias, especially acute lymphoblastic leukemia, can be targeted through CD19 (a protein expressed on the surface of malignant B-cells) using CAR T-cell therapies [60]. Furthermore, hepatocellular carcinoma (HCC), a type of liver cancer, was targeted by sorafenib (multikinase inhibitor) [61]. Active targeting has been used in ovarian cancer, predominantly with therapies that target folate receptor alpha [62]. Active targeting can also help improve the transport of chemotherapeutic drugs to the pancreatic adenocarcinoma [63].

5.7 LIGANDS USED IN ACTIVE TARGETING

There are many types of ligands that are used in active targeting, which help in drug delivery systems. These ligands can enhance the specificity of therapeutic agents to positive receptor cells or tissues. Furthermore, they can improve the treatment's efficacy and diminish the side effects. These ligands particularly bind to target receptors on the surface of cancer cells, which are more precisely to their proposed sites [16].

5.7.1 ANTIBODIES AND THEIR FRAGMENTS

Antibodies are among the most common types of ligands utilized in active targeting. Monoclonal antibodies can be engineered to identify specific antigens overexpressed on the surfaces of certain cancerous cells. Trastuzumab is an example of an antibody that can target the HER2, which is prevalent in some cancers of the breast. By attaching pharmaceutical materials such as chemotherapeutic agents or radioactive isotopes to these antibodies, treatment can be tailored to cancer cells while limiting damage to normal cells [64] (Figure 5.5I).

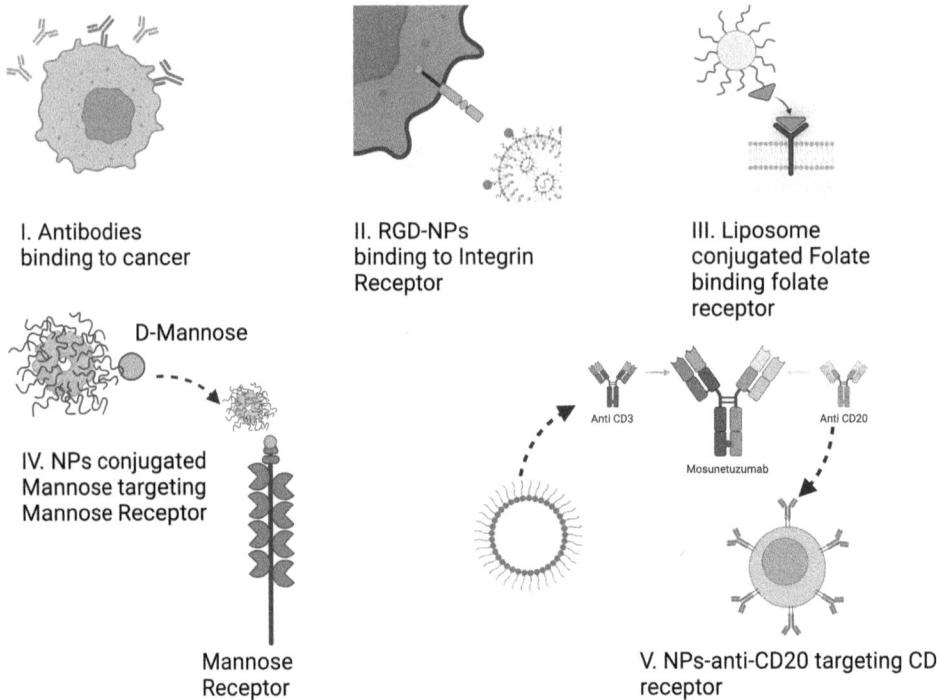

FIGURE 5.5 Illustrating different types of ligands targeting different receptors, this figure was created by Biorender with a license. Ligands are a potential advance to improve the specificity of NPs to target sites. Ligands such as antibodies, peptides, folate, sugar ligands, and cell-specific ligands (Created in BioRender. Abdellatif, A. (2025) https://BioRender.com/x40k791).

5.7.2 Peptides

Another notable example is the use of RGD peptides, which can target integrin receptors generally overexpressed in cancerous cells and angiogenic endothelium. RGD peptides were conjugated with drug-loaded NPs [65] (Figure 5.5II).

5.7.3 Folate

Folate receptors are commonly overexpressed in numerous cancerous types, including breast and ovarian cancer. Folate acts as a ligand that can be conjugated to NPs. For example, folate-conjugated liposomes were encapsulated chemotherapeutic agents, permitting special uptake by cancerous cells expressing folate receptors [66] (Figure 5.5III).

5.7.4 Sugar Ligands

Sugar or carbohydrate ligands can be conjugated with NPs for active targeting through binding to specific receptors on the cancerous cells. Mannose-modified NPs are an example of sugar ligands that are often used to deliver therapeutic agents to immune cells. This is especially considered advantageous in the vaccine-formulated NPs development [67,68] (Figure 5.5IV).

5.7.5 Cell-Specific Ligands

Specific ligands can also target definite cell types based on distinctive surface structures. The anti-CD20 monoclonal antibody is an example of a cell-specific ligand that is used to treat B-cell malignancies. Anti-CD20 was coupled with numerous cytotoxic drugs for selective targeting of B cells, with no interaction of other cell types [69]. Likewise, ligands that interact with chemokine receptors can be designed to target specific immune cells, enabling more precise administration of drugs in inflammatory conditions [70] (Figure 5.5V).

5.8 MECHANISM OF NP INTERNALIZATION OF ACTIVE TARGETING

The NPs' size has a substantial impact on passive targeting facilities. NPs ranging in size from 10 to 200 nm can enter the endothelium of cancerous cells' blood arteries. These average sizes permit them to easily circulate in the bloodstream for a longer time, giving a good chance of reaching the tumor location. Furthermore, passive targeting is influenced by NPs' composition and surface structures. Hydrophilic NPs can circulate for a longer time than hydrophobic NPs, which cluster and are quickly removed by the body. Additionally, surface changes can increase interactions with serum proteins, therefore increasing circulation times [20,71].

In contrast to passive targeting, active targeting incorporates the determined alteration of NPs to increase their attraction for positive receptor cell types. This approach often involves the coupling of targeting ligands with NPs to easily bind to receptors expressed on target cells [15]. The NPS can be internalized through active targeting and is often internalized through receptor-mediated endocytosis. Once the NPs attach to their target receptors on the cell surface, they pass by several endocytic routes, including clathrin-mediated endocytosis and caveolae-mediated endocytosis [15,31] (Figure 5.6).

Recent advances in nanoscale design have prompted the development of hybrid techniques that mix passive and active targeting mechanisms. This can enhance NP formulations, taking advantage of passive accumulation via the EPR effect and increasing specificity using targeted ligands.

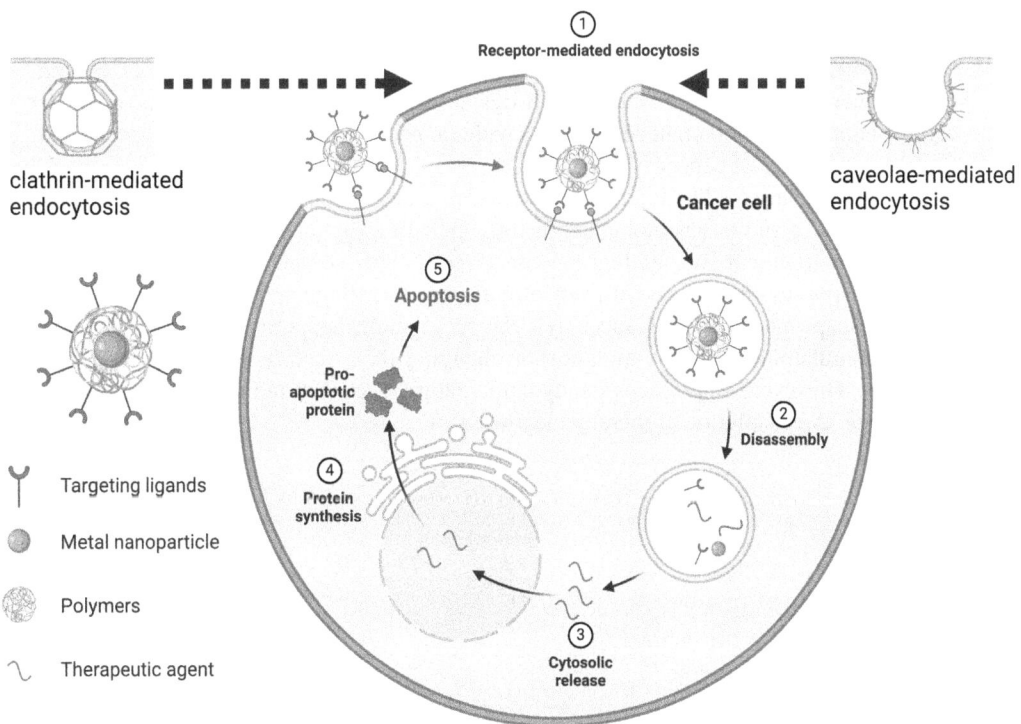

FIGURE 5.6 Nanoparticles internalized via receptor-mediated endocytosis for death-induced gene therapy in cancer (Created in BioRender. Abdellatif, A. (2025) https://BioRender.com/m04n179).

This dual-targeting approach can improve the therapeutic outcomes, as it not only increases the number of NPs at the target site but also augments their specificity, thereby minimizing off-target effects [25,26].

5.9 CHALLENGES IN PASSIVE AND ACTIVE TARGETING OF NPs TO CANCER CELLS

Passive targeting has the following challenges (Figure 5.7).:

1. Variability in a tumor microenvironment: Not every tumor has the same vascular features. The EPR impact varies greatly between tumor types and even within persons. This variety makes it impossible to anticipate NP behavior and accumulation in the tumor, resulting in variations in therapy efficacy.
2. NP Size and composition: The size, shape, and surface features of NPs all influence their capacity to use the EPR effect. While smaller NPs are more likely to reach tumor arteries, they may also be removed from the circulation too fast. Larger particles, on the other hand, may persist longer but are less efficient at penetrating the tumor. Optimizing these parameters to attain the ideal balance remains a major task.
3. Biological Obstacles: Once NPs enter the circulation, they may meet a variety of biological obstacles that prevent them from reaching the tumor location. These include interactions with serum proteins, immune cell recognition, and a variety of physiological variables that might influence circulation time.

Active targeting, on the other hand, is the modification of NPs with ligands that can bind to cancer cell receptors that are overexpressed. This technique tries to improve the selectivity and absorption of NPs. However, active targeting has its own set of challenges (Figure 5.7).:

1. Receptor heterogeneity: Tumor cells frequently express diverse surface receptors, complicating the creation of targeted therapies. A ligand that efficiently targets one kind of cancer cell may not target another, necessitating a wide range of targeting ligands to meet the different topographies of cancer types.
2. Biological complexity: The interactions between targeting ligands and cell receptors can be influenced by numerous biological factors, including the receptor's concentration and the microenvironment surrounding the cancer cells. This complexity can lead to unpredictable responses and reduced efficacy of the active targeting approach.
3. Antigen escape mechanisms: Cancer cells can occasionally adapt to resist targeted therapy by downregulating receptor expression or changing their surface properties to prevent detection. This event emphasizes the dynamic nature of cancer and the need for ongoing monitoring and adaptation of therapy measures.

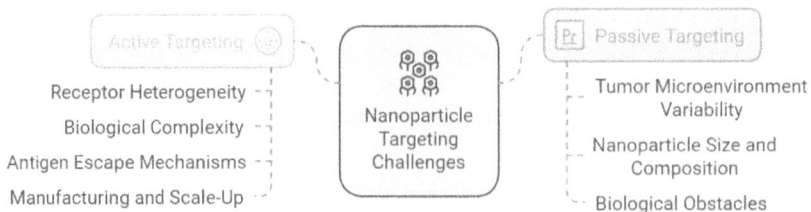

FIGURE 5.7 Challenges in passive and active targeting of nanoparticles to cancer cells (Graph created using napkin.ai).

4. Manufacturing and scale-up challenges: The production of actively targeted NPs introduces additional complications related to reproducibility, stability, and scalability. Ensuring that the modified NPs maintain their targeting abilities during clinical production remains a significant hurdle.

5.10 CONCLUSIONS

The selection between passive and active strategies depends on body cell characteristics and the drug's physical and chemical properties. Drugs easily penetrate biological barriers with no issues; simple encapsulation is a good choice with a high chance of reaching within the target cell, especially tumors, accompanied by reduced toxicity towards organs with high blood flow like the liver and heart. From the other side, drugs have issues with their penetration of the biological barriers that may damage normal tissues; active targeting is the preferred choice. Finally, carrier selection plays a pivotal role in coherent strategy and advancement for different disease formulations.

REFERENCES

1. Attia M. *Contrast Agent Based on Nano-Emulsion for Targeted Biomedical Imaging.* Universite de Strasbourg; 2016.
2. Attia MF, Anton N, Wallyn J, et al. An overview of active and passive targeting strategies to improve the nanocarriers efficiency to tumour sites. *J Pharm Pharmacol.* 2019;71(8):1185–1198. doi: 10.1111/jphp.13098.
3. Abdellatif AAH, Scagnetti G, Younis MA, et al. Non-coding RNA-directed therapeutics in lung cancer: Delivery technologies and clinical applications. *Colloids Surf B Biointerfaces.* 2023;229:113466. doi: 10.1016/j.colsurfb.2023.113466.
4. Torchilin V. Tumor delivery of macromolecular drugs based on the EPR effect. *Adv Drug Deliv Rev.* 2011;63(3):131–135. doi: 10.1016/j.addr.2010.03.011.
5. Matsumura Y, Maeda H. A new concept for macromolecular therapeutics in cancer chemotherapy: Mechanism of tumoritropic accumulation of proteins and the antitumor agent smancs. *Cancer Res.* 1986;46(12_Part_1):6387–6392.
6. Patel H. Serum opsonins and liposomes: Their interaction and opsonophagocytosis. *Crit Rev Ther Drug Carrier Syst.* 1992;9(1):39–90.
7. Noble GT, Stefanick JF, Ashley JD, et al. Ligand-targeted liposome design: Challenges and fundamental considerations. *Trends Biotechnol.* 2014;32(1):32–45.
8. Sebak AA, El-Shenawy BM, El-Safy S, et al. From passive targeting to personalized nanomedicine: Multidimensional insights on nanoparticles' interaction with the tumor microenvironment. *Curr Pharm Biotechnol.* 2021;22(11):1444–1465. doi: 10.2174/1389201021666201211103856.
9. Chen J, Yao Y, Mao X, et al. Liver-targeted delivery based on prodrug: Passive and active approaches. *J Drug Target.* 2024;32(10):1155–1168. doi: 10.1080/1061186X.2024.2386416.
10. Liu D, Lu N, Zang F, et al. Magnetic resonance imaging-based radiogenomic analysis reveals genomic determinants for nanoparticle delivery into tumors. *ACS Nano.* 2024. doi: 10.1021/acsnano.4c09387.
11. Gunjkar S, Gupta U, Nair R, et al. The neoteric paradigm of biomolecule-functionalized albumin-based targeted cancer therapeutics. *AAPS PharmSciTech.* 2024;25(8):265. doi: 10.1208/s12249-024-02977-6.
12. Al-Shadidi J, Al-Shammari S, Al-Mutairi D, et al. Chitosan nanoparticles for targeted cancer therapy: A review of stimuli-responsive, passive, and active targeting strategies. *Int J Nanomedicine.* 2024;19:8373–8400. doi: 10.2147/IJN.S472433.
13. Abdellatif AAH. A plausible way for excretion of metal nanoparticles via active targeting. *Drug Dev Ind Pharm.* 2020;46(5):744–750. doi: 10.1080/03639045.2020.1752710.
14. Abdellatif AAH, Ibrahim MA, Amin MA, et al. Cetuximab conjugated with octreotide and entrapped calcium alginate-beads for targeting somatostatin receptors. *Sci Rep.* 2020;10(1):4736. doi: 10.1038/s41598-020-61605-y.
15. Abdellatif AAH, Aldalaen SM, Faisal W, et al. Somatostatin receptors as a new active targeting sites for nanoparticles. *Saudi Pharm J.* 2018;26(7):1051–1059. doi: 10.1016/j.jsps.2018.05.014.
16. Abdellatif AA, Zayed G, El-Bakry A, et al. Novel gold nanoparticles coated with somatostatin as a potential delivery system for targeting somatostatin receptors. *Drug Dev Ind Pharm.* 2016;42(11):1782–1791. doi: 10.3109/03639045.2016.1173052.

17. Manikkath J, Manikkath A, Lad H, et al. Nanoparticle-mediated active and passive drug targeting in oral squamous cell carcinoma: Current trends and advances. *Nanomedicine.* 2023;18(27):2061–2080. doi: 10.2217/nnm-2023–0247.

18. Limongi T, Susa F, Dumontel B, et al. Extracellular vesicles tropism: A comparative study between passive innate tropism and the active engineered targeting capability of lymphocyte-derived EVs. *Membranes.* 2021;11(11). doi: 10.3390/membranes11110886.

19. Abdellatif AAH, Abou-Taleb HA, Abd El Ghany AA, et al. Targeting of somatostatin receptors expressed in blood cells using quantum dots coated with vapreotide. *Saudi Pharm J.* 2018;26(8):1162–1169. doi: 10.1016/j.jsps.2018.07.004.

20. Younis MA, Alsogaihi MA, Abdellatif AAH, et al. Nanoformulations in the treatment of lung cancer: Current status and clinical potential. *Drug Dev Ind Pharm.* 2024:1–17. doi: 10.1080/03639045.2024.2437562.

21. Das RP, Gandhi VV, Singh BG, et al. Passive and active drug targeting: Role of nanocarriers in rational design of anticancer formulations. *Curr Pharm Des.* 2019;25(28):3034–3056. doi: 10.2174/1381612825666190830155319.

22. Alavi M, Hamidi M. Passive and active targeting in cancer therapy by liposomes and lipid nanoparticles. *Drug Metab Pers Ther.* 2019;34(1). doi: 10.1515/dmpt-2018-0032.

23. Singh S, Sharma N, Behl T, et al. Promising strategies of colloidal drug delivery-based approaches in psoriasis management [Review]. *Pharmaceutics.* 2021;13(11). doi: 10.3390/pharmaceutics13111978.

24. Abdellatif AAH, Bouazzaoui A, Tawfeek HM, et al. MCT4 knockdown by tumor microenvironment-responsive nanoparticles remodels the cytokine profile and eradicates aggressive breast cancer cells. *Colloids Surf B Biointerfaces.* 2024;238:113930. doi: 10.1016/j.colsurfb.2024.113930.

25. Abdellatif AAH, Younis MA, Alsharidah M, et al. Biomedical applications of quantum dots: Overview, challenges, and clinical potential. *Int J Nanomedicine.* 2022;17:1951–1970. doi: 10.2147/IJN.S357980.

26. Younis MA, Tawfeek HM, Abdellatif AAH, et al. Clinical translation of nanomedicines: Challenges, opportunities, and keys. *Adv Drug Deliv Rev.* 2022;181:114083. doi: 10.1016/j.addr.2021.114083.

27. Abdellatif AAH. Identification of somatostatin receptors using labeled PEGylated octreotide, as an active internalization. *Drug Dev Ind Pharm.* 2019;45(10):1707–1715. doi: 10.1080/03639045.2019.1656735.

28. Zhang R, Zhang Y, Zhang Y, et al. Ratiometric delivery of doxorubicin and berberine by liposome enables superior therapeutic index than Doxil®. *Asian J Pharm Sci.* 2020;15(3):385–396. doi: 10.1016/j.ajps.2019.04.007.

29. Aldughaim MS, Muthana M, Alsaffar F, et al. Specific targeting of PEGylated liposomal doxorubicin (Doxil®) to tumour cells using a novel TIMP3 peptide. *Molecules.* 2020;26(1). doi: 10.3390/molecules26010100.

30. Abdellatif AAH, Tolba NS, Alsharidah M, et al. PEG-4000 formed polymeric nanoparticles loaded with cetuximab downregulate p21 & stathmin-1 gene expression in cancer cell lines. *Life Sci.* 2022;295:120403. doi: 10.1016/j.lfs.2022.120403.

31. Abdellatif AAH, Alshubrumi AS, Younis MA. Targeted nanoparticles: The smart way for the treatment of colorectal cancer. *AAPS PharmSciTech.* 2024;25(1):23. doi: 10.1208/s12249-024-02734-9.

32. Abdellatif AAH, Younis MA, Alsowinea AF, et al. Lipid nanoparticles technology in vaccines: Shaping the future of prophylactic medicine. *Colloids Surf B Biointerfaces.* 2023;222:113111. doi: 10.1016/j.colsurfb.2022.113111.

33. Beach MA, Nayanathara U, Gao Y, et al. Polymeric nanoparticles for drug delivery. *Chem Rev.* 2024;124(9):5505–5616. doi: 10.1021/acs.chemrev.3c00705.

34. Sueyoshi S, Vitor Silva J, Guizze F, et al. Dendrimers as drug delivery systems for oncotherapy: Current status of promising applications. *Int J Pharm.* 2024;663:124573. doi: 10.1016/j.ijpharm.2024.124573.

35. Sanka SM, Ramar K. Examining the effectiveness of polyamidoamine (PAMAM) dendrimers for enamel lesion remineralization: A systematic review. *Cureus.* 2024;16(7):e64490. doi: 10.7759/cureus.64490.

36. Abdellatif AAH, Mostafa MAH, Konno H, et al. Exploring the green synthesis of silver nanoparticles using natural extracts and their potential for cancer treatment. *3 Biotech.* 2024;14(11):274. doi: 10.1007/s13205-024-04118-z.

37. Heidari R, Assadollahi V, Shakib Manesh MH, et al. Recent advances in mesoporous silica nanoparticles formulations and drug delivery for wound healing. *Int J Pharm.* 2024;665:124654. doi: 10.1016/j.ijpharm.2024.124654.

38. Vines JB, Yoon JH, Ryu NE, et al. Gold nanoparticles for photothermal cancer therapy. *Front Chem.* 2019;7:167. doi: 10.3389/fchem.2019.00167.

39. Dheyab MA, Aziz AA, Khaniabadi PM, et al. Gold nanoparticles-based photothermal therapy for breast cancer. *Photodiagnosis Photodyn Ther.* 2023;42:103312. doi: 10.1016/j.pdpdt.2023.103312.

40. Hossen MN, Kajimoto K, Akita H, et al. Vascular-targeted nanotherapy for obesity: Unexpected passive targeting mechanism to obese fat for the enhancement of active drug delivery. *J Control Release.* 2012;163(2):101–110. doi: 10.1016/j.jconrel.2012.09.002.

41. Venkatesan J, Murugan D, Lakshminarayanan K, et al. Powering up targeted protein degradation through active and passive tumour-targeting strategies: Current and future scopes. *Pharmacol Ther.* 2024;263:108725. doi: 10.1016/j.pharmthera.2024.108725.

42. Al-Allaf FA, Abduljaleel Z, Taher MM, et al. Molecular dynamics simulation reveals exposed residues in the ligand-binding domain of the low-density lipoprotein receptor that interacts with vesicular stomatitis virus-G envelope [Article]. *Viruses.* 2019;11(11). doi: 10.3390/v11111063.

43. Ahad A, Aftab F, Michel A, et al. Development of immunoliposomes containing cytotoxic gold payloads against HER2-positive breast cancers. *RSC Med Chem.* 2024;15(1):139–150. doi: 10.1039/d3md00334e.

44. Villar-Alvarez E, Golan-Cancela I, Pardo A, et al. Inhibiting HER3 hyperphosphorylation in HER2-overexpressing breast cancer through multimodal therapy with branched gold nanoshells. *Small.* 2023;19(50):e2303934. doi: 10.1002/smll.202303934.

45. Marverti G, Gozzi G, Maretti E, et al. A peptidic thymidylate-synthase inhibitor loaded on pegylated liposomes enhances the antitumour effect of chemotherapy drugs in human ovarian cancer cells. *Int J Mol Sci.* 2020;21(12). doi: 10.3390/ijms21124452.

46. Lee J, Ahn HJ. PEGylated DC-Chol/DOPE cationic liposomes containing KSP siRNA as a systemic siRNA delivery carrier for ovarian cancer therapy. *Biochem Biophys Res Commun.* 2018;503(3):1716–1722. doi: 10.1016/j.bbrc.2018.07.104.

47. Essa D, Kondiah PPD, Kumar P, et al. Design of chitosan-coated, quercetin-loaded PLGA nanoparticles for enhanced PSMA-specific activity on LnCap prostate cancer cells. *Biomedicines.* 2023;11(4). doi: 10.3390/biomedicines11041201.

48. Caputo TM, Cusano AM, Principe S, et al. Sorafenib-loaded PLGA carriers for enhanced drug delivery and cellular uptake in liver cancer cells. *Int J Nanomedicine.* 2023;18:4121–4142. doi: 10.2147/IJN.S415968.

49. Cruz-Nova P, Gibbens-Bandala B, Ancira-Cortez A, et al. Chemo-radiotherapy with (177) Lu-PLGA(RGF)-CXCR4L for the targeted treatment of colorectal cancer. *Front Med.* 2023;10:1191315. doi: 10.3389/fmed.2023.1191315.

50. Cui Y, Xu Q, Chow PK, et al. Transferrin-conjugated magnetic silica PLGA nanoparticles loaded with doxorubicin and paclitaxel for brain glioma treatment. *Biomaterials.* 2013;34(33):8511–8520. doi: 10.1016/j.biomaterials.2013.07.075.

51. Hashemzadeh N, Dolatkhah M, Aghanejad A, et al. Folate receptor-mediated delivery of 1-MDT-loaded mesoporous silica magnetic nanoparticles to target breast cancer cells. *Nanomedicine.* 2021;16(24):2137–2154. doi: 10.2217/nnm-2021–0176.

52. Renner AM, Ilyas S, Schlosser HA, et al. Receptor-mediated in vivo targeting of breast cancer cells with 17alpha-ethynylestradiol-conjugated silica-coated gold nanoparticles. *Langmuir.* 2020;36(48):14819–14828. doi: 10.1021/acs.langmuir.0c02820.

53. Awad MG, Hanafy NAN, Ali RA, et al. Exploring the therapeutic applications of nano-therapy of encapsulated cisplatin and anthocyanin-loaded multiwalled carbon nanotubes coated with chitosan-conjugated folic acid in targeting breast and liver cancers. *Int J Biol Macromol.* 2024;280(Pt 2):135854. doi: 10.1016/j.ijbiomac.2024.135854.

54. Singh R, Kumar S. Cancer targeting and diagnosis: Recent trends with carbon nanotubes. *Nanomaterials.* 2022;12(13). doi: 10.3390/nano12132283.

55. Ziegengeist JL, Tan AR. A clinical review of subcutaneous trastuzumab and the fixed-dose combination of pertuzumab and trastuzumab for subcutaneous injection in the treatment of HER2-positive breast cancer. *Clin Breast Cancer.* 2024. doi: 10.1016/j.clbc.2024.10.005.

56. Zhang W, Zhang X, Zhao W, et al. What is the optimal first-line regimen for advanced non-small cell lung cancer patients with epidermal growth factor receptor mutation: A systematic review and network meta-analysis. *BMC Pulm Med.* 2024;24(1):620. doi: 10.1186/s12890-024-03438-3.

57. Zhao M, Zhang J, Gao J, et al. Osimertinib efficacy and safety in treating epidermal growth factor receptor mutation-positive advanced non-small-cell lung cancer: A meta-analysis. *Clin Pharmacol Drug Dev.* 2024. doi: 10.1002/cpdd.1483.

58. Yoshino T, Hooda N, Younan D, et al. A meta-analysis of efficacy and safety data from head-to-head first-line trials of epidermal growth factor receptor inhibitors versus bevacizumab in adult patients with RAS wild-type metastatic colorectal cancer by sidedness. *Eur J Cancer.* 2024;202:113975. doi: 10.1016/j.ejca.2024.113975.

59. Puik JR, Le C, Kazemier G, et al. Prostate-specific membrane antigen as target for vasculature-directed therapeutic strategies in solid tumors. *Crit Rev Oncol Hematol*. 2025;205:104556. doi: 10.1016/j. critrevonc.2024.104556.

60. Cao XY, Li JJ, Lu PH, et al. Efficacy and safety of CD19 CAR-T cell therapy for acute lymphoblastic leukemia patients relapsed after allogeneic hematopoietic stem cell transplantation. *Int J Hematol*. 2022;116(3):315–329. doi: 10.1007/s12185-022-03398-6.

61. Basu A, Namporn T, Ruenraroengsak P. Critical review in designing plant-based anticancer nanoparticles against hepatocellular carcinoma. *Pharmaceutics*. 2023;15(6). doi: 10.3390/pharmaceutics15061611.

62. Ndong C, Toraya-Brown S, Kekalo K, et al. Antibody-mediated targeting of iron oxide nanoparticles to the folate receptor alpha increases tumor cell association in vitro and in vivo. *Int J Nanomedicine*. 2015;10:2595–2617. doi: 10.2147/IJN.S79367.

63. Li H, Wang P, Deng Y, et al. Combination of active targeting, enzyme-triggered release and fluorescent dye into gold nanoclusters for endomicroscopy-guided photothermal/photodynamic therapy to pancreatic ductal adenocarcinoma. *Biomaterials*. 2017;139:30–38. doi: 10.1016/j.biomaterials.2017.05.030.

64. Lap CJ, Rajendran R, Martin JM, et al. Response of human epidermal growth factor receptor 2-expressing prostate cancer to trastuzumab deruxtecan. *Ann Intern Med*. 2024;177(12):1738–1741. doi: 10.7326/ANNALS-24-01409.

65. Lin RY, Dayananda K, Chen TJ, et al. Targeted RGD nanoparticles for highly sensitive in vivo integrin receptor imaging. *Contrast Media Mol Imaging*. 2012;7(1):7–18. doi: 10.1002/cmmi.457.

66. Christensen E, Henriksen JR, Jorgensen JT, et al. Folate receptor targeting of radiolabeled liposomes reduces intratumoral liposome accumulation in human KB carcinoma xenografts. *Int J Nanomedicine*. 2018;13:7647–7656. doi: 10.2147/IJN.S182579.

67. Kawakami S, Sato A, Nishikawa M, et al. Mannose receptor-mediated gene transfer into macrophages using novel mannosylated cationic liposomes. *Gene Ther*. 2000;7(4):292–299. doi: 10.1038/sj.gt.3301089.

68. Gao H, Goncalves C, Gallego T, et al. Comparative binding and uptake of liposomes decorated with mannose oligosaccharides by cells expressing the mannose receptor or DC-SIGN. *Carbohydr Res*. 2020;487:107877. doi: 10.1016/j.carres.2019.107877.

69. Adamo FM, Silva Barcelos EC, De Falco F, et al. Therapeutic targeting potential of novel silver nanoparticles coated with anti-CD20 antibody against chronic lymphocytic leukemia. *Cancers*. 2023;15(14). doi: 10.3390/cancers15143618.

70. Pisani A, Donno R, Valenti G, et al. Chemokine-decorated nanoparticles target specific subpopulations of primary blood mononuclear leukocytes. *Nanomaterials*. 2022;12(20). doi: 10.3390/nano12203560.

71. El Bahhaj F, Denis I, Pichavant L, et al. Histone deacetylase inhibitors delivery using nanoparticles with intrinsic passive tumor targeting properties for tumor therapy. *Theranostics*. 2016;6(6):795–807. doi: 10.7150/thno.13725.

6 Nanomedicine-Based Imaging Techniques in Cancer Diagnosis

Ahmed A.H. Abdellatif, Hesham M. Tawfeek, and Imran Saleem

6.1 INTRODUCTION

The evolution of fluorescent nanoparticles (F-NPs) loaded with different types of ligands makes the science of nanomedicine a promising formulation for targeting different diseases, especially cancers. F-NPs have persisted through the influence of nanomedicines as a first line in medical care for detecting and bioimaging diseases [1]. F-NPs are NPs that emit light; they have fluorescent emission that helps in the detection of different types of diseases such as cancer. They emit fluorescence based on semiconductors with amazing optical properties [2].

F-NPs can be classified according to their chemical composition into 12 types based on their element's compositions in the periodic table. F-NPs, such as CdTe, PbSe, ZnSe, or CdS, can overcome surface deficiency and enhance fluorescent yield [3,4]. F-NPs were formulated as novel semiconducting NPs, as polymeric nanomaterials can emit light near the infrared region (~800 nm), accelerating exciting biological applications such as flow cytometry and bioimaging [5].

F-NPs, with their tunable composition, high semiconductor yield, brightness, and intermittent light emission [6–8]. F-NPs, despite their promising properties, are still largely unexplored in clinical trials. This article explores recent methods for preparation and characterization, biomedical applications, ongoing trials, challenges, and technical comparisons. It highlights promising applications and identifies essential considerations to improve their clinical outcomes.

6.2 EXAMPLES OF FLUORESCENT NANOPARTICLES

These semiconductors and fluorescent materials have various material compositions. They demonstrate exclusive electrical properties that can be tuned for specific uses. There are different types of F-NPs; one type of F-NP is chalcopyrites, such as Cu_2S, which are mostly known for their photovoltaic applications due to their effective absorption of sunlight. Another type of F-NPs is transition metal dichalcogenides (TMDs). This type has high intrinsic carrier mobilities, such as $TiSe_2$, TaS_2, and $MoSe_2$, making them appropriate for nanoelectronics and flexible strategies. F-NPs such as Nb_2C and Ti_3C_2 are considered very interesting due to their conductive properties and capability to form thin films. Furthermore, there are three other groups (Group IV and Group II-VI) that are considered good semiconductors, such as silicon, carbon, and graphene, which are foundational in the semiconductor industry. They have many functions in LEDs and laser diodes. Further, a group of III-V semiconductors such as GaSb, AlSb, AlP, AlAs, GaAs, InAs, and InP are the best semiconductors for optoelectronic devices. Also, CsPbI3 is considered a perovskite, having transformed solar energy conversion, and is a foundation for next-generation solar cells (Table 6.1).

TABLE 6.1

Classification of F-NPs According to Chemical Compositions

Types of Semiconductors	Examples	Reference(s)
Ionic compound	AgBr	[9]
Polymeric semiconductor	NIR800	[5]
Metal carbide	Nb_2C, Ti_3C_2	[10,11]
Perovskite	CsPbI3	[12]
Alternative semiconductors	Black phosphorus	[13
Chalcogenides	$CuInS_2$, $CuInSe_2$, AgInS	[4]
Chalcopyrites	Cu_2S	[4]
III-V Semiconductors	AlSb, AlAs, AlP, GaSb, GaAs, InAs, InP	[4]
Robust thermoelectric properties; the binary compounds	PbS, PbSe, PbTe	[14]
Sustainable and high-performance materials	C, Si, Graphene	[15]
Transition metal dichalcogenides (TMDs)	ZnSe, ZnS, ZnO, CdS, CdSe, CdTe, HgS	[16]
Transition metal dichalcogenides and beyond	$TiSe_2$, TaS_2, $MoSe_2$	[17]

6.3 PREPARATION AND CHARACTERIZATION OF F-NPs

6.3.1 PREPARATION OF F-NPs

Preparing F-NPs involves assembling precursors into nanocrystals using physical and chemical approaches. Four fundamental methods for uniformity, reproducibility, and scalability are colloidal, biotemplate-based, electrochemical assembly, and biogenic syntheses (Figure 6.1) [18]. Colloidal synthesis is a popular method for preparing F-NPs, involving high-temperature injection of precursor metals into a solvent, assembling them into nuclei, and crystallizing them to achieve the desired particle size [19].

For the preparation of F-NPs such as CdTe/CdS and PbS F-NPs, the solvents used are mostly organic [19–21], which produce F-NPs with highly monodispersed phase, quantum yield, and narrow peak emission, making them ideal for imaging [22]. This method produces F-NPs with hydrophobic ligands, requiring post-synthesis modifications like exchange with hydrophilic ligands, polymeric coating, or silica-based shells to improve aqueous solubility [4,23]. Modifications increase production complexity, impacting scalability and costs. Aqueous solvents have emerged as solutions to address these shortcomings [24]. Aqueous solvent-based synthesis was used to prepare various F-NPs, including $CuInS_2$ F-NPs, CdS F-NPs, $AgInS_2$–ZnS, and ZnSe F-NPs, resulting in smaller particles and eliminating post-synthesis solubilization [24], whereas the produced soluble F-NPs exhibit reduced optical properties in comparison to the organometallic preparation [4].

The biotemplate-based method of synthesis is considered highly novel for preparing pharmaceutical crystals like F-NPs, using biomacromolecules like DNA or RNA as biosurfaces [4,25]. Kasotakis and colleagues synthesized CdSe@ZnS core-shell F-NPs on peptide-based templates, resulting in a significant red shift in wavelength due to synergistic interaction [26].

Electrochemical synthesis uses electrochemical forces to assemble precursor molecules into F-NPs. The applied potential, redox time, and supporting electrolyte concentration control F-NP size. Kalita et al. made graphene F-NPs using high-grade graphene oxide at room temperature. The obtained F-NPs were prepared with 3–5 nm diameters and were used as soil moisture sensors [27]. An innovative biotechnology-based method for scalable F-NP generation is biogenic synthesis. In this method, *E. coli* are used as bioreactors to synthesize CdS F-NPs. The first step detoxifies harmful cadmium ions by conjugating them to cysteine-terminated peptides. After entering the bacteria, the detoxified ions mix with endogenous sulfide ions to form CdS F-NPs, which are then exported

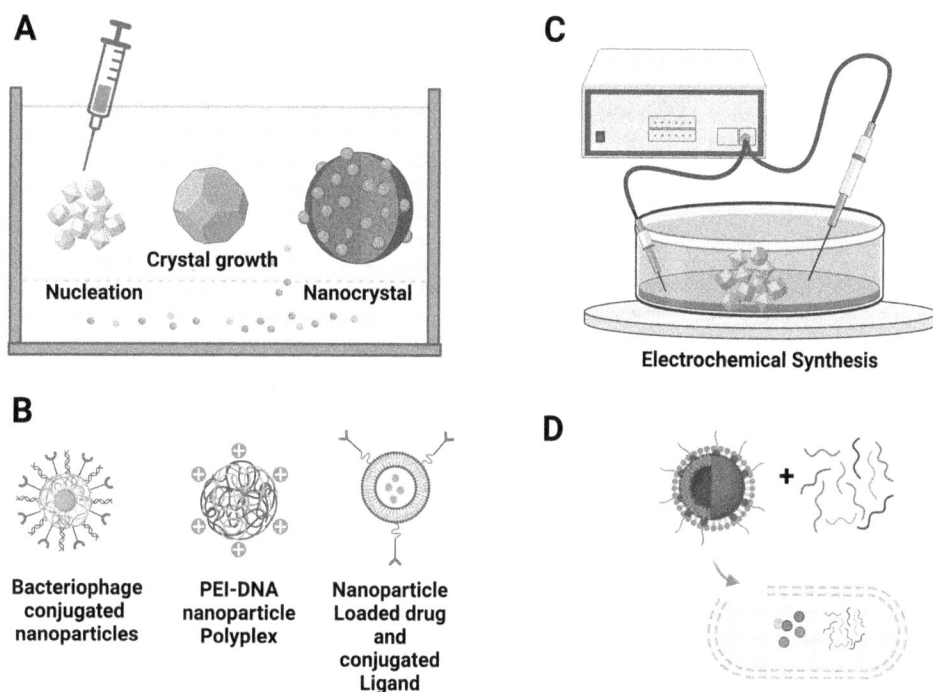

FIGURE 6.1 Different scheme preparation procedures. (a) The precursors are injected into the medium at elevated temperatures to generate F-NPs, which can be regulated via nucleation to nanocrystals. (b) Different types of bacteriophages conjugated with genetic viruses, DNA, or peptides. (c) An electrochemical method to form F-NPs from a starting precursor. (d) Biological synthesis, an example of *E. coli*, where they combine with endogenous co-precursors like sulfide ions to create F-NPs. (Created in BioRender. Abdellatif, A. (2025) https://BioRender.com/p62q522).

and recovered. Environmental tolerance, biosafety, economic production, and scalability are promised by this strategy. The detoxifying peptide composition can also affect F-NP physicochemical characteristics [28,29].

6.3.2 CHARACTERIZATION OF F-NPS

The optical characteristics of F-NPs depend on their particle size. Thus, reliable particle size determination is essential for F-NP characterization. DLS is a key method for nanomedicine characterization. F-NPs are ultra-fine; hence, more sensitive and dependable methods are needed. SEM, TEM, and AFM are frequently used to accurately determine F-NP particle size, shape, and internal composition [30]. Additionally, photoluminescence and Raman scattering spectroscopies can be used to study F-NP particle size and constitution [31]. This is because of the optical characteristics that these elements possess. Additionally, the hydrodynamic diameters of F-NPs are influenced by their surface chemistry. X-ray photoelectron spectroscopy, nuclear magnetic resonance spectroscopy, Rutherford backscattering, and analytical ultracentrifugation can be used to study F-NP surface chemistry [32]. Utilizing UV-VIS and photoluminescence spectroscopies allows for the determination of the excitation and emission spectra of F-NP, in addition to the quantum yield and brightness of the material [30]. F-NPs can be followed intracellularly using fluorescence microscopy and confocal laser scanning microscopy due to their excellent optical characteristics [33]. In addition, in vivo imaging devices can visualize F-NP biodistribution in experimental animals (Figure 6.2) [34].

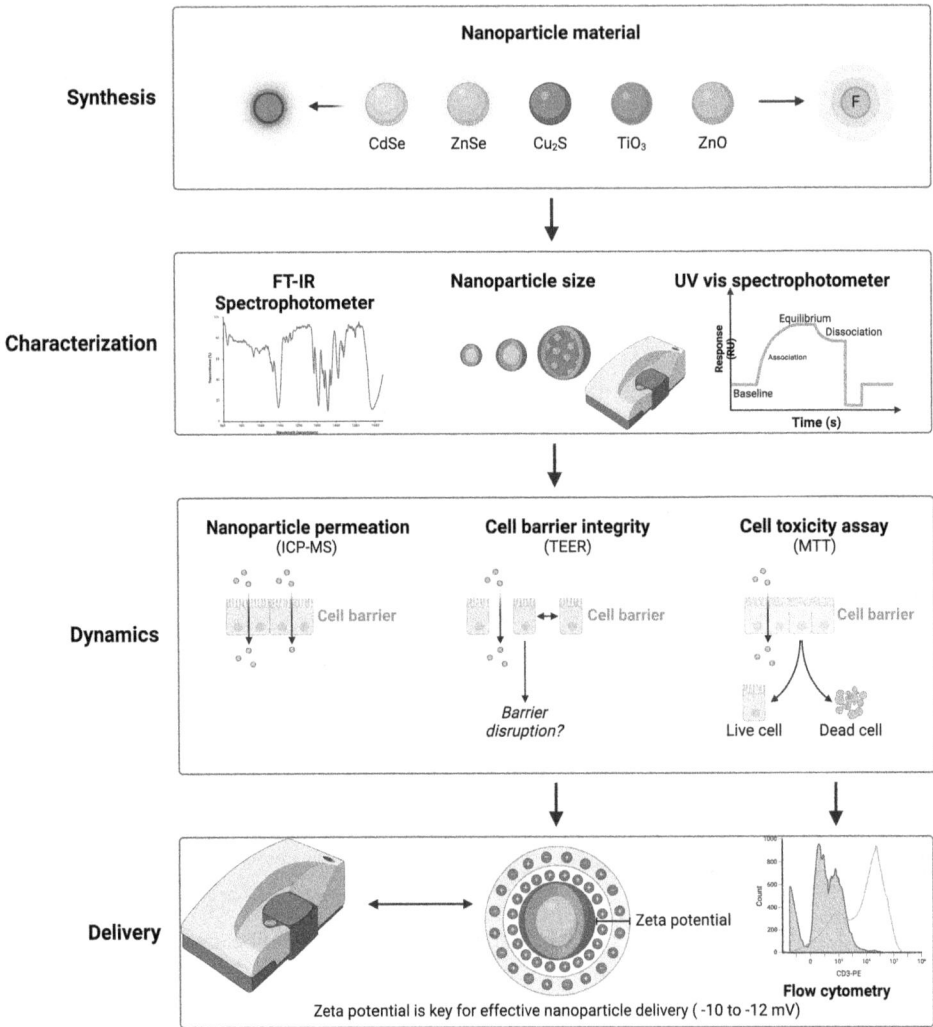

FIGURE 6.2 The names of nanoparticles, the steps of characterization, including the use of FTIR, size, zeta potential, UV-vis spectrum, and the cell toxicity assay for the fluorescent nanoparticles (Created in BioRender. Abdellatif, A. (2025) https://BioRender.com/b62c931).

6.4   BIOMEDICAL APPLICATIONS OF F-NPs

6.4.1   LIVE CELL IMAGING AND IN VIVO IMAGING

F-NP emission spectra are associated with particle diameters, allowing particle size manipulation to control optical characteristics [35]. By changing their particle size from 2 to 10 nm, CdSe F-NPs can glow at 400–600 nm. The average particle width directly affects the emission wavelength [36]. By manipulating core composition, F-NPs can have customized particle sizes and emission spectra. Depending on the particle size, for example, F-NPs having CdS cores have particle sizes ranging from 1 to 6 nm with emission spectra that fall between the ultraviolet and visible spectra. In contrast, F-NPs with InAs cores have similar particle sizes but different IR emission spectra. Table 6.2 shows other instances of how core composition affects F-NP particle size and emission spectra [37].

F-NPs are widely used as fluorescent probes for biomedical imaging due to their superior properties. F-NPs visualize intracellular components. Due to their tiny particle size, F-NPs are easily

TABLE 6.2

Effect of Fluorescent Nanoparticles Core Composition on the Particle Diameter and Emission Spectra [37]

Core Structure	Size Diameter (nm)	Emission Range
InAs	≃2.75–6.1	IR
SnTe	≃4.51–16	Mid-IR
PbSe	≃3–12	Near or mid-IR and size-dependable
GaInP$_2$	≃2–6.7	UV-Vis and size depend on
ZnSe: Mn	≃2.5–6.5	UV-Vis and size depend on
ZnSe	≃4–6	UV-Vis and size depend on
CdS	≃1–6	UV-Vis and size depend on
GaP	≃2–3	UV-Vis and size depend on
InP	≃2.6–6	UV-Vis, IR, and size-dependable
CdSe	≃1–25	Light visible
CdTe	≃1–8	Light visible

absorbed by target cells. After excitation, fluorescence microscopes or CLSM may identify the emission spectra. F-NPs flash to detect a single fluorescent peak and visualize subcellular components like proteins, unlike traditional fluorescent probes with continuous fluorescence emission [38]. After the injection of F-NPs functionalized with ligands to improve their affinity to organs and tissues, they are recruited for in vivo visualization [39]. Several uses of F-NPs in the field of biomedicine are shown in Figure 6.3. It is possible to conjugate F-NPs with a variety of different substances, including proteins, fluorescent probes, medicines, chelator-based radiolabeling, non-chelator-based radiolabeling, ScFVs, antibodies, small compounds, aptamers, and diabodies.

Their adaptability makes F-NPs useful biomedical research and application tools. This versatility allows multiple imaging, pharmacological, and treatment options. Here we discuss F-NP conjugation and medical therapy improvement. The ability to conjugate F-NPs with proteins is functional. Proteins are essential biological molecules needed for various physiological tasks. Increasing treatment efficiency can be improved by linking F-NPs to proteins to deliver drugs to cells or tissues through active targeting. This kind of conjugation can help precision-critical targeted drug delivery systems [40,41]. Real-time imaging of cellular activities with these sensors helps to understand cell dynamics and the nature of the mechanism. F-NPs with fluorescent probes enhance imaging sensitivity and specificity, indicating disease progression and treatment response [42]. F-NPs can be combined with proteins, fluorescence probes, and old and new medications. Drug solubility and stability alter with this conjugation, targeting diseased tissues. Anticancer drugs attached to F-NPs target tumor cells, decreasing side effects and improving efficacy [43]. Radiolabeling is another significant F-NP conjugation application. Chelator- or non-chelator-based radiolabeling exists. Stabilizing radioactive material with chelating agents' labels F-NPs with metal isotopes [44]. Cancer diagnosis and treatment often employ this technology to visualize aberrant processes and evaluate treatment. Non-chelator-based radiolabeling directly attaches radiolabels to F-NPs, simplifying the process.

F-NPs conjugated with ScFVs or antibodies can improve cell receptor targeting [45]. ScFVs are antibodies with high antigen affinity, making them ideal for cell targeting [46]. Conjugation of F-NPs with ScFVs can help in detecting therapeutic or diagnostic payloads. In addition, F-NPs conjugated with aptamers provide intriguing opportunities [47]. Small compounds can change nanoparticle surface properties to improve pharmacokinetics or biological system interactions. Small, single-stranded RNA or DNA molecules called aptamers bind targets with high affinity [48]. Their connection to F-NPs may improve targeted delivery system specificity. F-NPs conjugate two-antigen antibody fragments called diabodies. Dual targeting enhances therapeutic window and

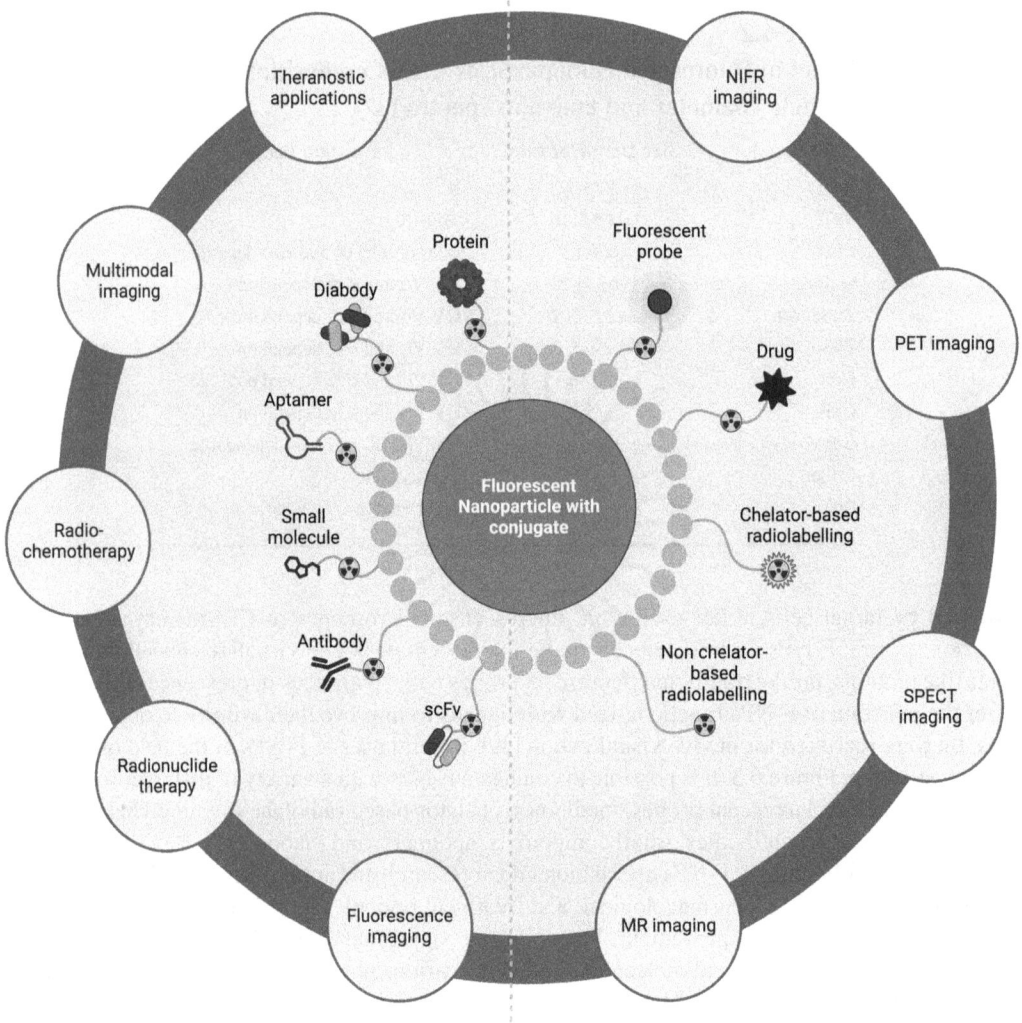

FIGURE 6.3 Several uses of F-NPs in the field of biomedicine. It is possible to conjugate F-NPs with a variety of different substances, including proteins, fluorescent probes, medicines, chelator-based radiolabeling, non-chelator-based radiolabeling, ScFVs, antibodies, small compounds, aptamers, and diabodies. Images taken inside the cell. For intracellular visibility via different bioimaging techniques that have been coated with specific targeting antibodies are utilized as fluorescent labels. (Created in BioRender. Abdellatif, A. (2025) https://BioRender.com/b62c931).

efficacy, especially in complex disorders like cancer. Combining F-NPs with chemicals affects biomedical applications. Targeted drugs, improved imaging, and new diagnostic techniques improve patient outcomes. More research will disclose medical breakthroughs [49].

6.4.2 FLUORESCENCE-ACTIVATED CELL SORTING (FACS)

FACS is employed in many biological applications, including drug delivery system uptake evaluation, cell population isolation, disease model characterizations, cellular marker identification, and immune cell mapping [50–52]. Due to numerous properties, F-NPs can be used as fluorescent markers in FACS. Compared to organic dyes, F-NPs have narrow emission spectra, decreasing overlap and allowing FACS polychromatic cell sorting with several labels. They have broad excitation spectra; therefore, one laser beam can excite many F-NPs probes, enhancing FACS equipment's capabilities [53]. F-NPs are brighter than most organic dyes, improving detection accuracy [54]. CdTe as a model of F-NPs was coupled to monoclonal antibodies to examine blood types, which was stable and efficient for 6 months [55]. If functionalized with targeting ligands to detect target markers, F-NPs could replace antibodies employed to stain cell surface markers due to their stability and low cost. F-NPs can be easily taken up into cells, unlike antibodies, allowing intracellular marker labeling without permeabilization buffers [56].

6.4.3 PHOTODYNAMIC THERAPY (PDT)

PDT shows promise for treating different types of malignancies, such as skin, tongue, and breast [57]. A photosensitizer is usually activated using light irradiation that helps in transferring energy intracellularly, generating reactive oxygen species (ROS) in situ, which indeed induce apoptosis in target tumor cells [58]. Additionally, F-NPs have the capability to be functionalized as photosensitizers or energy donors for other entities. In comparison to organic photosensitizers, F-NPs demonstrate excellent light absorption, emission, photostability, water solubility, tunable optical properties, and tissue accumulation. Additionally, the size and composition of F-NPs can be fine-tuned to enhance NIR emission, allowing for effective tissue penetration and treatment of deep-seated malignancies. Graphene F-NPs (GF-NPs) and graphene oxide F-NPs (GOF-NPs) serve as agents in cancer PDT. F-NPs exposed to 290 µW UV tube light successfully eliminate 90% of B16F10 melanoma and MCF-7 breast cancer cells in approximately 5 minutes, showcasing impressive effectiveness and capability [59]. Recent results suggest that carbon F-NPs could be employed in COVID-19 PDT via ROS production and type I interferon response stimulation [60,61].

6.4.4 TRACEABLE DRUG DELIVERY

The optical features of F-NPs make them useful in bioimaging, PDT, and medication administration. Due to their ease of fabrication, F-NPs have the ability to conjugate to many drugs, tuned physicochemical properties, and interesting optical properties [4,62]. F-NPs' sub-10 nm particle size makes them promising tumor-penetrating delivery vectors [63]. F-NPs can also be decorated with targeted ligands or functional coatings to alter biodistribution and pharmacokinetics. PEGylation is utilized to increase blood circulation retention time after intravenous delivery, which is necessary for tumor accumulation during cancer treatment [64,65]. Folic acid is commonly utilized to target cancer cell folate receptors that are overexpressed on cancer cells [66–68]. Recently, Cd/Se F-NPs were modified with vapreotide to target blood cell somatostatin receptors for the treatment of blood malignancies. A different study employed PEGylated MoS2 F-NPs for the delivery of doxorubicin to cancer cells. The F-NPs exhibited blue photoluminescence, pH-responsive drug release, and demonstrated high stability and biosafety in physiological conditions [69]. Table 6.3 contains a summary of other functional surface adjustments that have been made to F-NP.

The application of nanomedicine in treatment generally and diagnostic imaging specifically is considered a recent, effective approach for the diagnosis of different tumors compared to the traditional methods, such as surgery, MRI, and PET. These methods are expensive and cannot give a sharp diagnosis of tumors at early stages [85]. In addition, looking for effective and efficient techniques with acceptable sensitivity for early diagnosis and detection of cancer remains challenging for tumor diagnosis. Early detection of tumors as well as different types of cancers remains a challenge regarding their clinical investigation and diagnosis. Developed nanomaterials with unique properties, such as magnetic nanoparticles, liposomes, polymeric micelles, and carbon nanotubes, have shown a wide application nowadays in the field of cancer diagnosis, detection, and imaging [86,87]. New imaging probes for MRI, PET, SPECT, and ultrasound techniques can be considered types of detection. These probes showed a better contrast-amplified sensitivity, controlled in vivo biodistribution, and enhanced spatial imaging. Application of these newly synthesized probes with

TABLE 6.3

Different Functional Surface Modifications of F-NPs and Their Applications on Different Targeting Sites

Core/Multilayer Shell	Surface Modification	Targeting Receptor	References
Silica fluorescent nanoparticles	Anti-EGFR antibody	Anti-epidermal growth factor receptor	[70]
GF-NPs	Reduced graphene	Aryl hydrocarbon receptor	[71]
CdSe/ZnS	Peptide E5	Chemokine receptor 4	[72]
Graphene F-NPs (GF-NPs)	Anti-EGFR antibody	EGFR	[73]
Ag₂S F-NPs	ZEGFR1907 antibody	EGFR	[74]
Cd/Se	Aptamer 32	EGFRvIII	[75]
Gold F-NPs (AuF-NPs)	Cold atmospheric plasma	Fas/TRAIL-receptor	[76]
Silica-NPs	Folic a	Folate receptors (FO-RT)	[77]
CdSe/CdS/ZnS	Folic a	FO-RT	[66]
CdSe/ZnS	Dendrimer conjugated to PEG-folate	FO-RT	[67]
Cd/Se F-NPs	PEG-Folic acid	FO-RT	[68]
Gadolinium oxide	Bombesin gastrin-releasing peptide	GRP receptors	[78]
Cd/Se	Synaptic proteins	Neurotransmitter receptors	[79]
carbon F-NPs (CF-NPs)	Retinoic acid receptor responder protein 2	Retinoic acid receptor	[80]
Cd/Se	Vapreotide	Somatostatin receptors	[81,82]
Silica-NPs	Magnetic nanoparticles	Terpyridine receptor-immobilized	[83]
Cd/Se	Anti-VEGFR antibodies	VEGFR	[84]

Abbreviations: TRAIL, which stands for tumor necrosis factor-related apoptosis-inducing ligand; EGFR, which stands for epidermal growth factor receptor; PEG, which stands for polyethylene glycol.

their novel features offers numerous clinical advantages, like earlier tumor detection, assessment of disease state progression in real time, and personalized treatment, thanks to the unique properties of the nanomaterials, like optical, chemical, and magnetic properties, which participate in the synthesis of these imaging probes [88, 89]. Additionally, NPs can conjugate with certain cancer-specific antibodies for improved detection and imaging.

Different types of metallic NPs demonstrated a superior sensitivity in tumor tissue detection and could also conjugate with cancer-specific antibodies and penetrate cancer cells effectively. It has also been reported that nanocomposite hybrid nanomaterials, which contain more than one type of nanoparticles, are considered a better candidate for tumor detection and imaging [90]. Certain novel techniques can also perceive different biological and cellular processes through their improved and developed functions, such as bioimaging and biosensors. An example of these techniques is the 2D-graphene-based biosensor, which is considered a powerful tool in providing an efficient cancer diagnosis at the early stages in addition to monitoring and controlling the efficacy of the used treatments [91].

Langeveld S. et al. have developed phospholipid-coated targeted microbubbles for ultrasound molecular imaging. They delineated that this system is a powerful technique for studying disease progression, diagnostic imaging, and monitoring of therapeutic responses [92]. In radiotherapy, the use of inorganic compounds having a high atomic number could be useful to enhance the dose of the locally applied radiation and lower any possible adverse effects to normal tissues and cells. Gadolinium-based NPs showed a promising contrast material in MRI based on their proven potential as a radiation sensitizer [93]. Silver and gold NPs are also used effectively and efficiently for tumor diagnosis via radiotherapy. Gold NPs demonstrated good properties in cancer imaging and biosensing [94,95]. Núñez et al. utilized superparamagnetic quantum dots and gold NPs in tumor diagnosis, especially in MRI imaging; they also identified their characteristics and benefits regarding this issue [96]. Other promising materials are used for tumor diagnosis like silica, magnetics, and iron oxide. These materials have numerous advantages, such as lower cost and side effects as well as short tumor detection time when compared to the chemical methods and radiotherapy. In brain tumors, it was found that nanomaterials showed an enhanced permeation and retention in tumor tissues; hence, a better focused and sharp imaging of tumor tissues is obtained [97]. Superparamagnetic iron oxide NPs and ultra-superparamagnetic iron oxide NPs are used successfully for the detection and diagnosis of brain tumors, as they possess interesting physicochemical and magnetic properties as well as a proven stability [98]. Another interesting finding is the diagnosis and imaging of intracranial glioblastoma based on utilizing a multifunctional nanoprobe composed of PEGylated ultra-superparamagnetic iron oxide NPs conjugated with angiopep-2 to target glioblastoma tissues. Overexpressed materials found in tumor tissues like low-density lipoprotein could be considered an efficient target for angiopep-2 [99]. It has been found that AuNPs administered I.V. showed an enhanced fluorescence image rather than the produced image obtained with their direct administration in the tumor site. In glioma, good and enhanced resolution could be found with intravenous infusion of NPs [100]. One study demonstrated surface-decorated AuNPs for visualizing Glioblastoma multiforme, one of the most aggressive and lethal types of human brain cancer. AuNPs were attached to a small peptide, namely, CBP4, to target CD133 on the brain tumor cells. When the peptide-coated AuNPs localized in the cellular cytosol, it demonstrated a biocompatible stimulus-responsive fluorescence [101]. Another nanoprobe for glioma imaging was investigated, utilizing PEG-coated QDs based on CdSe/ZnS were developed. When a short peptide based on asparagines-glycine-arginine peptides is integrated into these QDs, it can effectively target the CD13 glycoprotein located on tumor tissues. Moreover, this system can also target the CD13 found in glioma tissues and shows good fluorescence upon its in vivo evaluation. This fluorescence is useful in the surgical removal of glioma [102]. A biosensor based on chitosan-modified reduced graphene oxide was fabricated for the vascular endothelial growth factor receptor 2 (VEGFR2) detection with acceptable susceptibility and a detection limit of 0.28 pM. The total amount of VEGFR2 could be measured by the proposed system in cell lysates. In addition, this fabricated nanocomposite can detect any changes in the expression upon administration of any inhibitor [103]. Kumar N. et al. proposed a novel electrochemical immunosensor for the detection of CD44 antigen found in breast tumor tissues. The developed immunosensor is composed of graphene quantum dots decorated with gold NPs (GQDs-GNPs). The authors suggested that the generated nanocomposite exhibited favorable properties for identifying various cancer biomarkers, particularly CD44 in the breast cancer [104]. Raj D et al. investigated a nanohybrid system based on innovative electrochemical biosensing for early-stage prostate cancer detection. The prepared gold graphene hybrid nanocomposite has the potential to detect prostate cancer early with enhanced treatment efficacy [105]. The flow cytometry is an effective tool used for studying the efficacy of nanocomposite-based systems. It is used for evaluating cell characteristics and facilitating the comprehension of the efficacy of novel cancer therapies and their response to varying environments [106]. It has been shown that it has an enhanced role in the diagnosis of cancer [107,108]. It has been investigated that flow cytometry is used successfully for accurate identification and measurement of cancer cells based on molybdenum disulfide nanocomposites [106]. The authors showed that the produced nanocomposite can target the tumor cells with little effect on normal cells, as well as enhanced diagnostic methods.

6.5 MECHANISMS OF NANOPARTICLE INTERNALIZATION

Functionalized F-NPs are commonly endocytosed by cells via receptors. Figure 6.4 shows three ways cells react with ligand-modified F-NPs during receptor-mediated endocytosis. F-NPs' endocytic fate depends on their ligand, target receptor, and composition [109,110]. Clathrin-mediated

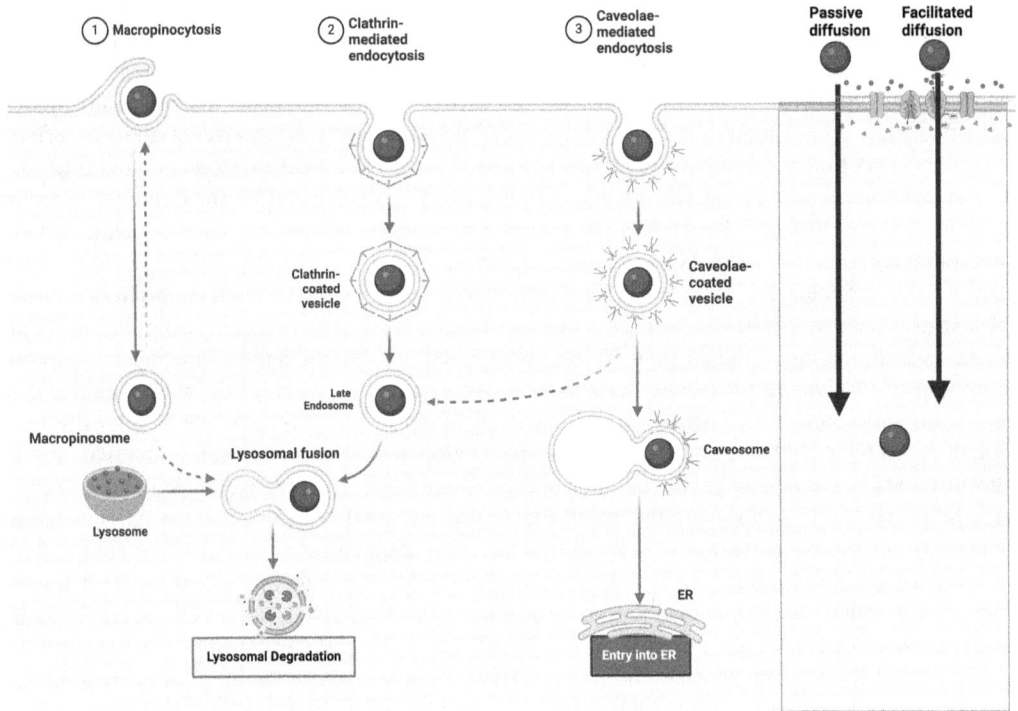

FIGURE 6.4 Nanoparticles with different labels can be employed in PET, SPECT, CT, MRI, US, and optical imaging (Reprinted, with permission under a CC BY 4.0 license, from C. H. Liu and P. Grodzinski et al. [1]).

endocytosis involves nanoparticles concentrating at the target receptor location and being absorbed by clathrin-coated cell membranes to create vesicles. The membrane protein dynamin dissociates these vesicles from the cell membrane into the cytoplasm, forming endosomes. If F-NPs leave the endosome, they can act and release their cargos in the cytosol. Lysosomal degradation will occur otherwise. After binding to their receptors, cholesterol-rich flask-shaped membrane invaginations, caveolae, import functionalized particles into target cells, and form caveosomes. Caveosomes improve medicine delivery to target cells because they are safer than endosomes. When ligands drive cell membrane ruffles to swallow particles into the cytoplasm, macropinosomes spill their contents [111].

6.6 CHALLENGES AND LIMITATIONS

Despite the usefulness of the widely investigated nanomaterials in bioimaging and cancer detection, they still have some concerns relating to their toxicity to tissues and biodistribution. It was found that ultra-small MRI contrast superparamagnetic iron oxide NPs could cause toxicity with high doses and repeated injection [112]. This effect can be minimized via surface coating of these NPs, as they possess a hydrophobic surface, so a proper surface coating like sugar or polymer could improve the particle stability, dispersion, and lower any proposed toxicity [113]. However, in certain cases, the coat could be released, exposing the iron oxide NPs to the biological system, which causes undesirable side effects. It was also found that iron-oxide NPs could cause neurotoxicity and trigger lung fibrosis as particles taken by alveolar macrophages [113]. In addition, they could affect the coagulation process through prolongation of prothrombin time [114]. The structure of QDs and their size and surface charge after modifications could significantly affect their toxicity. On the contrary, large-sized QDs, small QDs can induce more cell damage with chromosome condensation and membrane blebbing. In addition, they can also target nuclear histones and nucleoli after their active transport across the nuclear membrane, whereas the larger QDs can not [115].

Although the clinical translation of nanomedicines for treatment and diagnosis is a developing field, it offers promising opportunities to improve patient outcomes through innovative research. It also requires important consideration of different regulatory challenges. The safety and efficiency of NPs are important factors to consider, as are their pharmacokinetics, biodistribution, and clearance. The interaction with biological barriers like skin, organs, and cells has a pronounced effect on determining the biodistribution of the used nanomaterials [116]. Therefore, it is an important issue to study and characterize the interaction between nanotherapeutics and these barriers during medication transport, in addition to investigating the safety and efficacy of the nanotherapeutics during their transport and targeting. Other important issues are related to the regulatory agencies like the United States Food and Drug Administration (USFDA) and the European Medicine Agency (EMA), which have well-established criteria for evaluating and releasing these materials to the market. They also require the manufacturer to support documents with comprehensive data regarding the manufacturing processes, quality control, and characterization of these compounds as well as their stability [117]. One of the main duties of these regulatory bodies is to have a balance between the effectiveness of nanomaterials and their anticipated adverse effects. In terms of clinical trials, informed consent and ethical approvals are essential aspects to consider. For example, the nanocomposites regulations are complicated in clinical translation, but it is critical for guaranteeing the safety and efficacy of these materials [118]. Consolidation of regulatory standards is a major challenge, but it must be overcome to achieve efficient nanotherapeutic development.

Scale-up and cost of manufacturing are also crucial factors for nanomedicine clinical translation. The challenges of these materials to be included in the market are numerous due to their complexity. Small-scale production is useful in clinical trials and laboratory work; however, in large-scale production, any variations in the production process could significantly affect the physicochemical properties. Additionally, high costs prevent new nanotherapeutics from entering the market.

6.7 FUTURE DIRECTIONS AND INNOVATIONS

Nanomedicine and nanocomposites introduce novel approaches for cancer diagnosis, which seem to be more accurate and sensitive. One of these approaches holding enormous promise for early diagnosis is the nanocomposite biosensor, which is considered a non-invasive technique for tumor detection using blood or saliva as readily available fluids [90]. Microfluidic devices utilizing nanocomposites can be used efficiently for early diagnosis and detection of any tumor cells present, and also offer good monitoring of treatment [119]. One of the interesting approaches in cancer diagnosis is the functionalized DNA nanostructures. Biosensors and other diagnostic instruments can be used utilizing these nanomaterials, demonstrating the potential application in oncology [118].

AI and nanotechnology can now tailor cancer treatments to individual patients. Recent crossover between these domains is improving patient data collection and nanomaterial design for precision cancer therapy. AI can also help in the development and creation of novel materials as well as methods for effective treatments. Hence, it improves the early detection of cancer through high-resolution image analysis and could also personalize different treatment plans according to the available patient information [120]. It has been found that machine learning and AI are used in different medication aspects like medical imaging and gene expression analysis [121]. Upon using AI characteristics like pattern analysis and classification algorithms, AI can be used for improved diagnostic and therapeutic accuracy. Combining algorithm patterns obtained from machine learning with certain feature selection can be useful for detecting tumor-specific signatures. For example, a classification algorithm analyzed nearly 10,000 genes from 200 prostate cancer patients, recognizing 50 of them to be related to metastatic prostate cancer [122]. Applications of AI in research related to cancer offer significant potential approaches, but they also have some challenges, such as process validation and clinical manipulation, bias and fairness, data privacy and security, the quality of the obtained data and their quantity, and data interpretability.

6.8 CONCLUSION

Fluorescent nanoparticles show great promise for cancer bioimaging problems. Their unique properties make advanced imaging easier and enable precision medicine in new directions. Despite advances in F-NP preparation, application, and clinical testing, ongoing research is needed to overcome current challenges and fully realize their potential. F-NPs may help develop cancer diagnostics and therapies, improving patient outcomes. Nanomedicine is constantly changing, which explains this.

REFERENCES

1. Yan ZP, Yang M, Lai CL. COVID-19 vaccines: A review of the safety and efficacy of current clinical trials. *Pharmaceuticals*. 2021;14(5):406.
2. Ekimov AI, Efros AL, Onushchenko AA. Quantum size effect in semiconductor microcrystals. *Solid State Commun*. 1985;56(11):921–924.
3. Speranskaya ES, Beloglazova NV, Lenain P, De Saeger S, Wang Z, Zhang S, et al. Polymer-coated fluorescent CdSe-based quantum dots for application in immunoassay. *Biosens Bioelectron*. 2014;53:225–231.
4. Kargozar S, Hoseini SJ, Milan PB, Hooshmand S, Kim H-W, Mozafari M. Quantum dots: A review from concept to clinic. *Biotechnol J*. 2020;15(12):2000117.
5. Chen D, Wu IC, Liu Z, Tang Y, Chen H, Yu J, et al. Semiconducting polymer dots with bright narrow-band emission at 800 nm for biological applications. *Chem Sci*. 2017;8(5):3390–3398.
6. Abdellatif AA, Aldalaen SM, Faisal W, Tawfeek HM. Somatostatin receptors as a new active targeting sites for nanoparticles. *Saudi Pharm J*. 2018;26(7):1051–1059.
7. Jahangir MA, Gilani SJ, Muheem A, Jafar M, Aslam M, Ansari MT, et al. Quantum dots: Next generation of smart nano-systems. *Pharm Nanotechnol*. 2019;7(3):234–245.
8. Hu L, Zhao Q, Huang S, Zheng J, Guan X, Patterson R, et al. Flexible and efficient perovskite quantum dot solar cells via hybrid interfacial architecture. *Nat Commun*. 2021;12(1):466.
9. Wang D, Guo L, Zhen Y, Yue L, Xue G, Fu F. AgBr quantum dots decorated mesoporous Bi$_2$WO$_6$ architectures with enhanced photocatalytic activities for methylene blue. *J Mater Chem A*. 2014;2(30):11716–11727.
10. Xu Q, Ma J, Khan W, Zeng X, Li N, Cao Y, et al. Highly green fluorescent Nb$_2$C MXene quantum dots. *Chem Commun*. 2020;56(49):6648–6651.
11. Naguib M, Kurtoglu M, Presser V, Lu J, Niu J, Heon M, et al. Two-dimensional nanocrystals produced by exfoliation of Ti$_3$AlC$_2$. *Adv Mater*. 2011;23(37):4248–4253.
12. Xiao C, Zhao Q, Jiang C-S, Sun Y, Al-Jassim MM, Nanayakkara SU, et al. Perovskite quantum dot solar cells: Mapping interfacial energetics for improving charge separation. *Nano Energy*. 2020;78:105319.
13. Ren X, Yang X, Xie G, Luo J. Black phosphorus quantum dots in aqueous ethylene glycol for macroscale superlubricity. *ACS Appl Nano Mater*. 2020;3(5):4799–4809.
14. Zheng S, Chen J, Johansson EMJ, Zhang X. PbS colloidal quantum dot inks for infrared solar cells. *iScience*. 2020;23(11):101753.
15. Younis MR, He G, Lin J, Huang P. Recent advances on graphene quantum dots for bioimaging applications. *Front Chem*. 2020;8(424).
16. Babentsov V, Sizov F. Defects in quantum dots of IIB–VI semiconductors %. *J Opt-Electron Rev*. 2008;16(3):208–225.
17. Meng S, Zhang Y, Wang H, Wang L, Kong T, Zhang H, et al. Recent advances on TMDCs for medical diagnosis. *Biomaterials*. 2021;269:120471.
18. Wang J, Liu G, Leung KC, Loffroy R, Lu PX, Wang YX. Opportunities and challenges of fluorescent carbon dots in translational optical imaging. *Curr Pharm Des*. 2015;21(37):5401–5416.
19. Kim JY, Voznyy O, Zhitomirsky D, Sargent EH. 25th Anniversary article: Colloidal quantum dot materials and devices: A quarter-century of advances. *Adv Mater*. 2013;25(36):4986–5010.
20. Adegoke O, Montaseri H, Nsibande SA, Forbes PBC. Organometallic synthesis, structural and optical properties of CdSe quantum dots passivated with ternary AgZnS alloyed shell. *J Lumin*. 2021;235:118049.

21. Mozafari M, Moztarzadeh F, Seifalian AM, Tayebi L. Self-assembly of PbS hollow sphere quantum dots via gas–bubble technique for early cancer diagnosis. *J Lumin.* 2013;133:188–193.

22. Pu Y, Cai F, Wang D, Wang J-X, Chen J-F. Colloidal synthesis of semiconductor quantum dots toward large-scale production: A review. *Ind Eng Chem Res.* 2018;57(6):1790–1802.

23. Tyrakowski CM, Snee PT. A primer on the synthesis, water-solubilization, and functionalization of quantum dots, their use as biological sensing agents, and present status. *Phys Chem Chem Phys.* 2014;16(3):837–855.

24. Jing L, Kershaw SV, Li Y, Huang X, Li Y, Rogach AL, et al. Aqueous based semiconductor nanocrystals. *Chem Rev.* 2016;116(18):10623–10730.

25. Fernández-Fernández MR, Sot B, Valpuesta JM. Molecular chaperones: Functional mechanisms and nanotechnological applications. *Nanotechnology.* 2016;27(32):324004.

26. Kasotakis E, Kostopoulou A, Spuch-Calvar M, Androulidaki M, Pelekanos N, Kanaras AG, et al. Assembly of quantum dots on peptide nanostructures and their spectroscopic properties. *Appl Phys A.* 2014;116(3):977–985.

27. Kalita H, Palaparthy VS, Baghini MS, Aslam M. Electrochemical synthesis of graphene quantum dots from graphene oxide at room temperature and its soil moisture sensing properties. *Carbon.* 2020;165:9–17.

28. Mal J, Nancharaiah YV, van Hullebusch ED, Lens PNL. Metal chalcogenide quantum dots: Biotechnological synthesis and applications. *RSC Adv.* 2016;6(47):41477–41495.

29. Gallardo C, Monrás JP, Plaza DO, Collao B, Saona LA, Durán-Toro V, et al. Low-temperature biosynthesis of fluorescent semiconductor nanoparticles (CdS) by oxidative stress resistant Antarctic bacteria. *J Biotechnol.* 2014;187:108–115.

30. Drbohlavova J, Adam V, Kizek R, Hubalek J. Quantum dots - characterization, preparation and usage in biological systems. *Int J Mol Sci.* 2009;10(2):656–673.

31. Gu Y, Kuskovsky IL, Fung J, Robinson R, Herman IP, Neumark GF, et al. Determination of size and composition of optically active CdZnSe/ZnBeSe quantum dots. *Appl Phys Lett.* 2003;83(18):3779–3781.

32. Lees EE, Gunzburg MJ, Nguyen T-L, Howlett GJ, Rothacker J, Nice EC, et al. Experimental determination of quantum dot size distributions, ligand packing densities, and bioconjugation using analytical ultracentrifugation. *Nano Lett.* 2008;8(9):2883–2890.

33. Szymanski CJ, Yi H, Liu JL, Wright ER, Payne CK. Imaging intracellular quantum dots: Fluorescence microscopy and transmission electron microscopy. *Methods Mol Biol.* 2013;1026:21–33.

34. Texier I, Josser V. In vivo imaging of quantum dots. *Methods Mol Biol.* 2009;544:393–406.

35. Su G, Liu C, Deng Z, Zhao X, Zhou X. Size-dependent photoluminescence of PbS QDs embedded in silicate glasses. *Opt Mater Express.* 2017;7(7):2194–2207.

36. Bera D, Qian L, Tseng T-K, Holloway PH. Quantum dots and their multimodal applications: A review. *Materials (Basel).* 2010;3(4):2260–2345.

37. Murray CB, Kagan CR, Bawendi MG. Synthesis and characterization of monodisperse nanocrystals and close-packed nanocrystal assemblies. *Ann Rev Mater Sci.* 2000;30(1):545–610.

38. Schmidt R, Krasselt C, Göhler C, von Borczyskowski C. The fluorescence intermittency for quantum dots is not power-law distributed: A luminescence intensity resolved approach. *ACS Nano.* 2014;8(4):3506–3521.

39. Walling MA, Novak JA, Shepard JRE. Quantum dots for live cell and in vivo imaging. *Int J Mol Sci.* 2009;10(2):441–491.

40. Ma E, Fu Z, Chen K, Sun L, Zhang Y, Liu Z, et al. Smart protein-based fluorescent nanoparticles prepared by a continuous nanoprecipitation method for pesticides' precise delivery and tracing. *J Agric Food Chem.* 2023;71(22):8391–8399.

41. Liu X, Ren X, Chen L, Zou J, Li T, Tan L, et al. Fluorescent hollow ZrO_2@CdTe nanoparticles-based lateral flow assay for simultaneous detection of C-reactive protein and troponin T. *Mikrochim Acta.* 2021;188(6):209.

42. Yu BQ, Jin JC, Zou HF, Wang BB, He H, Jiang P, et al. Fluorescent protein nanoparticles: Synthesis and recognition of cellular oxidation damage. *Colloids Surf B Biointerfaces.* 2019;177:219–227.

43. Ravera M, Perin E, Gabano E, Zanellato I, Panzarasa G, Sparnacci K, et al. Functional fluorescent nonporous silica nanoparticles as carriers for Pt(IV) anticancer prodrugs. *J Inorg Biochem.* 2015;151:132–142.

44. Martinez Martinez T, Garcia Aliaga A, Lopez-Gonzalez I, Abella Tarazona A, Ibanez Ibanez MJ, Cenis JL, et al. Fluorescent DTPA-silk fibroin nanoparticles radiolabeled with [111]In: A dual tool for biodistribution and stability studies. *ACS Biomater Sci Eng.* 2020;6(6):3299–3309.

45. Tateo S, Shinchi H, Matsumoto H, Nagata N, Hashimoto M, Wakao M, et al. Optimized immobilization of single chain variable fragment antibody onto non-toxic fluorescent nanoparticles for efficient preparation of a bioprobe. *Colloids Surf B Biointerfaces.* 2023;224:113192.

46. Colombo M, Sommaruga S, Mazzucchelli S, Polito L, Verderio P, Galeffi P, et al. Site-specific conjugation of ScFvs antibodies to nanoparticles by bioorthogonal strain-promoted alkyne-nitrone cycloaddition. *Angew Chem Int Ed Engl.* 2012;51(2):496–499.

47. Gupta R, Gupta P, Wang S, Melnykov A, Jiang Q, Seth A, et al. Ultrasensitive lateral-flow assays via plasmonically active antibody-conjugated fluorescent nanoparticles. *Nat Biomed Eng.* 2023;7(12):1556–1570.

48. Xu R, Ouyang L, Chen H, Zhang G, Zhe J. Recent advances in biomolecular detection based on aptamers and nanoparticles. *Biosensors.* 2023;13(4):474.

49. Subjakova V, Oravczova V, Hianik T. Polymer nanoparticles and nanomotors modified by DNA/RNA aptamers and antibodies in targeted therapy of cancer. *Polymers.* 2021;13(3):341.

50. Younis MA, Khalil IA, Abd Elwakil MM, Harashima H. A multifunctional lipid-based nanodevice for the highly specific codelivery of sorafenib and midkine siRNA to hepatic cancer cells. *Mol Pharm.* 2019;16(9):4031–4044.

51. Adan A, Alizada G, Kiraz Y, Baran Y, Nalbant A. Flow cytometry: Basic principles and applications. *Crit Rev Biotechnol.* 2017;37(2):163–176.

52. Gouttefangeas C, Walter S, Welters MJP, Ottensmeier C, van der Burg SH, Chan C. Flow cytometry in cancer immunotherapy: Applications, quality assurance, and future. In: Rezaei N, editor. *Cancer Immunology: A Translational Medicine Context.* Cham: Springer International Publishing; 2020. pp. 761–783.

53. Petryayeva E, Algar WR, Medintz IL. Quantum dots in bioanalysis: A review of applications across various platforms for fluorescence spectroscopy and imaging. *Appl Spectrosc.* 2013;67(3):215–252.

54. Wu X, Zhu W. Stability enhancement of fluorophores for lighting up practical application in bioimaging. *Chem Soc Rev.* 2015;44(13):4179–4184.

55. Cabral Filho PE, Pereira MIA, Fernandes HP, de Thomaz AA, Cesar CL, Santos BS, et al. Blood group antigen studies using CdTe quantum dots and flow cytometry. *Int J Nanomed.* 2015;10:4393–4404.

56. Chattopadhyay PK. Chapter 18 - Quantum dot technology in flow cytometry. In: Darzynkiewicz Z, Holden E, Orfao A, Telford W, Wlodkowic D, editors. *Methods in Cell Biology.* Vol. 102. Academic Press; 2011. pp. 463–477.

57. Dolmans DEJGJ, Fukumura D, Jain RK. Photodynamic therapy for cancer. *Nat Rev Cancer.* 2003;3(5):380–387.

58. Satrialdi, Munechika R, Biju V, Takano Y, Harashima H, Yamada Y. The optimization of cancer photodynamic therapy by utilization of a pi-extended porphyrin-type photosensitizer in combination with MITO-Porter. *Chem Commun.* 2020;56(7):1145–1148.

59. Ahirwar S, Mallick S, Bahadur D. Photodynamic therapy using graphene quantum dot derivatives. *J Solid State Chem.* 2020;282:121107.

60. Łoczechin A, Séron K, Barras A, Giovanelli E, Belouzard S, Chen YT, et al. Functional carbon quantum dots as medical countermeasures to human coronavirus. *ACS Appl Mater Interfaces.* 2019;11(46):42964–42974.

61. Sanchez de Araujo H, Ferreira F. Quantum dots and photodynamic therapy in COVID-19 treatment. *Quant Eng.* 2021;3(4):e78.

62. Bao W, Ma H, Wang N, He Z. pH-sensitive carbon quantum dots–doxorubicin nanoparticles for tumor cellular targeted drug delivery. *Polym Adv Technol.* 2019;30(11):2664–2673.

63. Iannazzo D, Pistone A, Celesti C, Triolo C, Patané S, Giofré SV, et al. A smart nanovector for cancer targeted drug delivery based on graphene quantum dots. *Nanomaterials (Basel).* 2019;9(2):282.

64. Ulusoy M, Jonczyk R, Walter J-G, Springer S, Lavrentieva A, Stahl F, et al. Aqueous synthesis of PEGylated quantum dots with increased colloidal stability and reduced cytotoxicity. *Bioconj Chem.* 2016;27(2):414–426.

65. Younis MA, Tawfeek HM, Abdellatif AAH, Abdel-Aleem JA, Harashima H. Clinical translation of nanomedicines: Challenges, opportunities, and keys. *Adv Drug Deliv Rev.* 2022;181:114083.

66. Mangeolle T, Yakavets I, Lequeux N, Pons T, Bezdetnaya L, Marchal F. The targeting ability of fluorescent quantum dots to the folate receptor rich tumors. *Photodiagn Photodyn Ther.* 2019;26:150–156.

67. Zhao Y, Liu S, Li Y, Jiang W, Chang Y, Pan S, et al. Synthesis and grafting of folate-PEG-PAMAM conjugates onto quantum dots for selective targeting of folate-receptor-positive tumor cells. *J Colloid Interface Sci.* 2010;350(1):44–50.

68. Song EQ, Zhang ZL, Luo QY, Lu W, Shi YB, Pang DW. Tumor cell targeting using folate-conjugated fluorescent quantum dots and receptor-mediated endocytosis. *Clin Chem.* 2009;55(5):955–963.

69. Liu L, Jiang H, Dong J, Zhang W, Dang G, Yang M, et al. PEGylated MoS$_{(2)}$ quantum dots for traceable and pH-responsive chemotherapeutic drug delivery. *Colloids Surf B Biointerfaces.* 2020;185:110590.

70. Hun X, Zhang Z. Anti-epidermal growth factor receptor (anti-EGFR) antibody conjugated fluorescent nanoparticles probe for breast cancer imaging. *Spectrochim Acta A Mol Biomol Spectrosc.* 2009;74(2):410–414.

71. Zhang JH, Sun T, Niu A, Tang YM, Deng S, Luo W, et al. Perturbation effect of reduced graphene oxide quantum dots (rGOQDs) on aryl hydrocarbon receptor (AhR) pathway in zebrafish. *Biomaterials.* 2017;133:49–59.

72. Zu R, Fang X, Lin Y, Xu S, Meng J, Xu H, et al. Peptide-enabled receptor-binding-quantum dots for enhanced detection and migration inhibition of cancer cells. *J Biomater Sci Polym Ed.* 2020;31(12):1604–1621.

73. Nasrollahi F, Koh YR, Chen P, Varshosaz J, Khodadadi AA, Lim S. Targeting graphene quantum dots to epidermal growth factor receptor for delivery of cisplatin and cellular imaging. *Mater Sci Eng C Mater Biol Appl.* 2019;94:247–257.

74. Zhang Y, Zhao N, Qin Y, Wu F, Xu Z, Lan T, et al. Affibody-functionalized Ag2S quantum dots for photoacoustic imaging of epidermal growth factor receptor overexpressed tumors. *Nanoscale.* 2018;10(35):16581–16590.

75. Tang J, Huang N, Zhang X, Zhou T, Tan Y, Pi J, et al. Aptamer-conjugated PEGylated quantum dots targeting epidermal growth factor receptor variant III for fluorescence imaging of glioma. *Int J Nanomedicine.* 2017;12:3899–3911.

76. Kaushik NK, Kaushik N, Wahab R, Bhartiya P, Linh NN, Khan F, et al. Cold atmospheric plasma and gold quantum dots exert dual cytotoxicity mediated by the cell receptor-activated apoptotic pathway in glioblastoma cells. *Cancers (Basel).* 2020;12(2):457.

77. Yang H, Lou C, Xu M, Wu C, Miyoshi H, Liu Y. Investigation of folate-conjugated fluorescent silica nanoparticles for targeting delivery to folate receptor-positive tumors and their internalization mechanism. *Int J Nanomedicine.* 2011;6:2023–2032.

78. Cui D, Lu X, Yan C, Liu X, Hou M, Xia Q, et al. Gastrin-releasing peptide receptor-targeted gadolinium oxide-based multifunctional nanoparticles for dual magnetic resonance/fluorescent molecular imaging of prostate cancer. *Int J Nanomedicine.* 2017;12:6787–6797.

79. Labrecque S, Sylvestre JP, Marcet S, Mangiarini F, Bourgoin B, Verhaegen M, et al. Hyperspectral multiplex single-particle tracking of different receptor subtypes labeled with quantum dots in live neurons. *J Biomed Opt.* 2016;21(4):46008.

80. Mazumdar A, Haddad Y, Milosavljevic V, Michalkova H, Guran R, Bhowmick S, et al. Peptide-carbon quantum dots conjugate, derived from human retinoic acid receptor responder protein 2, against antibiotic-resistant gram positive and gram negative pathogenic bacteria. *Nanomaterials.* 2020;10(2):325.

81. Abdellatif AAH, Abou-Taleb HA, Abd El Ghany AA, Lutz I, Bouazzaoui A. Targeting of somatostatin receptors expressed in blood cells using quantum dots coated with vapreotide. *Saudi Pharm J.* 2018;26(8):1162–1169.

82. Abdellatif AAH. A plausible way for excretion of metal nanoparticles via active targeting. *Drug Dev Ind Pharm.* 2020;46(5):744–750.

83. Cho EJ, Jung S, Lee K, Lee HJ, Nam KC, Bae HJ. Fluorescent receptor-immobilized silica-coated magnetic nanoparticles as a general binding agent for histidine-tagged proteins. *Chem Commun.* 2010;46(35):6557–6559.

84. Chen S, Imoukhuede PI. Multiplexing angiogenic receptor quantification via quantum dots. *Anal Chem.* 2019;91(12):7603–7612.

85. Mukhtar M, Bilal M, Rahdar A, Barani M, Arshad R, Behl T, et al. Nanomaterials for diagnosis and treatment of brain cancer: Recent updates. *Chemosensors.* 2020;8(4):117.

86. Pandey PC. Properties, applications and toxicities of organotrialkoxysilane-derived functional metal nanoparticles and their multimetallic analogues. *Materials.* 2023;16(5):2052.

87. Raza F, Evans L, Motallebi M, Zafar H, Pereira-Silva M, Saleem K, et al. Liposome-based diagnostic and therapeutic applications for pancreatic cancer. *Acta Biomater.* 2023;157:1–23.

88. Chapman S, Dobrovolskaia M, Farahani K, Goodwin A, Joshi A, Lee H, et al. Nanoparticles for cancer imaging: The good, the bad, and the promise. *Nano Today.* 2013;8(5):454–460.

89. Liu CH, Grodzinski P. Nanotechnology for cancer imaging: Advances, challenges, and clinical opportunities. *Radiol Imaging Cancer.* 2021;3(3):e200052.

90. Alrushaid N, Khan FA, Al-Suhaimi EA, Elaissari A. Nanotechnology in cancer diagnosis and treatment. *Pharmaceutics.* 2023;15(3):1025.

91. Ashfaq M, Talreja N, Chauhan D, Afreen S, Sultana A, Srituravanich W. Two-dimensional (2D) hybrid nanomaterials for diagnosis and treatment of cancer. *J Drug Deliv Sci Technol.* 2022;70:103268.

92. Langeveld SAG, Meijlink B, Kooiman K. Phospholipid-coated targeted microbubbles for ultrasound molecular imaging and therapy. *Curr Opin Chem Biol.* 2021;63:171–179.

93. Sancey L, Lux F, Kotb S, Roux S, Dufort S, Bianchi A, et al. The use of theranostic gadolinium-based nanoprobes to improve radiotherapy efficacy. *Br J Radiol.* 2014;87(1041):20140134.

94. Yang J, Wang F, Yuan H, Zhang L, Jiang Y, Zhang X, et al. Recent advances in ultra-small fluorescent Au nanoclusters toward oncological research. *Nanoscale.* 2019;11(39):17967–17980.

95. Bromma K, Chithrani DB. Advances in gold nanoparticle-based combined cancer therapy. *Nanomaterials.* 2020;10(9):1671.

96. Nunez C, Estevez SV, Del Pilar Chantada M. Inorganic nanoparticles in diagnosis and treatment of breast cancer. *J Biol Inorg Chem.* 2018;23(3):331–345.

97. Meyers JD, Doane T, Burda C, Basilion JP. Nanoparticles for imaging and treating brain cancer. *Nanomedicine.* 2013;8(1):123–143.

98. Sonali, Viswanadh MK, Singh RP, Agrawal P, Mehata AK, Pawde DM, et al. Nanotheranostics: Emerging strategies for early diagnosis and therapy of brain cancer. *Nanotheranostics.* 2018;2(1):70–86.

99. Du C, Liu X, Hu H, Li H, Yu L, Geng D, et al. Dual-targeting and excretable ultrasmall SPIONs for T(1)-weighted positive MR imaging of intracranial glioblastoma cells by targeting the lipoprotein receptor-related protein. *J Mater Chem B.* 2020;8(11):2296–2306.

100. Smilowitz HM, Meyers A, Rahman K, Dyment NA, Sasso D, Xue C, et al. Intravenously-injected gold nanoparticles (AuNPs) access intracerebral F98 rat gliomas better than AuNPs infused directly into the tumor site by convection enhanced delivery. *Int J Nanomedicine.* 2018;13:3937–3948.

101. Lee C, Kim GR, Yoon J, Kim SE, Yoo JS, Piao Y. In vivo delineation of glioblastoma by targeting tumor-associated macrophages with near-infrared fluorescent silica coated iron oxide nanoparticles in orthotopic xenografts for surgical guidance. *Sci Rep.* 2018;8(1):11122.

102. Huang N, Cheng S, Zhang X, Tian Q, Pi J, Tang J, et al. Efficacy of NGR peptide-modified PEGylated quantum dots for crossing the blood-brain barrier and targeted fluorescence imaging of glioma and tumor vasculature. *Nanomedicine.* 2017;13(1):83–93.

103. Wei T, Tu W, Zhao B, Lan Y, Bao J, Dai Z. Electrochemical monitoring of an important biomarker and target protein: VEGFR2 in cell lysates. *Sci Rep.* 2014;4:3982.

104. Kumar N, Sadique MA, Khan R, Gowri VS, Kumar S, Ashiq M, et al. Immunosensor for breast cancer CD44 biomarker detection based on exfoliated graphene quantum dots integrated gold nanoparticles. *Hybrid Adv.* 2023;3:100065.

105. Raj D, Kumar A, Kumar D, Kant K, Mathur A. Gold-graphene quantum dot hybrid nanoparticle for smart diagnostics of prostate cancer. *Biosensors.* 2024;14(11):534.

106. Wang J, Sui L, Huang J, Miao L, Nie Y, Wang K, et al. $MoS_{(2)}$-based nanocomposites for cancer diagnosis and therapy. *Bioact Mater.* 2021;6(11):4209–4242.

107. Bouazzaoui A, Abdellatif AAH. Vaccine delivery systems and administration routes: Advanced biotechnological techniques to improve the immunization efficacy. *Vaccine X.* 2024;19:100500.

108. El-Readi MZ, Abdulkarim MA, Abdellatif AAH, Elzubeir ME, Refaat B, Althubiti M, et al. Doxorubicin-sanguinarine nanoparticles: Formulation and evaluation of breast cancer cell apoptosis and cell cycle. *Drug Dev Ind Pharm.* 2024;1:1–15.

109. Zhang LW, Monteiro-Riviere NA. Mechanisms of quantum dot nanoparticle cellular uptake. *Toxicol Sci.* 2009;110(1):138–155.

110. Zhang LW, Bäumer W, Monteiro-Riviere NA. Cellular uptake mechanisms and toxicity of quantum dots in dendritic cells. *Nanomedicine.* 2011;6(5):777–791.

111. Khalil IA, Kogure K, Akita H, Harashima H. Uptake pathways and subsequent intracellular trafficking in nonviral gene delivery. *Pharmacol Rev.* 2006;58(1):32–45.

112. Bourrinet P, Bengele HH, Bonnemain B, Dencausse A, Idee JM, Jacobs PM, et al. Preclinical safety and pharmacokinetic profile of ferumoxtran-10, an ultrasmall superparamagnetic iron oxide magnetic resonance contrast agent. *Invest Radiol.* 2006;41(3):313–324.

113. Li J, Chang X, Chen X, Gu Z, Zhao F, Chai Z, et al. Toxicity of inorganic nanomaterials in biomedical imaging. *Biotechnol Adv.* 2014;32(4):727–743.

114. Zhu MT, Feng WY, Wang B, Wang TC, Gu YQ, Wang M, et al. Comparative study of pulmonary responses to nano- and submicron-sized ferric oxide in rats. *Toxicology.* 2008;247(2–3):102–111.

115. Lovric J, Bazzi HS, Cuie Y, Fortin GR, Winnik FM, Maysinger D. Differences in subcellular distribution and toxicity of green and red emitting CdTe quantum dots. *J Mol Med.* 2005;83(5):377–385.

116. Sanhai WR, Sakamoto JH, Canady R, Ferrari M. Seven challenges for nanomedicine. *Nat Nanotechnol.* 2008;3(5):242–244.

117. Metselaar JM, Lammers T. Challenges in nanomedicine clinical translation. *Drug Deliv Transl Res.* 2020;10(3):721–725.
118. Andoh V, Ocansey DKW, Naveed H, Wang N, Chen L, Chen K, et al. The advancing role of nanocomposites in cancer diagnosis and treatment. *Int J Nanomedicine.* 2024;19:6099–6126.
119. Tiwari H, Rai N, Singh S, Gupta P, Verma A, Singh AK, et al. Recent advances in nanomaterials-based targeted drug delivery for preclinical cancer diagnosis and therapeutics. *Bioengineering.* 2023;10(7):760.
120. Weerarathna IN, Kamble AR, Luharia A. Artificial intelligence applications for biomedical cancer research: A review. *Cureus.* 2023;15(11):e48307.
121. Min Y, Caster JM, Eblan MJ, Wang AZ. Clinical translation of nanomedicine. *Chem Rev.* 2015;115(19):11147–11190.
122. Ren X, Wang Y, Chen L, Zhang XS, Jin Q. EllipsoidFN: A tool for identifying a heterogeneous set of cancer biomarkers based on gene expressions. *Nucleic Acids Res.* 2013;41(4):e53.

7 Immunotherapy of Cancer

Ikram A. Burney, Ishtiaq Ahmad Khan, and Samra Khan

7.1 IMMUNOTHERAPY AND CANCER

The relationship between immunity and cancer is complex and integral to carcinogenesis, cancer progression as well as potential treatments. The immune system plays a twofold role in inhibiting and encouraging tumor development. This idea is referred to as immunoediting [1,2]. The immunoediting hypothesis, suggested by Dunn and colleagues, consists of three phases [3].

- Elimination: In this phase, the nascent tumor cells are recognized and attacked by the immune system. Cytotoxic T lymphocytes (CTLs), Natural Killer lymphocytes, and macrophages collaborate to identify and eradicate abnormal cells.
- Equilibrium: During this phase, surviving tumor cells become dormant. This period might last for years. The immune pressure selects for genetically stable tumor variants. At this point, the immune system can only prevent the tumor from spreading further.
- Escape: In this phase, tumor cells are genetically or epigenetically modified; thus, they are unrecognizable by the immune system and can escape.

A key characteristic of cancer is its ability to evade recognition by the immune system [4]. While the immune system is anticipated to target the tumor-associated antigens (TAAs), the tumor microenvironment (TME) does not maintain an immunogenic state. Malignant cells can evade immune surveillance through mechanisms that inhibit immune responses, such as downregulating surface antigens and producing immunosuppressive molecules [4,5]. Certain cells, like tumor-associated macrophages (TAMs), can facilitate tumor growth by promoting angiogenesis and remodeling of the tissue. These cells can also fight the tumor [6]. Other cells, including lymphocytes, mast cells, monocytes, neutrophils, and macrophages, try to fight off cancer cells [7].

Over the last decade, there has been significant development in using the immune system to attack cancer cells. Cancer immunotherapy works by targeting regulatory networks that cancer cells use to dampen immune responses. Checkpoint inhibitors, such as Programmed Death Receptor 1 and Programmed-Death Receptor Ligand 1 (PD-1/PD-L1) blockers, essentially turn off the immune system's "brakes," allowing T lymphocytes to better target and destroy cancer cells [8]. Alternative immunotherapy modalities, including CAR-T cell treatment and cancer vaccines, improve the immune system's ability to identify cancer cells [9].

7.2 BRIEF HISTORY OF IMMUNOTHERAPY OF CANCER

The prospect of immunotherapy first emerged in the late 19th century, owing to the pioneering study of Dr. William Coley, known as the "father of immunotherapy." Dr. Coley noted that some cancer patients had tumor regression after contracting bacterial infections [10]. He hypothesized that the immune system might be targeting cancer cells and developed a blend of dead bacteria, known as "Coley's toxins," to provoke an immune response in cancer patients. While Coley's results met with skepticism at that time, his approach laid the groundwork for comprehending the possibilities of utilizing the immune system in cancer therapy.

Since the initial recognition of tumor antigenicity in 1957, immunotherapy has been a viable cancer therapeutic option [11]. Later on in the 20th century, the immune surveillance hypothesis

DOI: 10.1201/9781003517870-7

proposed by Thomas and Burnet further cemented the idea of immunotherapy. As per the hypothesis, our immune system constantly patrols the body for any unnatural cells like cancer cells to destroy them in the nascent stage [12]. The study of "tumor immunology" emerged as a result of a deeper comprehension of the immune system.

The second half of the 20th century was crucial for progress in tumor immunology. This was the time when TAAs were discovered. These TAAs act as the sites for the immune system to identify malignant cells so they can eliminate them [13]. TAAs offered the basis for the eventual creation of targeted immunotherapies. At the beginning of the 21st century, Dunn and colleagues proposed the hypothesis of immune editing, which laid the groundwork for our present comprehension of the intricate interaction between the immune system and cancer, including immunotherapy [3].

7.3 TYPES OF IMMUNOTHERAPY

Broadly speaking, immunotherapy is classified as either non-specific immunotherapy or targeted/personalized immunotherapy (Figure 7.1)

Both interferon and interleukin modulate the immune system stimulation and activation of T-lymphocytes and release of cytokines, which act in a non-specific manner. With the advent of better and safer forms of immunotherapy, these are obsolete in clinical practice. Bacille Calmette-Guérin (BCG) has limited indications and will not be discussed in this chapter. However, the interested reader could explore some recent reviews [14,15]. Furthermore, there is a separate chapter on CAR-T cell therapies (see chapter 9). For this chapter, the following will be discussed in detail:

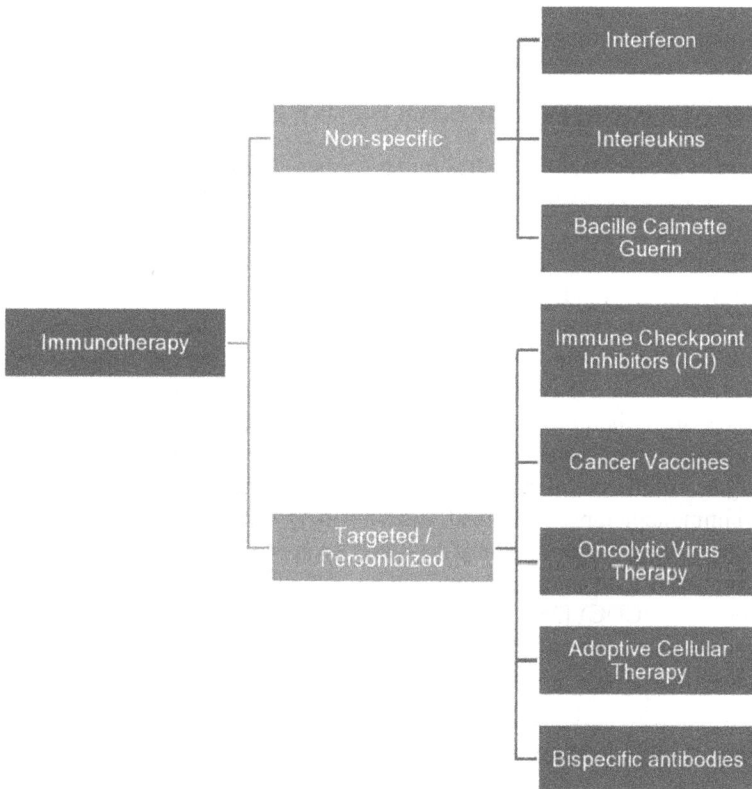

FIGURE 7.1 Types of immunotherapies.

- Immune Checkpoint Inhibitors: ICIs target proteins that act as "checkpoints" on immune cells.

    These proteins act as shields for normal cells to protect them from the immune system. Cancer cells exploit these proteins to their benefit, allowing them to avoid being recognized by the immune system. By focusing on these proteins, the immune system is capable of attacking the tumor cells. In 2011, the FDA approved Ipilimumab, the inaugural checkpoint inhibitor, for metastatic melanoma, which enhanced the outlook for patients with recurrent, metastatic advanced disease [16]. This spurred the development of other ICIs aiming at the PD-1/PD-L1 pathway, resulting in a broad application of this therapy across multiple cancers. ICIs have since demonstrated remarkable efficacy, leading to complete remission in some cases [17]. In 2018, Dr. James Allison and Dr. Tasuku Honjo received the Nobel Prize in Physiology or Medicine for discovering the Cytotoxic T-lymphocyte-associated protein 4 (CTLA-4) and PD-1 pathways [18]. Their work solidified immune checkpoint inhibition as a cancer therapeutic breakthrough.
- Cancer Vaccines: Cancer vaccines utilize the immune system to treat and prevent cancer [19]. Therapeutic vaccines, like Sipuleucel-T, are used to treat already-developed tumors. Since its approval in 2010, Sipuleucel-T has been employed in the treatment of castrate-resistant prostate cancer. Meanwhile, preventive vaccines are used to protect the body against oncogenic viruses like HPV [20].
- Oncolytic Virus Therapy: This approach utilizes either naturally occurring or laboratory-made viruses to fight off cancerous cells. These viruses specifically target tumor cells for replication, causing them to be destroyed while leaving healthy tissues unharmed [21]. OVT has the additional advantage of boosting anti-tumor immunity. The first ever oncolytic viral therapy to get regulatory approval was Talimogene Laherparepvec (T-VEC), which is authorized to treat metastatic melanoma [22].

7.4   WHAT IS CLASSICALLY NOT IMMUNOTHERAPY?

The development of monoclonal antibodies (mAbs) was seen as a significant advancement in cancer immunotherapy. mAbs trigger antibody-dependent complement-mediated cytotoxicity (ADCC) by attaching to particular tumor cell's antigens. The first major clinical success in this field was Rituximab, which is used to treat B-cell malignancies by targeting the CD20 receptors [23]. Since then, the antibody treatment industry has seen a massive rise, resulting in the approval of over 150 antibody therapeutics [24]. However, mAbs are often not classified as "immunotherapy" in the traditional sense [25]. At its core, immunotherapy involves utilizing the body's immune system to recognize and eliminate cancer cells by triggering an immune response. While mAbs do interact with the immune system, their mechanism often involves targeting specific cancer cell antigens and disrupting their signaling pathways [26]. This mechanism does not involve immune cell activation but rather a targeted inhibition of cancer cell growth, which classifies it closer to targeted therapy than traditional immunotherapy. Additionally, certain mAbs like bevacizumab exert their effects by either neutralizing growth factors or inducing direct apoptosis without eliciting a broader immune response [27]. Some mAbs do engage the immune system indirectly through ADCC and complement-dependent cytotoxicity (CDC) [28]. However, ADCC and CDC are secondary mechanisms. This difference in mechanism is one of the primary reasons mAbs are viewed distinctly from classical immunotherapies, such as ICIs or cancer vaccines.

7.5   IMMUNE CHECKPOINT INHIBITORS

ICIs have revolutionized cancer treatment by targeting specific pathways that tumors exploit to evade immune recognition. The rationale for using these inhibitors is grounded in understanding how cancer cells manipulate immune checkpoints (ICPs) [29]. ICPs are natural regulatory pathways that maintain self-tolerance and prevent autoimmunity. By blocking the ICPs, ICIs activate immune cells, particularly T-lymphocytes, enabling them to recognize and eliminate tumors more effectively.

7.5.1 Immune Checkpoints

ICP proteins operate like a "brake" in a healthy immune system. This allows them to prevent excessive immune system activation while keeping the healthy tissues safe. The most common examples of ICPs are PD-1 and CTLA-4 [30]. When there is an infection or occurrence of tumor cells, these proteins engage with the T-lymphocyte ligands, signaling the T-cells to reduce activity. Malignant cells overexpress checkpoint ligands like PD-L1, evading immune destruction. ICIs target these checkpoint pathways [31]. By binding to either the checkpoint proteins or their ligands, ICIs prevent their interaction and effectively unleash the immune system. The active immune system in turn facilitates T-cells in destroying cancer cells. Since several cancers exploit these pathways, ICIs dismantle this defense and promote the body's ability to combat cancer. By acting as inhibitory regulators, ICPs are also essential in limiting excessive immunological activation. Some inhibitory ICP molecules include T Cell Immunoglobulin and Mucin-Domain Containing-3 (TIM-3), Lymphocyte Activation Gene-3 (LAG-3), T cell immunoglobulin and ITIM domain (TIGIT), V domain Ig suppressor of T cell activation, CTLA-4, and PD-1/PD-L1. There are also stimulatory ICPs that positively regulate immune activity. For example, T cells express CD28 and ICOS (inducible T cell co-stimulator), which support effector T lymphocyte and Treg survival, growth, and activity [32]. Currently, other stimulatory ICPs' potential as therapeutic targets is being investigated.

One of the core advantages of ICIs is their potential to elicit sustained responses even at advanced stages and relapsed disease. Unlike chemotherapy, ICIs provide a more selective approach by targeting specific immune pathways exploited by cancer. Moreover, by activating the immune system's memory capabilities, some patients experience prolonged remissions even after the discontinuation of therapy, suggesting that their immune systems retain the ability to identify and control tumor growth long term.

7.5.2 Pivotal Phase I and Phase II Trials

The role of ICIs in cancer treatment was established by pivotal Phase I and II clinical trials. These trials provide substantial evidence regarding the safe use and possible mechanisms of ICIs to restart immune responses. Table 7.1 provides a brief description of some of the landmark trials.

These studies underscore the significant role of ICIs in various cancer types. The sustained survival benefits and response rates observed in these trials were transformative, leading to new approvals, phase III trials, combination therapies, and ongoing research.

7.5.3 Pivotal Phase III Trials

Phase III clinical studies are crucial for assessing the safety and effectiveness of novel agents in comparison to standard-of-care therapies and pave the way for getting regulatory approval. Some of the most studied ICIs like pembrolizumab, nivolumab, atezolizumab, durvalumab, and ipilimumab have been studied across various cancer types, establishing their role in current cancer therapy. Here is a summary of key trials in Table 7.2:

7.5.4 Selecting Patients for Treatment with ICIs

The effectiveness of ICIs varies across patient populations. Therefore, selecting appropriate patients for treatment with ICIs is essential to maximize benefits and minimize potential toxicities and costs. The selection process typically involves evaluating specific biomarkers, tumor characteristics, and patient-related factors [49].

TABLE 7.1

Pivotal Phase I and Phase II Clinical Trials of Immune Checkpoint Inhibitors

Reference	Study	Drug	Key Findings
MDX-010-20 [33]	CTLA-4 Inhibition in Melanoma (Phase II)	Ipilimumab	Improved overall survival (OS) and long-term disease control in metastatic melanoma; first immunotherapy to show survival benefits in melanoma.
CheckMate-037 [34]	PD-1 Inhibition in Advanced Melanoma (Phase II)	Nivolumab	Durable responses in advanced melanoma patients with fewer side effects compared to chemotherapy.
CheckMate-017 and -057 [35]	PD-1 Inhibition in Non-Small Cell Lung Cancer (NSCLC) (Phase II/III)	Nivolumab	Significant OS improvements in previously treated advanced NSCLC compared to docetaxel; granted FDA approval in NSCLC.
KEYNOTE-001 [36]	PD-1 Inhibition in Multiple Tumors (Phase I)	Pembrolizumab	Strong responses and durable survival across multiple tumor types, including melanoma and NSCLC, led to FDA approval in melanoma.
KEYNOTE-006 [37]	Pembrolizumab vs. Ipilimumab in Melanoma (Phase II)	Pembrolizumab	Improved survival and less toxicity in advanced melanoma as compared to ipilimumab established pembrolizumab as a frontline therapy.
IMvigor210 [38]	PD-L1 Inhibition in Urothelial Cancer (Phase II)	Atezolizumab	Meaningful response rates in metastatic urothelial carcinoma ineligible for cisplatin led to FDA approval.
FDA Approval Summary: [39]	Pembrolizumab in MMR-Deficient Tumors	Pembrolizumab	Objective response rate (ORR) of 39.6% and complete response (CR) rate of 7% across 15 tumor types with MMR deficiency; responses were durable with prolonged progression-free survival (PFS) and OS; led to the first tumor-agnostic FDA approval.

TABLE 7.2

Pivotal Phase III Clinical Trials of Immune Checkpoint Inhibitors

Study	Drug	Comparator	Key Findings
CheckMate-141 [40]	Nivolumab	Standard therapies	Extended OS and manageable safety profile in recurrent/metastatic head and neck squamous cell carcinoma.
KEYNOTE-024 [41]	Pembrolizumab	Chemotherapy	Improved OS and PFS in advanced NSCLC with high PD-L1 expression led to FDA approval.
CheckMate-067 [42]	Nivolumab + Ipilimumab	Monotherapies	Superior OS in the combination arm for advanced melanoma led to approval of dual ICI therapy for melanoma.
KEYNOTE-426 [43]	Pembrolizumab + Axitinib	None	Enhanced PFS and OS in advanced renal cell carcinoma resulted in combination therapy approval.
CheckMate-649 [44]	Nivolumab + Chemotherapy	None	OS improvement in advanced gastric and esophageal adenocarcinoma with PD-L1 expression influenced gastrointestinal cancer guidelines.
KEYNOTE-048 [45]	Pembrolizumab ± Chemotherapy	Chemotherapy	OS improvements in recurrent/metastatic head and neck squamous cell carcinoma prompted approvals in first-line settings.
JAVELIN Bladder 100 [46]	Avelumab	No maintenance therapy	Improved OS in metastatic urothelial carcinoma as maintenance therapy post-chemotherapy established its role in this setting.
PACIFIC Trial [47]	Durvalumab	No consolidation therapy	Improved OS as consolidation therapy post-chemoradiation in unresectable stage III NSCLC set a new standard of care.
KEYNOTE-590 [48]	Pembrolizumab + Chemotherapy	Chemotherapy	Significant OS benefits in esophageal carcinoma, particularly for PD-L1-positive tumors, led to regulatory approvals.

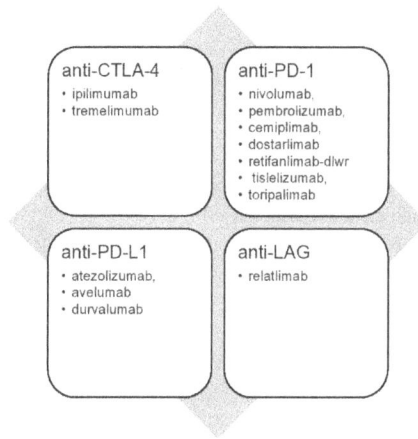

FIGURE 7.2   FDA-approved immune checkpoint inhibitors (ICIs).

The selection of patients for ICI treatment is a complex process that involves evaluating tumor-specific and patient-related factors. Biomarkers like PD-1/PD-L1, or Tumor Proportion Score/ Combined Positive Score (TPS/CPS), mismatch repair-deficiency dMMR/Microsatellite instability-high (MSI-H), and high tumor mutational burden (TMB) have provided useful guidance, but more refined predictors are needed to optimize outcomes [49]. Ultimately, a comprehensive assessment integrating clinical and molecular features, along with ongoing monitoring for efficacy and toxicity, is critical for the successful use of ICIs.

7.5.5   FDA-Approved ICIs

At the time of writing, the US FDA has authorized 13 ICIs that inhibit four separate ICPs (Figure 7.2).

7.5.6   Monitoring Response to ICIs

While ICIs have demonstrated significant efficacy across various cancers, their unique mechanisms of action have introduced challenges in response assessment. Traditionally, response to systemic treatment is evaluated by standardized methods such as the Response Evaluation Criteria in Solid Tumors (RECIST) [50]. However, classic RECIST criteria may fail to completely reflect the fluctuations of tumor response in the context of ICIs. Phenomena like pseudo-progression and hyper-progression have emerged as key considerations [51].

A key feature of pseudo-progression is a temporary increase in tumor size or the emergence of new lesions on imaging, which is followed by a subsequent tumor decrease or stabilization. 5%–10% of patients on ICIs have this condition, which has been often seen in melanoma, NSCLC, and renal cell carcinoma (RCC) [52]. This phenomenon challenges traditional RECIST criteria, which classify tumor growth as a progressive disease. To address this, immune-related response criteria (irRC) and immune RECIST (iRECIST) were developed [53]. These guidelines account for initial increases in tumor burden and recommend confirmatory imaging 4–8 weeks later to distinguish true progression from pseudo-progression. Biomarkers such as circulating tumor DNA (ctDNA) may be employed to differentiate pseudo-progression from actual tumor growth in some cases [54]. In contrast, hyper-progression represents a paradoxical acceleration of tumor growth in response to ICIs. It is characterized by a marked increase in tumor burden and rapid clinical deterioration, often associated with poor outcomes. Hyper-progression occurs in approximately 10% of ICI-treated patients, though its incidence varies across cancer types [55]. The development of PET/CT with immune-specific tracers can assess immune-related responses.

TABLE 7.3

Toxicity of Immune Checkpoint Inhibitors, Grades, and Management Strategies

Grade	General Description	Management Strategy
Grade 1	Patient is asymptomatic or has a minimal skin rash. The irAE has been detected on a routine blood test or radiological investigation	Continue ICI and manage side effects symptomatically, if any.
Grade 2	The patient is minimally symptomatic	Withhold ICIs and delay treatment till the toxicity returns to grade 1
Grade 3	The patient is symptomatic and requires intervention to treat the symptom	Withhold ICIs and treat with corticosteroids. If toxicity returns to grade 1, rechallenge can be given carefully in selected patients
Grade 4	Life-threatening toxicity, requiring HDU/ICU transfer	Withhold and permanently discontinue treatment with ICIs

Additionally, integrating liquid biopsies, particularly ctDNA provides complementary insights. ctDNA dynamics have shown promise also in differentiating pseudo-progression, stable disease, and true progression [56,57].

7.5.7 TOXICITY AND SIDE EFFECTS

ICIs promote enduring responses across various cancers by stimulating the immune system. Nonetheless, this reactivation frequently leads to immune-related adverse events (irAEs). Although ICIs have a distinct side effect profile than traditional cytotoxic chemotherapy, irAEs can vary from negligible to lethal and affect almost every organ system [58,59]. Understanding and managing these side effects is crucial to optimizing ICI therapy. Factors leading to irAEs are being studied and include certain genetic polymorphisms and the presence of autoantibodies. For example, patients with pre-existing autoimmune diseases often experience more severe irAEs when treated with ICIs. The management of irAEs involves early detection and grading toxicity, often requiring discontinuation of ICIs and prompt intervention. The American Society of Clinical Oncology (ASCO) [60] and the World Health Organization (WHO), as shown in Table 7.3, have suggested a general scheme of management of irAEs:

7.5.8 MECHANISMS OF RESISTANCE

A major problem with ICI resistance is that either many patients do not react at all (primary resistance) or relapse after responding initially (acquired resistance). The main culprits behind these resistance mechanisms are the TME, tumor intrinsic factors, and the adaptive immune system. Comprehending these systems has resulted in tactics intended to surmount resistance and improve the effectiveness of ICIs [61–63].

7.5.9 IMPACT OF ICIs ON CANCER RESEARCH LANDSCAPE

ICIs have improved the response rates, disease-free survival, progression-free survival, and overall survival (OS) across various cancers. The success of ICIs has spurred research into their use in several areas, such as reporting clinical studies, clinical trials, combination with other therapies, mitigating irAEs, and overcoming resistance. Figure 7.3 shows the publication trends of ICIs extracted from SCOPUS.

Number of Publications

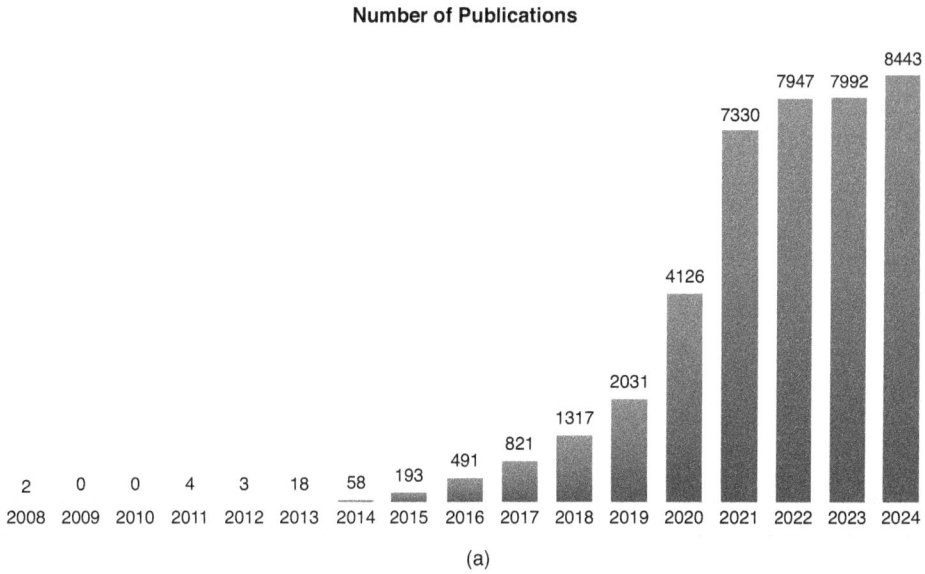

(a)

40,776 document results

Select year range to analyze: 2008 ☑ to 2024 ☑ Analyze

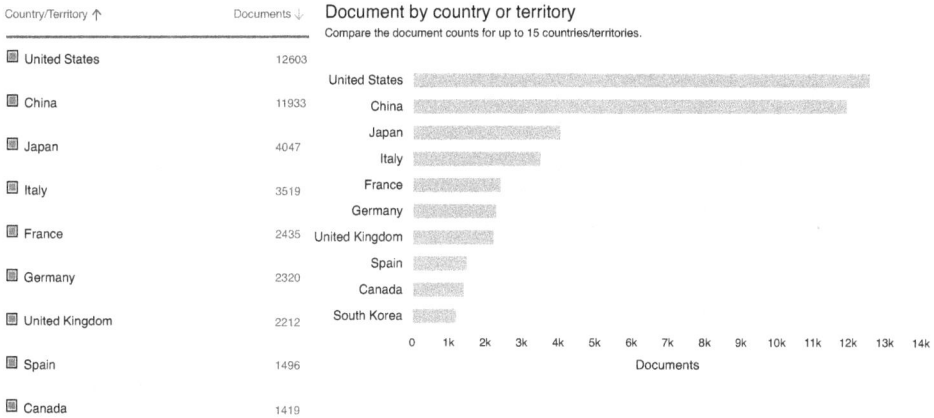

Document by country or territory
Compare the document counts for up to 15 countries/territories.

(b)

FIGURE 7.3 (a and b): Publication trends of ICIs from Scopus.

7.6 CANCER VACCINES

Cancer vaccines are proving to be an effective approach to cancer prevention and treatment. Preventive vaccines target oncogenic viruses, and therapeutic vaccines stimulate the immune system to attack existing tumors. These approaches leverage various mechanisms, including whole cells, DNA, and, more recently, mRNA technology [64].

7.6.1 PREVENTIVE CANCER VACCINES

The most successful preventive cancer vaccines to date target virus-associated malignancies, particularly the hepatitis B virus (HBV) and HPV. HBV vaccination significantly reduces the risk of liver cancer, while HPV vaccines have shown effectiveness in preventing cervical, anal, and other HPV-related cancers. The widespread use of HPV vaccine has resulted in a considerable reduction in HPV infections, benign tumors, and cervical cancer incidence worldwide. HPV vaccination can reduce HPV infections by up to 90% and HPV-related malignancies by more than 70% in vaccinated populations, highlighting the promise of preventative immunization regimens [65]. A Scottish population-based observational study found that women inoculated at 12 or 13 years old did not develop invasive cervical cancer, regardless of the number of doses received [66].

7.6.2 Therapeutic Cancer Vaccines

Designed to provoke an immune reaction, therapeutic cancer vaccines help the immune system identify tumor-specific antigens. The most noteworthy example is Sipuleucel-T, which got initial approval in 2010 for treating prostate cancer. The vaccine notably extended the median survival time for men with metastatic castration-resistant prostate cancer from 21.7 to 25.8 months [67]. This vaccine involves priming the patient's immune cells with a fusion protein to trigger an anti-cancer immune response. However, like other therapeutic vaccines, Sipuleucel-T face challenges such as tumor heterogeneity and immunosuppressive microenvironments that limit efficacy [68], and despite being the first therapeutic vaccine to be approved, is not used routinely in clinical settings.

After the success of COVID-19 mRNA vaccines, new developments in mRNA technology have led to investigations of mRNA vaccines for cancer therapies. The aim of mRNA cancer vaccines, such as those developed by BioNTech and Moderna, is to encode tumor antigens within mRNA sequences to elicit an immune response targeting those antigens. Personalized mRNA vaccines customized to specific tumor patterns have demonstrated encouraging results in early-phase studies for melanoma, head and neck cancer, and other malignancies [69]. For example, an ongoing trial using a personalized mRNA vaccine combined with ICIs in head and neck cancer patients showed significant tumor reduction in most participants, indicating mRNA vaccines' potential to target even advanced cancers effectively [70].

A promising strategy in cancer immunotherapy is the combination of therapeutic vaccines with ICIs. Combining these therapies has shown synergistic effects in increasing the efficacy of cancer vaccines, especially in tumors with an immunosuppressive microenvironment. Clinical trials combining checkpoint inhibitors with therapeutic vaccines showed encouraging outcomes in melanoma and NSCLC [71].

Neoantigen-based vaccines are highly personalized, using genetic information from the patient's tumor to create a vaccine tailored to specific mutations within the cancer cells. By focusing on unique tumor neoantigens, these vaccines minimize off-target effects on normal tissues. Studies involving melanoma and glioblastoma patients have highlighted the therapeutic potential of neoantigen vaccines, although logistical challenges, such as the need for rapid vaccine production and comprehensive sequencing, must be addressed [72].

However, while cancer vaccines have made significant strides, challenges remain. Tumor heterogeneity, immune escape mechanisms, and the TME can all affect vaccine efficacy. Furthermore, clinical translation requires robust manufacturing and delivery systems, especially for individualized vaccines like mRNA and neoantigen-based approaches. Ongoing research is exploring ways to improve vaccine delivery, including nanoparticle carriers and intra-tumoral injections, which may enhance vaccine efficacy and stability [73].

In conclusion, cancer vaccines are a promising frontier in cancer treatment and prevention. While there are hurdles to overcome, the continued development of these vaccines may transform cancer treatment paradigms, potentially making personalized, effective, and targeted cancer therapies widely accessible in the coming decade.

7.7 ONCOLYTIC VIRUS THERAPY

OVT represents a cutting-edge approach to cancer treatment that utilizes either naturally occurring or genetically modified viruses to selectively attack and eliminate cancer cells [74]. This strategy harnesses the unique properties of viruses to replicate within tumor cells, inducing cell lysis while sparing normal tissues. Moreover, OVT enhances anti-tumor immune responses, presenting a dual mode of action.

The foundation of OVT lies in the selective replication of viruses within cancer cells [21]. Cancer cells, due to their genetic and metabolic abnormalities, often lack robust antiviral defenses, rendering them more susceptible to viral infection and replication. When the virus enters the tumor cell, it multiplies, causing cell rupture and releasing offspring virions that infect nearby cancer cells. This oncolysis causes the release of TAAs, which stimulate both intrinsic and adaptive immune responses. The immune system, which is stimulated by viral infection and TAAs, can target both infected and uninfected cancer cells, leading to systemic anti-tumor activity.

At present, only a few oncolytic viruses have received regulatory clearance. For example T-VEC is a genetically altered herpes simplex virus used to treat metastatic melanoma [75]. Patients given this intralesional therapy achieved longer OS (23.3 months vs. 18.9 months) in comparison to the standard therapy. The FDA authorized the medicine in 2015, which is currently used together with pembrolizumab.

Beyond T-VEC, a variety of oncolytic viruses, including adenoviruses, reoviruses, and vaccinia viruses, are in various stages of clinical trials for multiple cancers. Three other oncolytic viruses have received FDA approval to date; in 2022, a non-oncolytic virus was authorized for non-muscle invasive bladder cancer [76]. Despite promising results, OVT faces several challenges. The immunogenicity of viral vectors can lead to neutralization by the host immune system, reducing therapeutic efficacy. Strategies such as repeated dosing, immune-evasive viral designs, and combination therapies with ICI are under investigation to overcome these hurdles [77]. Additionally, the TME's complexity, including hypoxia and immune suppression, necessitates further optimization of virus delivery and function. OVT exemplifies the potential of combining virology and immunotherapy in the treatment of cancer and is poised to become a cornerstone of precision oncology.

7.8 CONCLUSIONS

ICIs have transformed cancer treatment by providing new therapeutic alternatives to patients. A new cancer therapy paradigm has emerged, focusing on reactivating the immune system rather than cancer cells. This technique has resulted in considerable improvements in patient outcomes, particularly for tumors that were previously challenging to treat, such as melanoma, NSCLC, and RCC. The standard treatment protocol for locally advanced fully resected melanoma, metastatic malignant melanoma, metastatic RCC, and recurrent endometrial cancer is to use ICIs either alone, alongside another ICI, or in conjunction with a tyrosine kinase inhibitor (TKI). ICIs' transformational influence stems from their capacity to provide long-term responses and the possibility of survival. Furthermore, the development of combination therapies, where ICIs are paired with other treatments like chemotherapy, targeted therapies, or radiation, is often more effective than mono therapy, leading to even greater clinical benefits and broadening the use of ICIs to additional cancer types. For example, ICIs commonly combined with chemotherapy, which has established itself as a standard treatment for metastatic NSCLC due to significantly improved survival outcomes compared to chemotherapy alone. Data are emerging to suggest that for selected patients, it is possible to treat cancers without the need for surgery, radiotherapy, and attendant morbidity. dMMR rectal cancer is one example.

Today, 13 different ICIs, blocking 4 different types of ICPs, are approved for the treatment of almost 20 cancer sites either as monotherapy or with chemotherapy, radiotherapy, TKI, anti-antigenic therapy, other ICIs, with BsAbs, and OVT. ICIs are used for the treatment of advanced metastatic disease, second-line treatment after failure of primary treatment of metastatic disease, in the maintenance setting, neo-adjuvant setting, adjuvant setting, and concomitantly with radiotherapy. New ICIs, such as anti-LAG-3 and anti-TIGIT antibodies, have been approved over the last 2 years only.

Assessing response to ICIs requires novel tools and methods due to the unique immune-mediated mechanisms of tumor response. Combining advanced imaging, biomarker analysis, and irRC enhances the precision of response assessment. Despite their promise, only a subset of patients respond to these therapies. Research on biomarkers, including tests for MMR deficiency, PD-L1 expression, and measuring TMB, has been beneficial in identifying patients who are most likely to respond positively to ICIs, though some of these biomarkers are not yet universally reliable.

ICIs have a distinct toxicity profile from standard cytotoxic chemotherapy, TKIs, and mAbs. In rare cases, ICIs can cause irAEs that might be minor to fatal, making patient treatment challenging.

Cancer vaccines are a promising frontier in cancer treatment and prevention. While there are hurdles to overcome, the continued development of these vaccines may transform cancer treatment paradigms, potentially making personalized cancer therapies accessible over the next few years. OVT exemplifies the potential of combining virology and immunotherapy in the treatment of cancer and is poised to become a cornerstone of precision oncology.

While challenges remain, ICIs have undeniably set a new standard in cancer care, and arguably their introduction is considered a quantum leap in efforts to overcome cancer. We are living in the era of immunotherapy, spearheaded by ICIs.

REFERENCES

1. Schreiber RD, Old LJ, Smyth MJ. Cancer immunoediting: integrating immunity's roles in cancer suppression and promotion. *Science*. 2011;331(6024):1565–1570.
2. Vesely MD, Kershaw MH, Schreiber RD, et al. Natural innate and adaptive immunity to cancer. *Annual Review of Immunology*. 2011;29(1):235–271.
3. Dunn GP, Old LJ, Schreiber RD. The immunobiology of cancer immunosurveillance and immunoediting. *Immunity*. 2004;21(2):137–148.
4. Hanahan D, Weinberg RA. Hallmarks of cancer: the next generation. *Cell*. 2011;144(5):646–674.
5. Fridman WH, Zitvogel L, Sautès-Fridman C, et al. The immune contexture in cancer prognosis and treatment. *Nature Reviews Clinical Oncology*. 2017;14(12):717–734.
6. Quail DF, Joyce JA. Microenvironmental regulation of tumor progression and metastasis. *Nature Medicine*. 2013;19(11):1423–1437.
7. Wang R, Jaw JJ, Stutzman NC, et al. Natural killer cell-produced IFN-γ and TNF-α induce target cell cytolysis through up-regulation of ICAM-1. *Journal of Leukocyte Biology*. 2012;91(2):299–309.
8. Topalian SL, Drake CG, Pardoll DM. Immune checkpoint blockade: a common denominator approach to cancer therapy. *Cancer Cell*. 2015;27(4):450–461.
9. June CH, O'Connor RS, Kawalekar OU, et al. CAR T cell immunotherapy for human cancer. *Science*. 2018;359(6382):1361–1365.
10. McCarthy EF. The toxins of William B. Coley and the treatment of bone and soft-tissue sarcomas. *The Iowa Orthopaedic Journal*. 2006;26:154.
11. Burney IA. Immunotherapy in cancer: incremental gain or a quantum leap? *Oman Medical Journal*. 2017;32(1):1.
12. Burnet F. The concept of immunological surveillance. *Progress in Experimental Tumor Research*. 1970;13:1–27.
13. Durmaz AA, Karaca E, Demkow U, et al. Evolution of genetic techniques: past, present, and beyond. *BioMed Research International*. 2015;2015(1):461524.
14. Mukherjee N, Wheeler KM, Svatek RS. Bacillus Calmette–Guérin treatment of bladder cancer: a systematic review and commentary on recent publications. *Current Opinion in Urology*. 2019;29(3):181–188.
15. Jiang S, Redelman-Sidi G. BCG in bladder cancer immunotherapy. *Cancers*. 2022;14(13):3073.
16. Culver ME, Gatesman ML, Mancl EE, et al. Ipilimumab: a novel treatment for metastatic melanoma. *Annals of Pharmacotherapy*. 2011;45(4):510–519.

17. Callahan MK, Postow MA, Wolchok JD. CTLA-4 and PD-1 pathway blockade: combinations in the clinic. *Frontiers in Oncology.* 2015;4:385.
18. Tanriverdi O, Tasar M, Yilmaz M, et al. Important milestones for cancer at the Nobel prize. *Indian Journal of Cancer.* 2020;57(4):370–375.
19. Butterfield LH. Cancer vaccines. *BMJ.* 2015;350:h988.
20. Cheever MA, Higano CS. PROVENGE (Sipuleucel-T) in prostate cancer: the first FDA-approved therapeutic cancer vaccine. *Clinical Cancer Research.* 2011;17(11):3520–3526.
21. Fukuhara H, Ino Y, Todo T. Oncolytic virus therapy: a new era of cancer treatment at dawn. *Cancer Science.* 2016;107(10):1373–1379.
22. Pol J, Kroemer G, Galluzzi L. First oncolytic virus approved for melanoma immunotherapy. *Oncoimmunology.* 2015;5(1):e1115641.
23. Pierpont TM, Limper CB, Richards KL. Past, present, and future of rituximab—the world's first oncology monoclonal antibody therapy. *Frontiers in Oncology.* 2018;8:163.
24. Lyu X, Zhao Q, Hui J, et al. The global landscape of approved antibody therapies. *Antibody Therapeutics.* 2022;5(4):233–257.
25. Galluzzi L, Vacchelli E, Bravo-San Pedro J-M, et al. Classification of current anticancer immunotherapies. *Oncotarget.* 2014;5(24):12472.
26. Scott AM, Wolchok JD, Old LJ. Antibody therapy of cancer. *Nature Reviews Cancer.* 2012;12(4):278–287.
27. Mansfield AS, Nevala WK, Lieser EAT, et al. The immunomodulatory effects of bevacizumab on systemic immunity in patients with metastatic melanoma. *Oncoimmunology.* 2013;2(5):e24436.
28. Zahavi D, Weiner L. Monoclonal antibodies in cancer therapy. *Antibodies.* 2020;9(3):34.
29. Postow MA, Callahan MK, Wolchok JD. Immune checkpoint blockade in cancer therapy. *Journal of Clinical Oncology.* 2015;33(17):1974–1982.
30. Toor SM, Nair VS, Decock J, et al., editors. Immune checkpoints in the tumor microenvironment. In *Seminars in Cancer Biology*, 2020;65:1–12..
31. Reschke R, Sullivan RJ, Lipson EJ, et al. Targeting molecular pathways to control immune checkpoint inhibitor toxicities. *Trends in Immunology.* 2025;46(1):61–73.
32. Van Berkel ME, Oosterwegel MA. CD28 and ICOS: similar or separate costimulators of T cells? *Immunology Letters.* 2006;105(2):115–122.
33. Hodi FS, O'Day SJ, McDermott DF, et al. Improved survival with ipilimumab in patients with metastatic melanoma. *New England Journal of Medicine.* 2010;363(8):711–723.
34. Topalian SL, Hodi FS, Brahmer JR, et al. Safety, activity, and immune correlates of anti–PD-1 antibody in cancer. *New England Journal of Medicine.* 2012;366(26):2443–2454.
35. Borghaei H, Paz-Ares L, Horn L, et al. Nivolumab versus docetaxel in advanced nonsquamous non–small-cell lung cancer. *New England Journal of Medicine.* 2015;373(17):1627–1639.
36. Garon EB, Rizvi NA, Hui R, et al. Pembrolizumab for the treatment of non–small-cell lung cancer. *New England Journal of Medicine.* 2015;372(21):2018–2028.
37. Robert C, Schachter J, Long GV, et al. Pembrolizumab versus ipilimumab in advanced melanoma. *New England Journal of Medicine.* 2015;372(26):2521–2532.
38. Rosenberg JE, Hoffman-Censits J, Powles T, et al. Atezolizumab in patients with locally advanced and metastatic urothelial carcinoma who have progressed following treatment with platinum-based chemotherapy: a single-arm, multicentre, phase 2 trial. *The Lancet.* 2016;387(10031):1909–1920.
39. Lemery S, Keegan P, Pazdur R. First FDA approval agnostic of cancer site—when a biomarker defines the indication. *New England Journal of Medicine.* 2017;377(15):1409–1412.
40. Ferris RL, Blumenschein Jr G, Fayette J, et al. Nivolumab for recurrent squamous-cell carcinoma of the head and neck. *New England Journal of Medicine.* 2016;375(19):1856–1867.
41. Reck M, Rodríguez-Abreu D, Robinson AG, et al. Pembrolizumab versus chemotherapy for PD-L1–positive non–small-cell lung cancer. *New England Journal of Medicine.* 2016;375(19):1823–1833.
42. Larkin J, Chiarion-Sileni V, Gonzalez R, et al. Combined nivolumab and ipilimumab or monotherapy in untreated melanoma. *New England Journal of Medicine.* 2015;373(1):23–34.
43. Rini BI, Plimack ER, Stus V, et al. Pembrolizumab plus axitinib versus sunitinib for advanced renal-cell carcinoma. *New England Journal of Medicine.* 2019;380(12):1116–1127.
44. Janjigian YY, Shitara K, Moehler M, et al. First-line nivolumab plus chemotherapy versus chemotherapy alone for advanced gastric, gastro-oesophageal junction, and oesophageal adenocarcinoma (CheckMate 649): a randomised, open-label, phase 3 trial. *The Lancet.* 2021;398(10294):27–40.
45. Burtness B, Harrington KJ, Greil R, et al. Pembrolizumab alone or with chemotherapy versus cetuximab with chemotherapy for recurrent or metastatic squamous cell carcinoma of the head and neck (KEYNOTE-048): a randomised, open-label, phase 3 study. *The Lancet.* 2019;394(10212):1915–1928.

46. Powles T, Park SH, Caserta C, et al. Avelumab first-line maintenance for advanced urothelial carcinoma: results from the JAVELIN Bladder 100 trial after≥ 2 years of follow-up. *Journal of Clinical Oncology*. 2023;41(19):3486–3492.

47. Antonia SJ, Villegas A, Daniel D, et al. Durvalumab after chemoradiotherapy in stage III non–small-cell lung cancer. *New England Journal of Medicine*. 2017;377(20):1919–1929.

48. Kato K, Shah MA, Enzinger P, et al. KEYNOTE-590: phase III study of first-line chemotherapy with or without pembrolizumab for advanced esophageal cancer. *Future Oncology*. 2019;15(10):1057–1066.

49. Yao Y, Chen YF, Zhang Q. Optimized patient-specific immune checkpoint inhibitor therapies for cancer treatment based on tumor immune microenvironment modeling. *Briefings in Bioinformatics*. 2024;25(6):bbae547.

50. Eisenhauer EA, Therasse P, Bogaerts J, et al. New response evaluation criteria in solid tumours: revised RECIST guideline (version 1.1). *European Journal of Cancer*. 2009;45(2):228–247.

51. Borcoman E, Nandikolla A, Long G, et al. Patterns of response and progression to immunotherapy. *American Society of Clinical Oncology Educational Book*. 2018;38(38):169–178.

52. Waxman ES, Gerber DL. Pseudoprogression and immunotherapy phenomena. *Journal of the Advanced Practitioner in Oncology*. 2020;11(7):723.

53. Seymour L, Bogaerts J, Perrone A, et al. iRECIST: guidelines for response criteria for use in trials testing immunotherapeutics. *The Lancet Oncology*. 2017;18(3):e143–e152.

54. Pessoa LS, Heringer M, Ferrer VP. ctDNA as a cancer biomarker: a broad overview. *Critical Reviews in Oncology/Hematology*. 2020;155:103109.

55. Toki MI, Syrigos N, Syrigos K. Hyperprogressive disease: a distinct pattern of progression to immune checkpoint inhibitors. *International Journal of Cancer*. 2021;149(2):277–286.

56. Hegi-Johnson F, Rudd S, Hicks RJ, et al. Imaging immunity in patients with cancer using positron emission tomography. *NPJ Precision Oncology*. 2022;6(1):24.

57. Jang A, Lanka SM, Jaeger EB, et al. Longitudinal monitoring of circulating tumor DNA to assess the efficacy of immune checkpoint inhibitors in patients with advanced genitourinary malignancies. *JCO Precision Oncology*. 2023;7:e2300131.

58. Postow MA, Sidlow R, Hellmann MD. Immune-related adverse events associated with immune checkpoint blockade. *New England Journal of Medicine*. 2018;378(2):158–168.

59. Salman B, AlWard NM, Al-Hashami Z, et al. The prevalence and patterns of toxicity with immune checkpoint inhibitors in solid tumors: a real-world experience from a Tertiary Care Center in Oman. *Cureus*. 2023;15(10):e47050.

60. Brahmer JR, Lacchetti C, Schneider BJ, et al. Management of immune-related adverse events in patients treated with immune checkpoint inhibitor therapy: American Society of Clinical Oncology Clinical Practice Guideline. *Journal of Clinical Oncology*. 2018;36(17):1714–1768.

61. Sharma P, Hu-Lieskovan S, Wargo JA, et al. Primary, adaptive, and acquired resistance to cancer immunotherapy. *Cell*. 2017;168(4):707–723.

62. Litchfield K, Reading JL, Puttick C, et al. Meta-analysis of tumor-and T cell-intrinsic mechanisms of sensitization to checkpoint inhibition. *Cell*. 2021;184(3):596–614. e14.

63. Wei J, Li W, Zhang P, et al. Current trends in sensitizing immune checkpoint inhibitors for cancer treatment. *Molecular Cancer*. 2024;23(1):279.

64. Gupta M, Wahi A, Sharma P, et al. Recent advances in cancer vaccines: challenges, achievements, and futuristic prospects. *Vaccines*. 2022;10(12):2011.

65. Jit M, Prem K, Benard E, et al. From cervical cancer elimination to eradication of vaccine-type human papillomavirus: feasibility, public health strategies and cost-effectiveness. *Preventive Medicine*. 2021;144:106354.

66. Palmer TJ, Kavanagh K, Cuschieri K, et al. Invasive cervical cancer incidence following bivalent human papillomavirus vaccination: a population-based observational study of age at immunization, dose, and deprivation. *JNCI: Journal of the National Cancer Institute*. 2024;116(6):857–865.

67. Kantoff PW, Higano CS, Shore ND, et al. Sipuleucel-T immunotherapy for castration-resistant prostate cancer. *New England Journal of Medicine*. 2010;363(5):411–422.

68. Palucka AK, Coussens LM. The basis of oncoimmunology. *Cell*. 2016;164(6):1233–1247.

69. Yaremenko AV, Khan MM, Zhen X, et al. Clinical advances of mRNA vaccines for cancer immunotherapy. *Med*. 2025;6(1):100562.

70. Shibata H, Zhou L, Xu N, et al. Personalized cancer vaccination in head and neck cancer. *Cancer Science*. 2021;112(3):978–988.

71. Collins JM, Redman JM, Gulley JL. Combining vaccines and immune checkpoint inhibitors to prime, expand, and facilitate effective tumor immunotherapy. *Expert Review of Vaccines*. 2018;17(8):697–705.

72. Ott PA, Hu Z, Keskin DB, et al. An immunogenic personal neoantigen vaccine for patients with melanoma. *Nature.* 2017;547(7662):217–221.
73. Lin Y, Chen X, Wang K, et al. An overview of nanoparticle-based delivery platforms for mRNA vaccines for treating cancer. *Vaccines.* 2024;12(7):727.
74. Russell SJ, Peng K-W, Bell JC. Oncolytic virotherapy. *Nature Biotechnology.* 2012;30(7):658–670.
75. Andtbacka RH, Kaufman HL, Collichio F, et al. Talimogene laherparepvec improves durable response rate in patients with advanced melanoma. *Journal of Clinical Oncology.* 2015;33(25):2780–2788.
76. Shalhout SZ, Miller DM, Emerick KS, et al. Therapy with oncolytic viruses: progress and challenges. *Nature Reviews Clinical Oncology.* 2023;20(3):160–177.
77. Zhao J-L, Lin B-L, Luo C, et al. Challenges and strategies toward oncolytic virotherapy for leptomeningeal metastasis. *Journal of Translational Medicine.* 2024;22(1):1000.

8 RNA Therapeutics for Cancer

Ella McGovern, Kehinde Ross, and Imran Saleem

8.1 RNA THERAPEUTICS

8.1.1 CRISPR/Cas9 Gene Editing

Clustered regularly interspaced short palindromic repeats (CRISPR) systems were originally found in *Escherichia coli* in 1987 and then in archaea, with investigations concluding they had an immunological defense role against foreign genetic elements such as bacteriophages and plasmids [1,2]. CRISPR-associated (Cas) proteins are a family of endonucleases and a primary component of the CRISPR systems, with the most researched protein being Cas9. The CRISPR/Cas9 system is largely known for genomic editing and has been the focus of many developing biotechnologies for gene therapy to treat hereditary diseases, hemoglobinopathies, cardiovascular diseases, and cancers, among other diseases [3]. Cas9 forms a complex with a single-guide RNA (sgRNA) that is specific to the target DNA sequence. The sgRNA binds to Cas9 via a loop section called the trans-activating CRISPR RNA (tracrRNA), and to the target gene by the CRISPR RNA (crRNA) section [4], and guides the endonuclease to the gene of interest via its complementary sequence, as well as its recognition of a protospacer-adjacent motif (PAM) [5]. The sgRNA binds to the target DNA sequence, and Cas9 induces a double-strand break, with the HNH (His-Asn-His) domain cleaving the complementary strand and the Ruv-C domain cleaving the non-complementary strand (Figure 8.1) [6,7].

This mechanism provides the opportunity for therapeutic action through knocking out disease-causing genes and knocking in disease-suppressing genes. Specifically in the case of cancer, this refers to the deletion of oncogenes and insertion of tumor-suppressor genes into target cells. Not only can this technique be used for cancer therapy, as outlined below, but it can also be used in cancer modeling to create a more accurate imitation of tumors to both understand the roles and importance of specific genes and to test therapeutics. Mou et al. simultaneously knocked in the oncogenic *KRAS-G12D* mutant allele and deleted the p53 tumor suppressor gene in mice, which subsequently formed intrahepatic cholangiocarcinomas [8]. By using CRISPR/Cas9, Cai et al. were able to identify gene mutations required for prostate cancer (PCa) progression and metastasis to the lungs. Inducing loss-of-function mutations in five tumor suppressor genes such as *PTEN* in mice caused initiation and progression of PCa in mice, and that loss-of-function mutation of *Kmt2c* in mouse models and human PCa cell lines was essential for metastasis, identifying that its loss resulted in upregulation of metastasis-linked Odam and Cabs1 [9]. Therefore, the use of this CRISPR/Cas9 model revealed essential pathways of PCa formation and identified potential tumor suppressor genes and oncogenes to target therapeutically.

8.1.2 Tumor Suppressor Genes

The CRISPR/Cas9 system was exploited to activate transcription of previously downregulated wild-type PTEN, a phosphatase that dampens the activation of oncogenic pathways such as PI3K/AKT, mTOR, and MAPK. Dead Cas9 (dCas9) was attached to transactivator VPR (VP64-p65-Rta), which was guided by sgRNA generated to complement the PTEN proximal promoter, subsequently increasing PTEN mRNA expression and reducing migration and resistance to B-Raf and PI3K/mTOR inhibitor in melanoma cell lines [10].

One proposed therapy that holds promise is a system to restore p53 function, a tumor-suppressor gene and the most altered gene in human cancers. Delivered via a viral vector, the model proposes

DOI: 10.1201/9781003517870-8

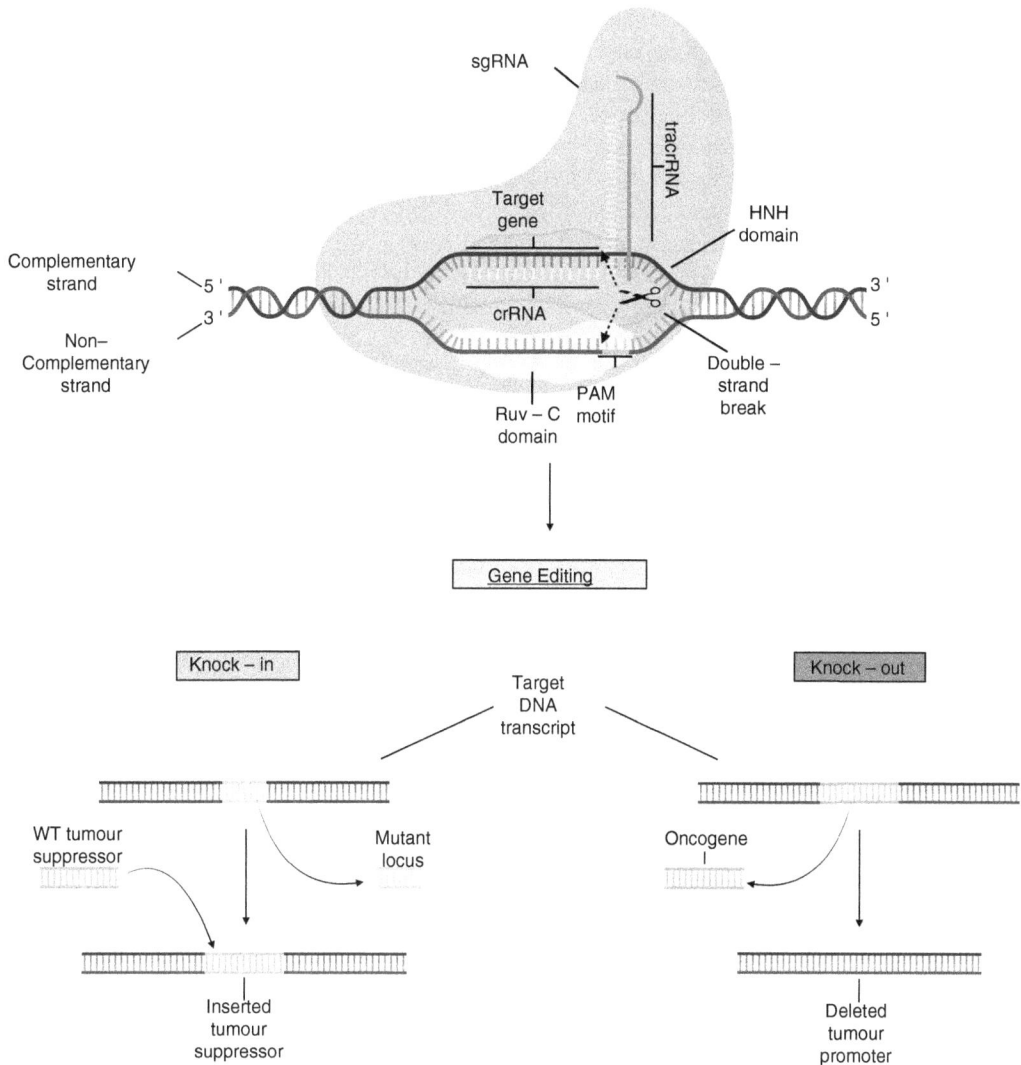

FIGURE 8.1 CRISPR/Cas9 mechanism of action. sgRNA and PAM recognition guides Cas9 to the target gene and induces cleavage. This can be utilized to knock out oncogenes or knock in tumor suppressor genes (Created in BioRender. McGovern, E. (2025) https://BioRender.com/kxim7m2).

a combination of two sgRNAs that bind to both sides of the mutant *TP53* locus, causing cleavage via an altered Cas9 (Cas9-HF). This cleavage is tumor-specific by the inclusion of the survivin promoter, the expression of which has been shown to be specific to cancer cells when compared to non-cancerous cells [11], within the viral vector. Homology-directed repair (HDR) causes the insertion of wild-type TP53 cDNA and upon stimulation by doxycycline results in transcription of the wild-type TP53, restoring its tumor suppressive function [12].

8.1.3 SPECIFIC ONCOGENE TARGETING

Depending on the type of tumor, mutated loci that have resulted in a gain-of-function to drive onco-genesis can be identified by knock-in/-out studies and patient data and genome-wide screening [13]. sgRNA is then designed to complement the mutated sequence, but not the non-mutant gene, thus allowing binding of Cas9 and cleavage of the mutant loci and the subsequent downregulation of the cancer-driving gene and its protein expression [14]. Ideally, mutant alleles that produce a PAM, downstream of the gene, should be targeted to increase specificity and reduce off-target effects where wild-type alleles do not contain a PAM sequence [15]. For instance, Koo et al. performed an *in vitro* study of non-small cell lung cancer (NSCLC), in which a single nucleotide missense mutation/substitution resulted in the formation of a PAM motif (CCG) within the EGFR allele, resulting in recognition by *Streptococcus pyogenes*-derived Cas9 and binding of the generated sgRNA [16]. The sgRNA complementary sequence was also immediately upstream of the PAM motif, which aids in effective cleavage and reduction of off-target effects [14,17].

Research is ongoing to develop a PAM-free system with the capability of targeting any sequence. The CRISPR/Cas9 system is currently limited to sequences upstream of a PAM motif, in the absence of which the sequence will be ignored by the nuclease [18]. Currently, work is focused on engineering nucleases to increase tolerance of mismatch variations within PAM [19]. In 2020, Martinez-Lage et al. used CRISPR/Cas9 to target fusion oncogenes, the fusion of two genes, forming a chimera. sgRNA was generated to target the most common fused loci of the *EWSRI* and *FLI1* genes (EF). An adenovirus vector (Ad) was chosen to deliver the Cas9/EF vectors, and when this was administered to A673-xenografted mice significantly reduced tumor growth. When administered to patient-derived xenograft mice models, AdCas9_EF significantly reduced tumor volume and increased survival to 50% after 70 days, compared to 0% at day 40 in the control group. This increased to 66% survival past day 80 with a combined therapy with doxorubicin [20].

It should be noted that the editing of the genome does not come without risks and can, in some cases, result in carcinogenesis, largely due to off-target effects [21]. Mutagenesis may also be induced by CRISPR/Cas9 by large-scale base deletions and other lesions to both on-target and off-target alleles [22]. An alternative strategy is to use Cas13a in place of Cas9 to target the mRNA of transcribed oncogenes. This has been shown to be effective in inducing apoptosis in pancreatic cancer cells by generating an sgRNA with a single mismatch within the crRNA domain, causing targeting of mutant KRAS mRNA (KRAS-G12D) with a knockdown efficiency of 94%. The complex was shown to be specific to KRAS-G12D, with no effect on wild type KRAS, while achieving silencing of KRAS-G12D mRNA expression at a rate of 70% as well as reducing tumor growth when delivered via intra-tumoral injection [23]

8.2   MRNA THERAPY

8.2.1   Protein Replacement Therapy

One common genetic alteration in cancers is a loss-of-function mutation, in which the DNA sequence of a tumor suppressor gene changes, resulting in the prevention of gene transcription or the translation into non-functional proteins. Without these tumor suppressor gene products, oncogenic functions such as apoptotic resistance and rapid cell replication can occur unchecked [24].

One way in which mRNA can be utilized for cancer therapy is protein replacement therapy to restore under-expressed proteins such as tumor suppressors. mRNA delivery holds an advantage over the direct delivery of proteins, having a longer half-life and less risk of degradation and immunogenicity [25].

8.2.2   Examples of Protein Replacement mRNA Therapy in Preclinical Studies

One study delivered p53 mRNA via lipid nanoparticles to *p53*-null hepatocellular carcinoma and NSCLC and found increased p53 expression and apoptosis after delivery. When Hep3B xenograft mice were treated with the p53-mRNA NPs, tumor size was reduced, and when used in combination with everolimus, an mTOR inhibitor, tumor size decreased even further. This shows that delivery of the mRNA restores p53 expression and aids in the targeting of oncogene mTOR to reduce cancer proliferation [26]. Another application of this method is to deliver mRNA encoding proteins that are toxic to cell survival and subsequently cause cell death. Hirschberger et al. aimed to achieve this by delivering abrin, a chemically modified mRNA, naturally expressed in *Abrus precatorius* and a known toxin, reducing tumor growth both *in* vitro and *in vivo* [27]. This technique, however, often causes toxicity to other organs such as the liver and spleen. One study sought to overcome this by generating mRNA with parts of its sequence complementary to microRNAs (miRNAs), specifically miR-122, typically expressed in healthy hepatocytes, and miR-142, typically expressed in antigen-presenting cells, to exploit the inhibitory function of the miRNA and avoid protein expression in healthy liver cells and reduce immune activation. This reduced tumor growth in mice while also reducing accumulation in the liver and spleen and liver damage [28].

8.2.3 mRNA Vaccine

Tumor-associated antigens (TAAs), while their expression can be found in other cells, are abnormally overexpressed on the surface of cancer cells. Tumor-specific antigens (TSAs), also referred to as neoantigens, are specific to tumor cells and hold promise to generate a high-level immune response that is targeted to tumors to reduce off-target effects [29]. It is the mutational nature of cancer cells that gives rise to aberrantly expressed TSAs, which occur due to alterations in the transcripts such as frameshift mutations, splicing and single-nucleotide variations, or due to expression of onco-fetal genes which are otherwise transcriptionally repressed in adults [30].

Tumor mRNA vaccines function by firstly activating the innate immune system through initial T cell activation. Antigen-presenting cells (APCs) such as dendritic cells (DCs) can recognize the mRNA delivered, activating toll-like receptors (TLRs), which recognize pathogen-associated molecular patterns within the mRNA sequence [31]. The TLRs then trigger the expression of pro-inflammatory cytokines and chemokines through signaling pathways that translocate into the nucleus and induce the transcription of molecules such as interleukins-2 and -6. These cytokines/chemokines stimulate the first load of T cells (Figure 8.2a) [32]. The mRNA vaccine also activates the adaptive immune system. When delivered to APCs, the mRNA is translated into its encoded antigen, which is then broken down into peptides. These are recognized and presented to CD8$^+$ T cells by MHC I, triggering the activation of cytotoxic T cells (Figure 8.2b). The antigen peptides may be incorporated into endosomes from the cytoplasm, which are delivered for presentation on MHC II and recognized by CD4$^+$ T cells. This pathway triggers the activation of helper T cells, which stimulate the clonal expansion of B cells and, therefore, the release of antibodies (Figure 8.2c) [33,34].

8.2.4 Examples of TAA mRNA Vaccine Studies

In ovarian carcinoma xenografted mice, TAAs CLDN6 and EpCAM were used to target cancer cells, allowing for the delivery of a bispecific antibody (bsAb) targeted to CD3, which is expressed on the surface of T-cells. This combination forms a bi-(scFv)$_2$ (CD3×CLDN6 and EpCAM×CD3), and treatment with it resulted in tumor regression [35]. BNT111 is a liposomal RNA therapy that encodes four TAAs and completed Phase I clinical trials (NCT02410733). Melanoma patients were given eight doses and received either monotherapy or co-therapy with anti-PD1 antibodies. Patients who received monotherapy had upregulated antigen-specific cytotoxic CD4$^+$ and CD8$^+$ T-cell activity and cytokine levels. Patients receiving co-therapy saw a higher rate of target lesion regression than those receiving monotherapy. Furthermore, some patients who were previously non-responsive to anti-PD1 therapy, after treatment with BNT111, were re-challenged with anti-PD1 and showed partial response and lesion regression [36]. BNT111 is currently in Phase II trials for monotherapy or co-therapy with cemiplimab (NCT04526899). Also, in melanoma, one study combined tyrosinase-related protein 2 (TRP2), a melanoma-specific antigen, encoding mRNA and PD-L1 targeting siRNA, loaded into a lipid-coated, PEGylated calcium phosphate NP for uptake by DCs. Mice treated with the drug showed raised antigen-specific T-cell

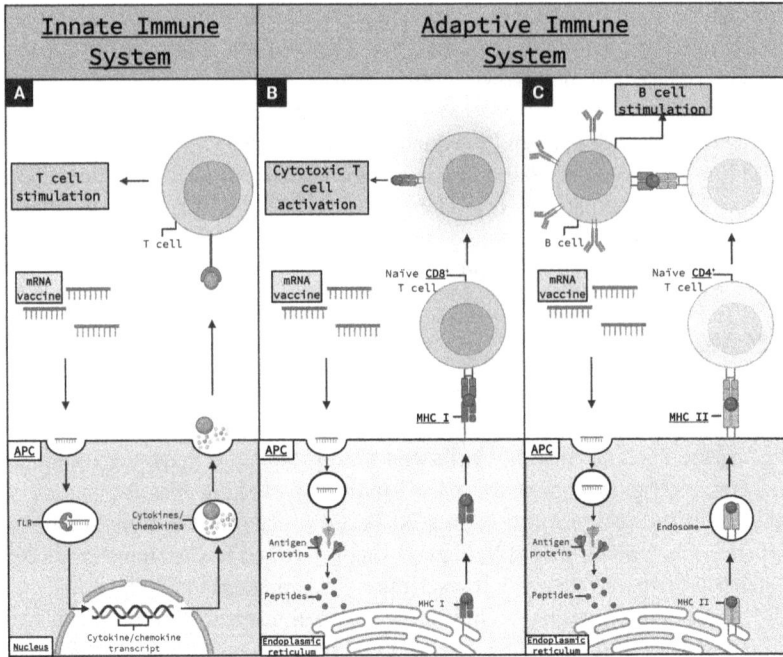

FIGURE 8.2 Illustration of how the mRNA vaccine activates and sensitizes the immune system to cancer cells. The mRNA vaccine is endocytosed into APCs. (a) TLRs trigger the expression and translation of pro-inflammatory cytokines and chemokines, which stimulate T cell production. (b) The mRNA is translated into the encoded antigens, which are broken down into peptides. The peptides bind to MHC I, which then present the antigen to naïve CD8+ T cells, which mature into cytotoxic T cells. (c) The mRNA is translated into the encoded antigens, which are broken down into peptides. The peptides bind to MHC II, which then present the antigen to naïve CD4+ T cells, which mature into T helper cells that stimulate B cell proliferation and antibody release (Created in BioRender. McGovern, E. (2025) https://BioRender.com/uwxxg31).

responses and increased tumor regression. This was further enhanced when combined with the PD-L1 siRNA [37].

There are issues, however, with mRNA vaccines, particularly those using TAAs. TAAs, unlike TSAs, are not specific to cancer cells, meaning, to some degree, they are expressed on the surface of normal cells. This means that the immune system may identify the TAA expressed on APCs as a "self-antigen" and, due to a system designed to tolerate self-tissue, may "ignore" the antigen and fail to activate a T cell response. These mechanisms are central and peripheral tolerance, which remove lymphocytes that have a strong recognition for self-antigens [38]. On the other hand, research is ongoing to develop TAA-based mRNA vaccines that "break" this tolerance [39].

8.2.5 Examples of TSAs mRNA Vaccines in Clinical Studies

TSAs, therefore, attract a lot of attention due to their specificity and low tolerance, but also for their potential in personalized therapy. For example, a completed Phase I clinical trial (NCT04161755) used individualized neoantigen mRNA vaccines to amplify T cell activity and stimulate naïve T cells following anti-PD-L1 treatment in pancreatic cancer. Patients had pancreatic tumors surgically resected and sequenced to identify neoantigens. Following treatment with atezolizumab, the mRNA vaccine "autogene cevumeran" with the individualized neoantigens was intravenously administered at a median time of 9.4 weeks after surgery, and around 3 weeks after that, mFOLFIRINOX, a chemotherapy regiment, was initiated. Results of the study showed a significant increase in neo-antigen-specific T cells up to two years post-vaccination in 50% (n=8) of patients. At an updated median follow-up at 3.2 years, the median recurrence-free survival for responders (n=8) was not yet reached, meaning the majority had not experienced a recurrence of disease, compared to a recurrence-free survival of 13.4 months in non-responders (n=8) [40,41]. This method of RNA cancer therapy holds promise for effective and personalized medicine and is currently the focus of numerous clinical trials [31,42].

8.3   RNA INTERFERENCE

RNA interference (RNAi) refers to the process through which oligonucleotides target mRNA in a post-transcriptional manner to dampen or silence protein translation. The effectors of RNAi include antisense oligonucleotides (ASOs), small interfering RNAs (siRNAs), and microRNAs (miRNAs).

8.3.1   microRNAs

miRNAs are endogenously expressed small RNAs, 19–25 nucleotides long, that regulate post-transcriptional expression, due to their ability to target mRNA and silence such genes. The importance of miRNAs spans molecular effects on cellular homeostasis through to roles in larger tissues and organs and maintenance of normal systematic function [43]. The impact of miRNA dysregulation extends from neurological disorders such as Parkinson's disease and schizophrenia, to cardiovascular disorders and cancer [44–46].

8.3.1.1   Biogenesis

To function as gene silencers, miRNAs must first be processed to produce mature miRNAs and the guide and passenger strands separated. Most miRNA genes are transcribed in the nucleus by RNA polymerase II into a long primary miRNA transcript (pri-miRNA) that folds into a hairpin RNA that is often capped and polyadenylated [47,48]. The pri-miRNA is cleaved by the Microprocessor complex, which consists of DiGeorge Syndrome Critical Region 8 (DGCR8) or Pasha, and Drosha (Ribonuclease III) [49,50], to form pre-miRNA. The pre-miRNA is then translocated into the cytoplasm by exportin5 (XPO5) and RAS-related nuclear protein-guanosine-5′-triphosphate (Ran-GTP) [51]. Dicer, an RNAase III endonuclease, then binds to pre-miRNA and cleaves the terminal loop, aided by the catalytic center made of two amino acid residues and the PAZ domain, producing a mature miRNA duplex [52,53].

The double-stranded miRNA is then loaded onto argonaute (Ago) to form the RNA-induced silencing complex (RISC) [54]. In mammals, while miRNA can load onto Ago 1–4, it is only Ago2 that is catalytically active [55]. It has been suggested that a RISC-loading complex (RLC) made up of Dicer and Transactivation Response Element RNA-binding protein (TRBP) is responsible for delivering and transferring the miRNA duplexes to RISC, and TRBP may also orient the miRNA for correct directional sequence positioning on Ago [56,57]. Assembly of RISC has been found to be ATP-dependent, requiring ATPases Hsp70 and Hsp90 to load the miRNA onto Ago2 [58], which opens to allow binding and forms pre-RISC. While previously thought otherwise, the unwinding of the duplex is ATP-independent. The "rubber band" model proposes the chaperones Hsp70 and -90 attach to Ago2, using ATP to open the protein or "stretch the rubber band", with only the guide strand binding to Ago2, and when taut the rubber band is released, that is, Ago shifts to a more closed position, causing the miRNA to unwind—the passenger strand is forced out of the complex, while the guide strand remains bound [59]. This forms the mature RISC or miRNA-Induced Silencing Complex (miRISC) alongside GW182 (TNRC6A) (Figure 8.3).

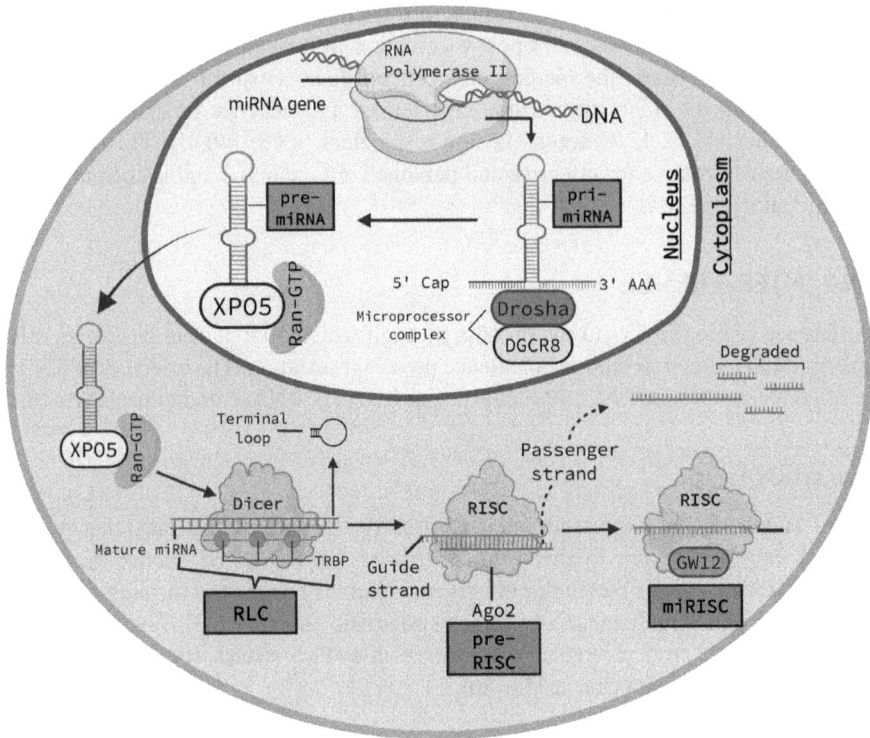

FIGURE 8.3 Biogenesis of miRNA and gene silencing of oncogenes. pri-miRNA is transcribed and converted into pre-miRNA by the microprocessor complex, which is transported into the cytoplasm. The RLC cleaves the terminal loop, and the pre-RISC ejects the passenger strand, leaving the guide strand ready to function on miRISC (Created in BioRender. McGovern, E. (2025) https://BioRender.com/6l82mu8).

8.3.1.2 Gene Silencing

Gene silencing by miRNAs involves mRNA decay or translational repression. Normally, mRNA strands are protected from degradation by eukaryotic initiation factors (eIFs), which interact with the 5′ m7G cap, and poly-A-binding proteins (PABPs), which associate with the 3′ poly-A-tail [60]. When miRISC interacts with target mRNA, GW182 recruits CCR4-NOT and PAN-2-PAN-3, which de-adenylate the 3′ poly-A-tail of the mRNA [61] and DCP2 removes the 5′ m7G cap [62]. The mRNA is then vulnerable to degradation by exonuclease XRN1 (Figure 8.4) [63]. Translational repression is a more complex and debated method; however, it allows for an inhibition of mRNA that is reversible, should protein synthesis be required to resume. It is suggested that the interruption of the eIF4F cap-binding complex at the 5′ end of the target mRNA by the miRISC is a major factor in suppressing translation (Figure 8.5a) [64]. Studies also suggest that disruption of ribosomal 43S PIC scanning of target mRNA occurs when correct scanning is prevented, that is, the start codon is not read, and the 80S ribosomal unit is not formed to initiate translation (Figure 8.5b) [65].

8.3.1.3 Examples of Tumor Suppressor miRNAs in Preclinical Studies

The let-7 family is a cluster of miRNAs that have tumor-suppressor effects; however, in some cancers, let-7 is overexpressed. Mimic delivery of let-7 has been found to be effective in reducing cell proliferation in lung cancer, breast cancer, and hepatocellular cancer (HCC) by silencing genes such as K-RAS, glycolytic enzyme HK2, and the PI3K/AKT/mTOR pathway [66–69]. The miRNA-34 family is also reported to be a potent tumor suppressor, made up of miR-34a, -b, and -c. Many studies have confirmed its anti-tumor effect, particularly miR-34a, reporting reduced cell proliferation,

mRNA Degradation

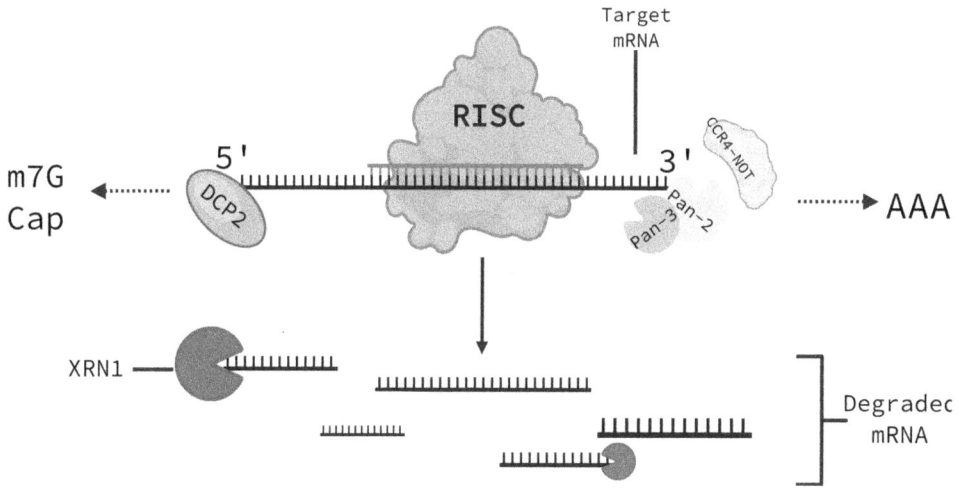

FIGURE 8.4 Gene silencing enacted through target mRNA degradation. Deadenylation and decapping of target mRNA leads to degradation by XRN1 (Created in BioRender. McGovern, E. (2025) https://BioRender.com/z5zhqx6).

Translation Repression

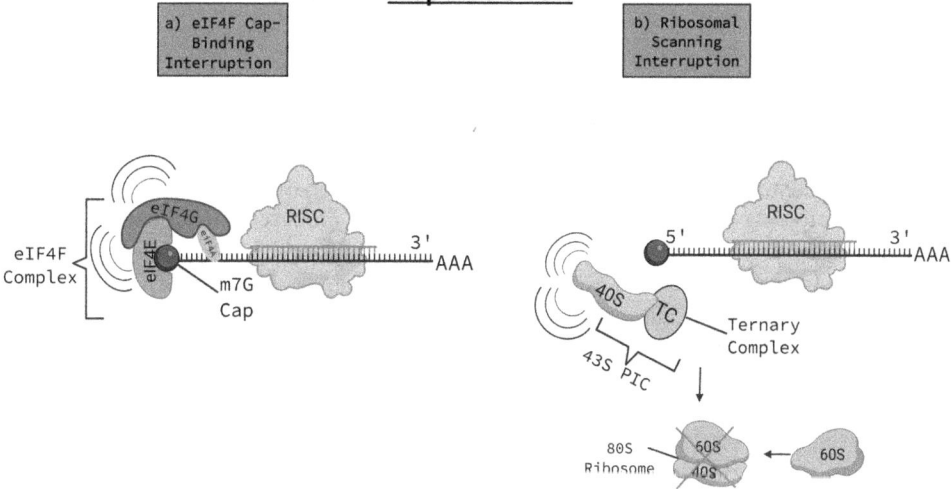

FIGURE 8.5 Gene silencing enacted through target translation repression. The eIF4F complex (a) and ribosomal scanning (b) are interrupted, preventing mRNA translation and protein expression (Created in BioRender. McGovern, E. (2025) https://BioRender.com/yjmdcuo).

metastasis, and epithelial-mesenchymal transition (EMT) due to inhibiting targets such as NOTCH1, SNAIL1, WNT1, and BCL-2 [70–73]. Notably, miR-34a (MRX34) was the first miRNA treatment to reach clinical trials but was terminated in 2016 due to immune-related adverse effects [74].

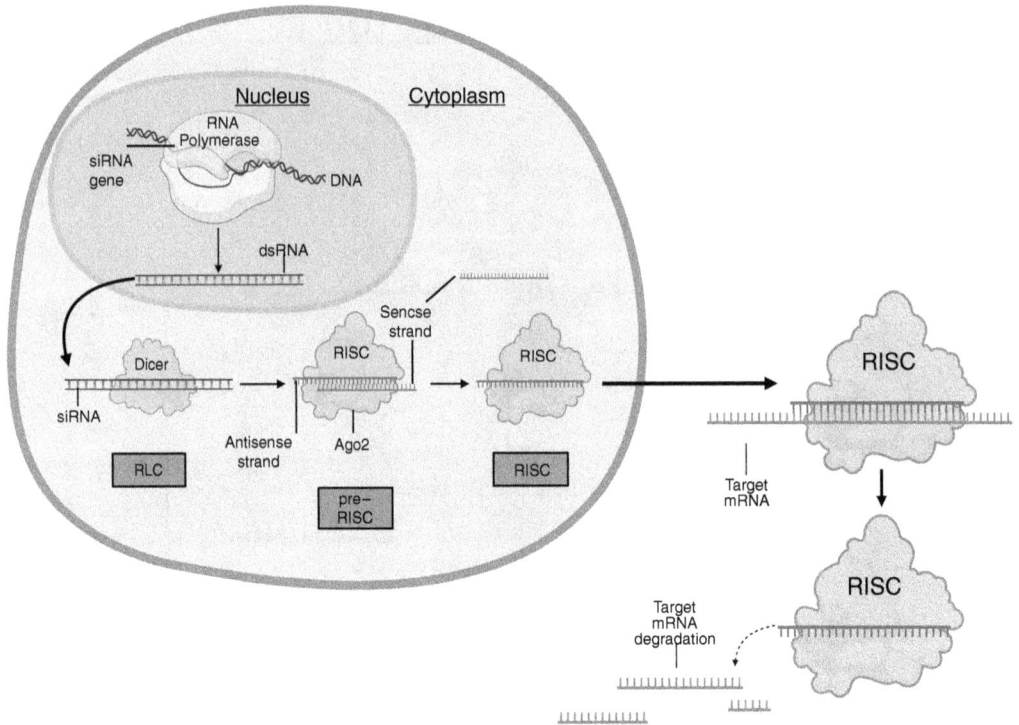

FIGURE 8.6 Biogenesis of siRNA and gene silencing of oncogene. dsRNA is converted into siRNA by the RLC in the cytoplasm. The sense strand is ejected when on pre-RISC. Mature RISC binds to the target mRNA and degrades it (Created in BioRender. McGovern, E. (2025) https://BioRender.com/1ok9w0o).

8.3.2 siRNA

siRNAs are similar to miRNAs, in that they promote the degradation of target mRNA; however, they differ in their targeting mechanism. While miRNAs have numerous targets with short regions that are complementary to short sequences on multiple mRNA, siRNAs are most often highly complementary to only one target mRNA [75]. Endogenously produced siRNA must be processed before functioning as a gene silencer. When transcribed from DNA, dsRNA is translocated into the cytosol, where dicer cleaves it into siRNA. The siRNA is then loaded onto RISC, where the sense strand is degraded by Ago-2. The antisense strand of siRNA guides RISC to the target mRNA and binds via base-paring due to its perfectly complementary sequence. The monophosphate 5′ end of the guide strand provides the start point for Ago-2 to cleave 10–11 nucleotides of the target mRNA in an upstream direction. Thus, the mRNA is degraded, translation is prevented, and the target gene is silenced (Figure 8.6) [76,77].

8.3.2.1 Examples of Tumor Suppressor siRNAs in Preclinical Studies

Modified siRNAs PnkRNA and nkRNA were synthesized to self-anneal, allowing them to be administered alone via inhalation to treat lung cancer. Tumor growth of A549-luc-C8 xenograft mice models was inhibited by PnkRNA and nkRNA by targeting the tumor promoter, RPN2, reducing its expression [78].

A toxic 6mer seed sequence of siRNA/shRNAs has been identified to selectively kill cancer cells via a mechanism termed Death Induced by Survival gene Elimination (DISE) [79]. The series of si-/shRNAs only require a 6-nucleotide (nt) seed sequence from nts 2–7 that is commonly high in G content to target C matches of the target mRNA [80]. To confirm it was the seed sequence that was responsible for cell death, CRISPR/Cas9 was used to knock down the target sites of the si-/shRNA, and toxicity was still induced irrespective of the designed target site binding [81].

8.3.3 Antisense Oligonucleotides

ASOs are single-stranded RNAs comprised of 15–20 nucleotides, which are also capable of tumor suppression by gene silencing. ASOs function in a similar way to siRNAs, by binding via Watson–Crick base pairing to complementary mRNA strands, and in some sources, siRNAs are included as a type of ASO. In others, however, they are separated due to the difference in processing, as siRNAs are double-stranded until cleavage by Ago-2, whereas ASOs are single-stranded [82]. ASOs function by inducing the cleavage of target mRNA via ribonuclease H1 (RNaseH1) or blocking mRNA translation [83]. mRNA translation can be blocked by steric hindrance, in which the ribosome is prevented from binding and thus translation cannot occur, or through splicing modulation [84,85]. ASOs modulate splicing of pre-mRNA by binding to *cis*-elements motifs for exon recognition, and disrupting target mRNA formation by causing spliceosomes to skip or include exons. This causes alterations in mRNA sequence, disrupting protein expression [86].

Gapmer ASOs are a promising type of ASOs that form a complex with strands on either side of an antisense DNA strand and bind to target mRNA. The formation of this DNA:RNA hybrid triggers recruitment of RNase H1, while the gapmer ASOs are resistant to RNase H and are free to form further complexes [87]. Gapmers were used by Morelli et al to activate RNAse H to down-regulate pri-mir-17–92, preventing the maturation of tumor promoter miRNAs in the miR-17–92 cluster. The MIR17PTi gapmer induced apoptosis in a C-MYC-dependent manner in multiple myeloma cells [88].

8.3.4 Aptamers

Aptamers are made up of a single strand of DNA or RNA and are capable of folding to form tertiary and quaternary structures, resembling antibodies, and have shown high selectivity and affinity for targets *in vitro* and *in vivo* [89]. Target mRNA and protein-specific aptamers can be determined using Systematic Evolution of Ligands by Exponential Enrichment (SELEX), in which a large library of RNA aptamers are incubated with the target protein, washed, and then amplified and analyzed through several rounds using reverse transcriptase-PCR to determine the aptamers with the highest affinity binding affinity [90]. When used as an anti-cancer drug, aptamers have several methods of functioning; one such method is their inhibitory function. Aptamers can bind proteins involved in carcinogenesis, thus inhibiting their function. Pegaptanib, FDA-approved and originally formulated to treat age-related macular degeneration, targets an isoform of vascular endothelial growth factor and has been shown to reduce angiogenesis and tumor formation in HUVECs and Cal27s, showing particular efficacy when delivered via tetrahedron DNA nanostructures [91].

8.4 NANOPARTICLES

RNA therapeutics are an effective way to target the drivers of carcinogenesis and treat cancer *in vitro*, but when advancing to *in vivo* studies, delivery is problematic. RNAs are readily degraded and face various issues with cellular uptake when systemically delivered, reducing the anti-tumor effect and increasing off-target effects [92].

In order to overcome this, nanoparticles can be used to carry the RNA therapy to the target site, ensuring stable delivery and maximizing the therapeutic effect (Figure 8.7). Some of the issues are discussed below in Table 8.1.

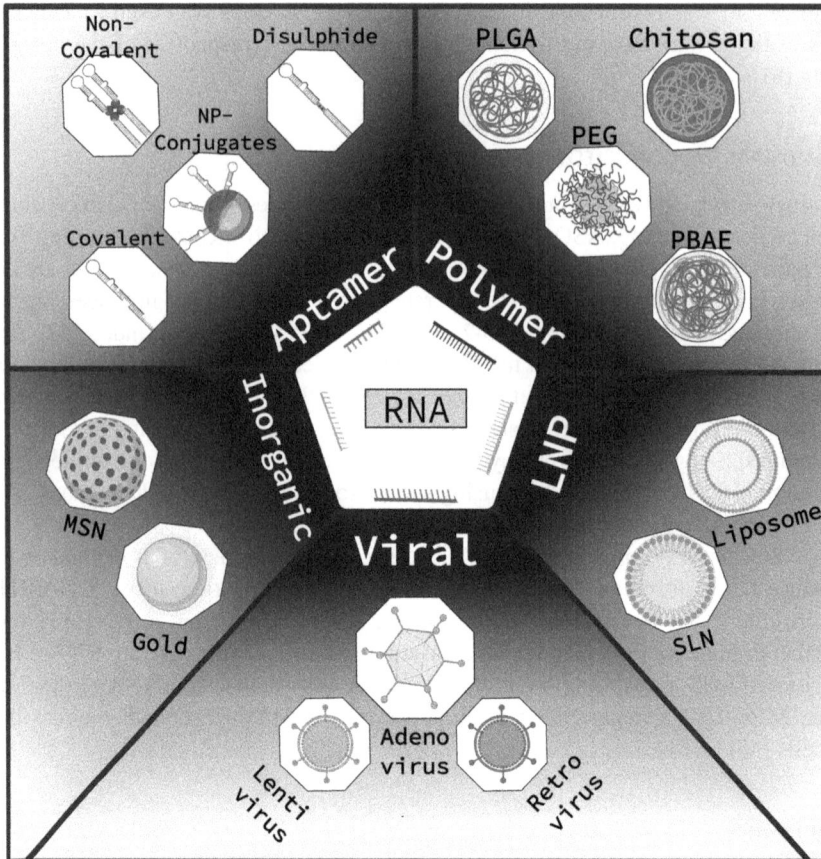

FIGURE 8.7   Schematic of nanoparticles to deliver RNA to cancer (Created in BioRender. McGovern, E. (2025) https://BioRender.com/k1to8dv).

TABLE 8.1

The Issues with RNA Delivery and How Nanoparticles Overcome Them

Issue	Description	Solution
Degradation	Multiple routes of administration such as intravenous and nasal delivery leave RNA vulnerable to degradation. Ribonucleases (RNases) include endonucleases, which cleave within the RNA sequence, and exonucleases that cleave from the 3′ and 5′ ends [93]. This causes degradation of the therapeutic RNA, reducing its half-life and preventing sufficient cellular uptake.	• Nanoparticles provide a physical barrier against RNases, preventing cleavage. • Modifications such as those to the 2′ ribose sugar like 2′-fluoro, 2′-O-methylation 2′-OME, and the locked nucleic acid (LNA) method, or those to the phosphorothioate backbone [94,95].
Size, charge, hydrophilicity	RNAs are negatively charged, hydrophilic, and often too large to be uptaken into the cell by endocytosis through the lipid bilayer [96].	• RNA can be carried by a positively charged nanoparticle. • Packaging into a nanoparticle condenses the size of an RNA molecule [92]. • Hydrophilic lipid nanoparticles can be used to increase uptake/ • miRNAs are smaller than other RNAs, allowing for easier internalization into cells.

(Continued)

TABLE 8.1 (*Continued*)
The Issues with RNA Delivery and How Nanoparticles Overcome Them

Issue	Description	Solution
Endosomal escape	Once taken into the cell, RNAs struggle to escape the endosome into the cytoplasm.	• pH-sensitive nanoparticles are designed to disassemble when the endosome enters the acidic environment of the cytoplasm and cause instability of the endosomal membrane, allowing release [97]. PEI and chloroquine can be used as a "proton sponge" to this effect [98,99].
Non-specific uptake	When naked RNAs are administered, they have poor targeting capabilities, leading to reduced uptake by target cells and increasing off-target effects by uptake into non-target cells and accumulation and subsequent toxicity in organs such as the liver and spleen.	• Conjugating ligands or peptides specific to the target cell allows for a significant increase in uptake and accumulation in tumors, and a reduction in off-target effects, hepatic accumulation, and toxicity and renal clearance [100].

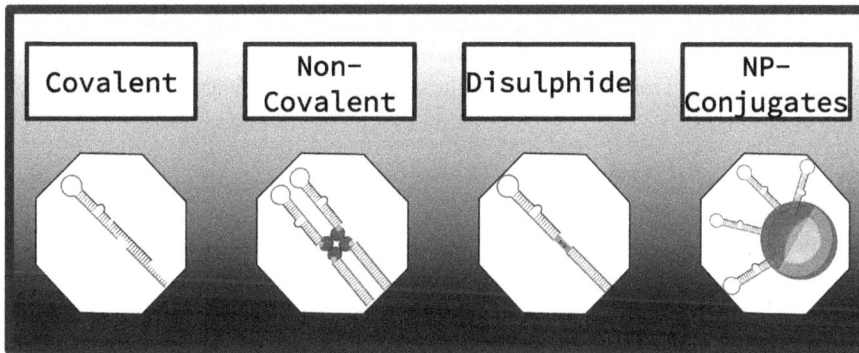

FIGURE 8.8   Schematic of aptamer-based delivery of RNA and the methods of conjugation to RNA (Created in BioRender. McGovern, E. (2025) https://BioRender.com/e4s2rwo).

8.4.1 APTAMERS

In addition to their cancer-suppressing abilities, aptamers may also be used as a targeted RNA delivery system through either aptamer-based nanoparticles or aptamer-RNA chimeras/conjugates (Figure 8.8) [101,102]. siRNAs are usually the basis for aptamer-RNA chimeras, combining the gene silencing of the siRNA and the targeted-binding capacity of the aptamers. SELEX allows for the selection of aptamers that target the cells/ligands of interest, which are then isolated and conjugated to the siRNA of interest. There are several methods of conjugation.

1. Covalent bonds: This approach uses bonding in the form of complementary base pairs.
   • The "sticky bridge" as developed by Rossi et al., a 16 nt "stick" is generated and annealed onto the 3′ end of the aptamer, which complements the RNA, forming a bridge and binding the two oligonucleotides together. A C3 linker may be added to reduce steric effects and to protect against improper folding of the aptamer and interference with Dicer processing [103].
   • An aptamer-siRNA chimera was developed to reduce polo-like kinase 1 (PLK1) expression and tumor growth in prostate cancer, using the A10 aptamer covalently bonded with the PLK1 targeting siRNA. The aptamer was modified with 2′-fluoro pyrimidines to prevent nuclease degradation [104]. This template was later used to sensitize HER2+ cells to cisplatin [105].

2. Non-covalent bonds: One method of conjugation is to exploit the strong non-covalent bonds between other structures, such as biotin and streptavidin. One study biotinylated both a 27mer siRNA and the A9 aptamer, which targets prostate-specific membrane antigen, and incubated them with streptavidin to form the aptamer/streptavidin/siRNA conjugates. Treatment of prostate cells with the conjugate inhibited expression of lamin A/C, a driver of prostate cancer [106].

3. Disulfide bond: A disulfide bond may be used as a linker to increase release into the cytoplasm once the chimera reaches its target. This may be achieved through the use of DMT, PEG, amidites, and/or thiol groups. It is the cleavage of the disulfide bond that releases the RNA from the chimera [107–109].

4. Aptamer-conjugated nanoparticles: Aptamers may be used to aid in the delivery of RNA drug-loaded nanoparticles. They can be conjugated onto the surface of various nanoparticles such as polymers and liposomes to combine the advantages of a nanocarrier such as protection and stability, the specificity of aptamers, and the tumor-suppressing RNAs.

A study on NSCLC used the AS1411 aptamer chimerized with PD-L1 targeted siRNA combined with a platform of polyethylenimine/glutamine/β-cyclodextrin and chemotherapy drug doxorubicin (Dox). The results showed the combination increased T cell activation and reduced tumor growth in an immune-dependent manner *in vivo* [110]. Aptamer-siRNA chimeras were also used by Jeong et al. to deliver and enhance the effect of Dox in multidrug-resistant cancer cells. Multivalent aptamers, Bcl-2 targeted siRNA, and Dox conjugates significantly reduced cell viability of multidrug-resistant MCF-7 cells *in vitro* [111].

8.4.2 POLYMERS

Polymeric NPs are advantageous nanocarriers due to their potential to be modified and their formulations altered to achieve target NP characteristics such as size and charge. Polymers may be natural such as cellulose and chitosan or synthetic such as polyethene glycol (PEG) and poly(lactic-co-glycolic) acid (PLGA) and can form nanocapsules or nanospheres. Their ability to encapsulate drugs makes them ideal for delivery to cancer, and their ability to encapsulate RNAs makes them ideal nanocarriers for RNA therapeutics (Figure 8.9) [112].

8.4.2.1 Poly(lactic-co-glycolic) Acid

PLGA is a polyester co-polymer synthesized from lactic acid and glycolic acid moieties. It is a popular option for NPs due to its biocompatibility, biodegradability, and solubility in solvents such as acetone, ethanol, and DMSO. However, it is insoluble in water due to the hydrophobic nature

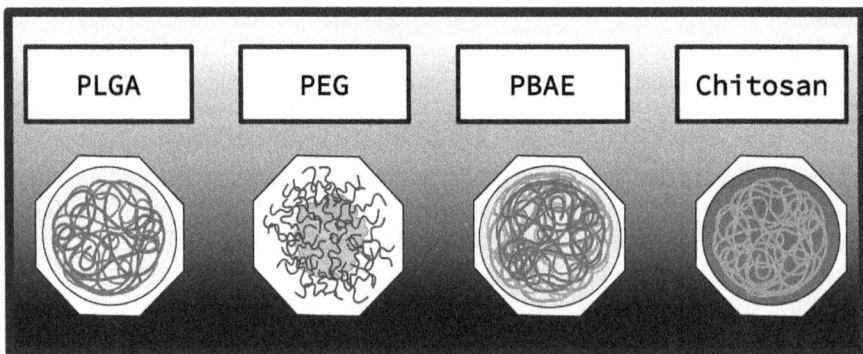

FIGURE 8.9 Schematic of polymeric nanoparticles to deliver RNA (Created in BioRender. McGovern, E. (2025) https://BioRender.com/6e6ta8v).

of lactic acid, yet it will degrade in water via hydrolysis of its ester linkages. The composition of PLGA may be altered through varying ratios of lactic and glycolic acid, with a higher proportion of glycolic acid increasing its rate of biodegradation due to the lack of a methyl side group [113,114]. One study incorporated poly-L-lysine (PLL) into PLGA NPs loaded with miR-34a and found that the combination NPs delivered more miRNA to triple negative breast cancer cells than PLL alone. Treatment with the PLGA-PLL-miRNA NPs suppressed Notch-1, Survivin, and Bcl-2 gene expression and inhibited cell cycle progression and cell proliferation [115]. Another study loaded PLGA NPs with Bcl-2 and MDR1 targeting siRNA to successfully reduce resistance to cisplatin and paclitaxel and increase cell death in ovarian cancer cells [116].

However, PLGA's anionic nature poses an issue for carrying negatively charged RNAs. To overcome this, cationic compounds may be incorporated in the NP such as hyaluronic acid (HA) or PEI. It was implemented by Camorani et al., who synthesized PLGA-b-PEG-COOH NPs, decorated with a triple negative breast cancer cell-specific aptamer, and loaded them with PD-L1 targeting siRNA, which resulted in effective cell-specific uptake and downregulation of PD-L1 [117].

8.4.2.2   Polyethylene Glycol (PEG)

Polymer and lipid nanoparticles are often coated with PEG in order to improve circulation time, biodistribution, and reduce toxicity [118], and this has been implemented in RNA therapeutics for cancer. PEG was incorporated into PDMAEMA nanoparticles decorated with a C-peptide to deliver miR-128-3p to colorectal cancer, resulting in the targeting of PI3K/AKT and MEK/ERK pathways and reducing tumor growth. The PEGylation of the nanoparticles served to increase circulation time and aid in target specificity by reducing interactions with other proteins/ligands [119]. PEGylation of a PEI nanoparticle, modified with disulfide linkages and a cell-permeable peptide, was implemented to increase biocompatibility and biodistribution. The study achieved delivery of miR-145 to prostate cancer in both cell and mouse models and reduced tumor growth [120].

Often PEG is selected due to its so-called "stealth" attributes, reducing uptake of the NPs by the mononuclear phagocyte system [121]. However, further research has uncovered anti-PEG antibodies both in patients treated with PEGylated drugs and those who have not received such treatment [122]. Therefore, caution should be exercised when considering PEG for RNA nanotherapeutics.

8.4.2.3   Poly(beta-amino ester)s (PBAEs)

PBAEs are most commonly formed by the Micheal addition of acrylates and amines. They are suited for RNA delivery due to their cationic properties, which also aid in the conjugation of targeted peptides [123]. PBAEs also facilitate endosomal escape. The addition of tertiary amines increases buffering capacity due to the proton sponge effect, increasing osmotic pressure, leading to endosomal rupture. Their low toxicity allows for a high ratio of PBAE to RNA to maximize encapsulation [124]. Dosta et al. investigated how modifications to PBAEs affected the encapsulation and transfection efficiency of siRNA, of which a GFP-targeted siRNA was used to knock down EGFP. Cchol-50 (a cholesterol-modified PBAE at 50% esterification) and C6–50 (a C6 polymer modified by 50% hexylamine) were the best performers for EGFP silencing and stability in the presence of serum proteins [125]. PBAEs have even been shown to selectively target cancer cells over non-cancerous cells. Also using GFP-targeted siRNA, Kozielski et al. found siRNA-mediated cell death of up to 97% in glioblastoma cells compared to 27% in fetal neural progenitor cells, suggesting targeted uptake of nanoparticles [126].

8.4.2.4   Chitosan (CS)

CS is a linear polysaccharide, and a derivative of chitin is a natural polymer and a popular basis for nanoparticles because of its low toxicity and high stability. To form the nanoparticle, CS is often cross-linked with negatively charged sodium tripolyphosphate (TPP) [127]. In one study, CS was modified by HA to form nanoparticles loaded with Bcl-2 targeting siRNA. The CS-HA-siRNA NPs significantly downregulated Bcl-2 expression and tumor growth when compared to naked siRNA

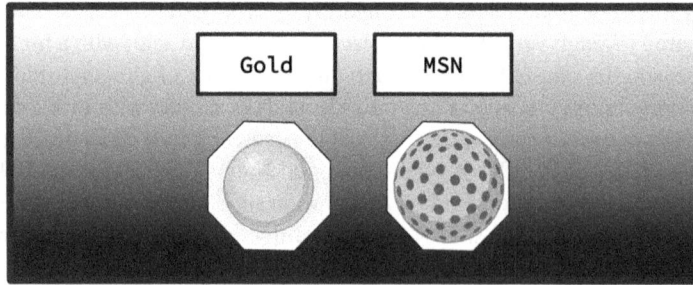

FIGURE 8.10 Schematic of inorganic nanoparticles to deliver RNA (Created in BioRender. McGovern, E. (2025) https://BioRender.com/ss1wpck).

[128]. HA was also combined with trimethyl chitosan (TMC) and siRNA to downregulate IL-6 and STAT3 in various CD44$^+$ cancer cell lines. Treatment resulted in target gene silencing, reduced cancer cell proliferation and angiogenesis, which was upheld in chicken chorioallantoic membrane models [129]. A similar study also using TMC, HA, and STAT3 siRNA, with the addition of thio-lated chitosan and HIV-1 derived TAT peptide, as well as PD-L1 siRNA, achieved silencing of PD-L1 and STAT3 and saw reduced cancer cell growth both *in vitro* and *in vivo* melanoma and breast cancer models [130]. The thiolation or trimethylation of CS, as mentioned above, is a method to modify CS to become water-soluble. The use of water-soluble CS holds an advantage over those soluble in acidic solution by allowing the loading of water-soluble drugs, eliminating the restriction of drug release only in acidic environments, and reducing toxicity, particularly for oral and nasal delivery routes [131].

8.4.3 INORGANIC NPs

Inorganic NPs use materials such as gold, silver, silica (Figure 8.10), and selenium to form nanocar-riers for RNA that, in the majority of NPs, can be conjugated onto the surface or, for porous materi-als, encapsulated within the core. In addition to acting as a delivery vehicle for RNA therapies, some inorganic NPs display anticancer capabilities themselves [132]. For example, silver NPs were shown to reduce cell viability in breast cancer cells [133].

8.4.3.1 Gold

Gold NPs were used to exploit their radio-sensitizing abilities [134] by Zhuang et al., conjugating specificity protein 1 (SP1)-specific siRNA. SP1 has been found to be overexpressed in lung can-cers and to be downregulated following radiotherapy, making it a viable candidate for targeting. Delivery of AuNPs-si-SP1 downregulated SP1 expression in NSCLC cells, which reversed GZMB suppression, resulting in decreased cell viability and increased radiosensitivity in cell culture and subcutaneous tumors [135]. Gold can also be utilized in photothermal therapy (PTT), in which light is converted into heat energy inducing hyperthermia and subsequently cell apoptosis. For gold NP therapies, this is often achieved by local delivery of NPs to the tumor site followed by exposure to near-infrared (NIR) laser light [136]. PTT was implemented by Cai et al., conjugating the iRGD peptide and CDK7 targeting siRNA onto Au@mPEG-PEI NPs. In lung adenocarcinoma cells, AuNP/siCDK7 successfully downregulated CDK7 expression, raised cellular temperature, and reduced cancer cell survival after laser irradiation. The efficacy of the treatment was extended to *in vivo* studies, in which the delivery of the NPs was assessed, showing significant accumulation in the tumor compared to lung, liver, and kidney tissue. The combination of the NP complexes and irradiation therapy resulted in prolonged overall survival, supporting the anticancer effects of the treatment [137].

8.4.3.2 Mesoporous Silica

Mesoporous silica nanoparticles (MSNs) are made up of amorphous silica with pores on the surface. They are a promising nanocarrier that holds an advantage due to the pores largely increasing the surface area, allowing for a higher level of loading/conjugating [138]. One study utilized a combination of chemotherapy drug, sorafenib, and CRISPR/Cas9 therapy to treat HCC. Sorafenib and two EGFR-specific sgRNAs with CRISPR/CAS9 plasmids were loaded into the hollow core of the NP, and the surface was conjugated with the dendrimer PAMAM and targeting DNA-based aptamers. In cell culture, inhibition of the EGFR/PI3K/AKT and angiogenesis was shown, and in mice treatment, it resulted in a significant reduction of tumor growth and active tumor targeting [139]. MSNs can also be used to load photosensitizing drugs, as Wang et al. did by loading ICG into the NP and exposing them to NIR, which led to the generation of reactive oxygen species to aid in endosomal escape of RNA into the cytosol. The study used both miRNA and siRNA (miR-200c and siPlk1) to inhibit Plk1 and EMT, an iRGD peptide on the surface was used to target breast tumor cells, and a lipid coating was added. The results showed preferential tumor targeting and significant reduction in cell viability *in vitro* and tumor growth *in vivo,* most significantly when treated with NIR [140].

8.4.4 Lipid-Based Nanoparticles

Lipid nanoparticles (LNPs) are an advantageous drug delivery platform due to the lipid composition, allowing for easier entry into the cell through the phospholipid bilayer. The incorporation of cationic and ionizable lipids is ideal for negatively charged RNA [141]. Liposomes are more commonly used for RNA delivery over solid lipid nanoparticles (SLN) due to the solid and lipophilic core making an inadequate environment to load RNA; however, it may be conjugated onto the surface (Figure 8.11) [142].

Zhang et al. developed a complex of the cholesterol-modified cationic peptide, DP7, liposome, and mRNA to serve as a neoantigen-specific mRNA vaccine. Results showed a significant increase in mature dendritic cells and cytotoxic lymphocytes alongside reduced tumor volume in lung tumor-bearing mice [143]. Commonly used LNPs for RNA delivery are cationic and ionizable lipids. Furthermore, the most popular cationic lipids are DOTAP and DOTMA [144]. In 2022, Gao et al. used DOTAP with polymers (mPEG and PCL) and cell-penetrating peptides (R9 and cRGD) to deliver mRNA encoding the suicide gene, *Bim,* resulting in induction of apoptosis in colonic cancer cells. When tested *in vivo* with mice, intratumoral delivery of the LNP/mRNA complexes to

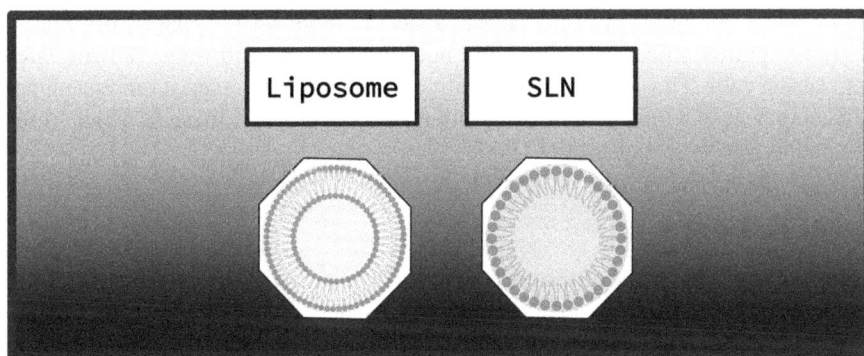

FIGURE 8.11 Schematic of lipid-based nanoparticles to deliver RNA. (Created in BioRender. McGovern, E. (2025) https://BioRender.com/84q918a).

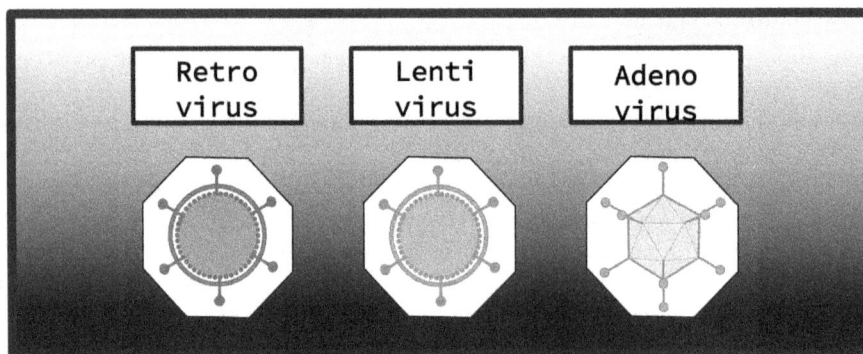

FIGURE 8.12   Schematic of viral vector nanoparticles to deliver RNA (Created in BioRender. McGovern, E. (2025) https://BioRender.com/ctufyvd).

abdominally xenografted colonic tumors and intravenous delivery to metastatic lung models were effective in reducing tumor growth and metastases [145].

8.4.5 Viral Vectors

As natural agents capable of effective delivery of genetic material to the cytoplasm, viral vectors pose a promising avenue for RNA therapies, particularly as viral tropism allows for cell-specific targeting. Viral vectors range from integrating vectors such as retroviruses and lentiviruses to non-integrating vectors like adenoviruses and adeno-associated viruses (Figure 8.12). However, caution should be exercised, particularly when using retroviruses, which have been linked to inducing mutagenesis [146].

Lou et al. altered an adenovirus to target cancer cells and limit replication in non-cancerous cells by inserting a tumor-specific survivin motif and an RGD motif to facilitate targeting to integrins on the tumor cell surface and deleting the E1B-55kDa (ZD55) region. The viral vector was constructed to deliver miR-143, which was found to target KRAS and suppress cell proliferation in colorectal cancer cell lines. The Ad-RGD-Survivin-ZD55-miR-143 complex successfully targeted colorectal cell lines *in vitro* and reduced tumor growth in xenografted mice [147]. Using viral vectors can not only act as a delivery platform for RNA therapeutics but also provide an opportunity to exploit the anti-cancer abilities of oncolytic viruses. Chen et al. inserted siRNA targeted to Bcl-2 and Survivin into herpes simplex virus type 1 (HSV-1) to form HSV010-BS. HSV010-BS was found to reduce cell viability in the breast cancer cell line, MCF-7, and reduce tumor growth in MCF-7 tumor-grafted mice; however, it did not affect normal cell lines, showing preferential targeting to cancer cells. The efficacy of the treatment is limited; however, it shows a higher degree of gene silencing in cells with high phosphorylation levels of protein kinase R, which limits HSV-1 replication and its oncolytic capabilities [148].

8.5   CLINICAL TRIALS

Preclinical trials have seen positive results both *in vitro* and *in vivo* across the various classes and combinations of RNA and nanoparticle therapies. Consequently, this has transitioned into ongoing clinical trials to develop a potent and effective drug and regimen for cancer treatment and to improve the survival rate and quality of life in patients. Examples of such clinical trials can be found in Table 8.2.

TABLE 8.2
Clinical Trials of Nanoparticle Encapsulated RNA Drugs for Cancer Therapy

RNA	Name	Nanoparticle	Description	Cancer	Outcomes	Phase	NTC Code
CRISPR/Cas9	TT52CAR19	Lentiviral vectors	Anti-CD19 CAR T cells (CAR19) generated with knockdown of TRAC and CD52	– Pediatric B cell acute lymphoblastic leukemia (B-ALL)	1/3 (n=6) patients entered complete remission and are alive and well following restorative allo-SCT [149]	I	NCT04557436
mRNA vaccine	mRNA-4157	LNP	Individualized neoantigen therapy (≤34 antigens) in combination with pembrolizumab	– Melanoma	RFS at 18 months was 79% compared to 62% with pembrolizumab monotherapy [150]	IIb	NCT03897881
					TBD	III	NCT05933577
				– NSCLC	TBD	III	NCT06077760
				– Cutaneous squamous cell carcinoma (cSCC)	TBD	II/III	NCT06295809
	FixVac (BNT111)	LNP	mRNA encoding 4 TAAs; NY-ESO-1, MAGE-A3, TPTE and tyrosinase	– Melanoma	Those treated with the combination of 100 μg FixVac and anti-PD1 showed the highest rate of partial response with a lesion regression rate of >35% [36]	I	NCT02410733
					TBD; to be trialled in three arms; BNT111+cemiplimab, BNT111 monotherapy and cemiplimab monotherapy	II	NCT04526899
miRNA	MRX34	LNP	miR-34a delivered via liposomal injection to target and suppress> 30 oncogenes	– Primary Liver Cancer – SCLC – NSCLC – Lymphoma – Melanoma – Multiple Myeloma – Renal Cell Carcinoma	19% (n=16) had stable disease after four cycles of treatment, and 4% (n=3) had partial regression. Terminated after five immune-related deaths [74,151]	I	NCT01829971
				– Various solid tumors	Withdrawn following five deaths during previous phase I study	I/II	NCT02862145

(Continued)

TABLE 8.2 (Continued)
Clinical Trials of Nanoparticle Encapsulated RNA Drugs for Cancer Therapy

RNA	Name	Nanoparticle	Description	Cancer	Outcomes	Phase	NTC Code
	INT-1B3 (miR-193a-3p)	LNP	Modified miR-193a-3p delivered via LNP to reduce tumor growth and modulate tumor microenvironment [152]	– Various solid tumors	Terminated; insufficient funding	I	NCT04675996
	MesomiR 1	TargomiRs	MiR-16 packaged by minicells (EDV) to target EGFR	– NSCLC – Malignant Pleural Mesothelioma	19% (n=5) survival at the median follow-up point of 6 months, with two patients surviving past 2 years. Median overall survival was 200 days [153]	I	NCT02369198
siRNA	CALAA-01	Cyclodextrin-based polymer (CDP) and PEG	Ligands conjugated onto the surface to target Tf receptors on cancer cells with RRM2-specific siRNA (siR2B+5)	– Solid tumors	Phase 1a saw delivery of NPs when assessing post-treatment biopsies; however, phase 1b found dose-limiting toxicities and was subsequently terminated [77,154]	I	NCT00689065
	LODER	Polymeric biodegradable matrix	KRAS G12D-targeted siRNA	– Non-Resectable Pancreatic Cancer	83% (n=10) had stable disease during phase I, and so far, phase II has seen a 9.3 months improvement in overall survival. Response rate is reported to be 56% in combination with chemotherapy in comparison to 20% with chemo-monotherapy	II	NCT01188785 NCT01676259
Aptamer	NOX-A12	PEG	Targets and inhibits CXCL12, linked to cancer signaling, adhesion, and invasion	– Metastatic Colorectal and Pancreatic Cancer	Stable disease was achieved in 25% of patients and 19% survival rate at 12 months when combined with pembrolizumab	I/II	NCT03168139
				– Glioblastoma	90% (n=9) showed response to radiotherapy and median overall survival was 389 days [155]	I/II	NCT04121455
				– Metastatic Pancreatic Cancer	TBD	II	NCT04901741

8.6 CONCLUSION

There is extensive evidence that RNA nanotherapeutics hold the potential to treat a range of cancers at various stages. Their capability to silence oncogenes, upregulate and insert tumor suppressor genes, and sensitize the immune system to tumor cells makes RNA an ideal candidate for clinical development. Although their instability *in vivo* raises an issue for drug delivery, nanoparticles offer an optimal solution to increase stability, cellular uptake, and endosomal escape, while decreasing immunogenicity and providing a platform for targeted therapy. Future directions involve advancing cell-specific therapies such as targeting peptides and aptamers to maximize the anti-cancer effect, while reducing off-target effects. Further development is needed in order to advance these therapies through clinical trials and into clinical practice to help cancer patients and their prognosis.

REFERENCES

1. Ishino Y, Shinagawa H, Makino K, et al. Nucleotide sequence of the iap gene, responsible for alkaline phosphatase isozyme conversion in *Escherichia coli*, and identification of the gene product. *Journal of Bacteriology*. 1987;169(12):5429–5433.
2. Janik E, Niemcewicz M, Ceremuga M, et al. Various aspects of a gene editing system-CRISPR-Cas9. *International Journal of Molecular Sciences*. 2020;21(24):9604.
3. Ravichandran M, Maddalo D. Applications of CRISPR-Cas9 for advancing precision medicine in oncology: From target discovery to disease modeling. *Frontiers in Genetics*. 2023;14:1273994.
4. Deltcheva E, Chylinski K, Sharma CM, et al. CRISPR RNA maturation by trans-encoded small RNA and host factor RNase III. *Nature*. 2011;471(7340):602–627.
5. Anders C, Niewoehner O, Duerst A, et al. Structural basis of PAM-dependent target DNA recognition by the Cas9 endonuclease. *Nature*. 2014;513(7519):569–573.
6. Gasiunas G, Barrangou R, Horvath P, et al. Cas9–crRNA ribonucleoprotein complex mediates specific DNA cleavage for adaptive immunity in bacteria. *Proceedings of the National Academy of Sciences – PNAS*. 2012;109(39):E2579–E2586.
7. Babu K, Kathiresan V, Kumari P, et al. Coordinated actions of Cas9 HNH and RuvC nuclease domains are regulated by the bridge helix and the target DNA sequence. *Biochemistry*. 2021;60(49):3783–3800.
8. Mou H, Ozata DM, Smith JL, et al. CRISPR-SONIC: Targeted somatic oncogene knock-in enables rapid in vivo cancer modeling. *Genome Medicine*. 2019;11(1):21.
9. Cai H, Zhang B, Ahrenfeldt J, et al. CRISPR/Cas9 model of prostate cancer identifies Kmt2c deficiency as a metastatic driver by Odam/Cabs1 gene cluster expression. *Nature Communications*. 2024;15(1):2088.
10. Moses C, Nugent F, Waryah CB, et al. Activating PTEN tumor suppressor expression with the CRISPR/dCas9 system. *Molecular Therapy Nucleic Acids*. 2019;14:287–300.
11. Chen J-S, Liu J-C, Shen L, et al. Cancer-specific activation of the survivin promoter and its potential use in gene therapy. *Cancer Gene Therapy*. 2004;11(11):740–747.
12. Chira S, Gulei D, Hajitou A, et al. Restoring the p53 'Guardian' phenotype in p53-deficient tumor cells with CRISPR/Cas9. *Trends in Biotechnology* (Regular ed). 2018;36(7):653–660.
13. Liu B, Saber A, Haisma HJ. CRISPR/Cas9: A powerful tool for identification of new targets for cancer treatment. *Drug Discovery Today*. 2019;24(4):955–970.
14. Oppel F, Schürmann M, Goon P, et al. Specific targeting of oncogenes using CRISPR technology. *Cancer Research*. 2018;78(19):5506–5512.
15. Jinek M, Chylinski K, Fonfara I, et al. A programmable dual-RNA–guided DNA endonuclease in adaptive bacterial immunity. *Science*. 2012;337(6096):816–821.
16. Koo T, Yoon AR, Cho H-Y, et al. Selective disruption of an oncogenic mutant allele by CRISPR/Cas9 induces efficient tumor regression. *Nucleic Acids Research*. 2017;45(13):7897–7908.
17. Hsu PD, Scott DA, Weinstein JA, et al. DNA targeting specificity of RNA-guided Cas9 nucleases. *Nature Biotechnology*. 2013;31(9):827–832.
18. Collias D, Beisel CL. CRISPR technologies and the search for the PAM-free nuclease. *Nature Communications*. 2021;12(1):555–555.
19. Walton RT, Christie KA, Whittaker MN, et al. Unconstrained genome targeting with near-PAMless engineered CRISPR-Cas9 variants. *Science (American Association for the Advancement of Science)*. 2020;368(6488):290–296.

20. Martinez-Lage M, Torres-Ruiz R, Puig-Serra P, et al. In vivo CRISPR/Cas9 targeting of fusion onco-genes for selective elimination of cancer cells. *Nature Communications*. 2020;11(1):5060.

21. Sarkar E, Khan A. Erratic journey of CRISPR/Cas9 in oncology from bench-work to successful-clinical therapy. *Cancer Treatment and Research Communications*. 2021;27:100289.

22. Kosicki M, Tomberg K, Bradley A. Repair of double-strand breaks induced by CRISPR–Cas9 leads to large deletions and complex rearrangements. *Nature Biotechnology*. 2018;36(8):765–771.

23. Zhao X, Liu L, Lang J, et al. A CRISPR-Cas13a system for efficient and specific therapeutic targeting of mutant KRAS for pancreatic cancer treatment. *Cancer Letters*. 2018;431:171–181.

24. Dakal TC, Dhabhai B, Pant A, et al. Oncogenes and tumor suppressor genes: Functions and roles in cancers. *MedComm (2020)*. 2024;5(6):e582-n/a.

25. Vavilis T, Stamoula E, Ainatzoglou A, et al. mRNA in the context of protein replacement therapy. *Pharmaceutics*. 2023;15(1):166.

26. Kong N, Tao W, Ling X, et al. Synthetic mRNA nanoparticle-mediated restoration of p53 tumor sup-pressor sensitizes. *Science Translational Medicine*. 2019;11(523):eaaw1565.

27. Hirschberger K, Jarzebinska A, Kessel E, et al. Exploring cytotoxic mRNAs as a novel class of anti-cancer biotherapeutics. *Molecular Therapy Methods & Clinical Development*. 2018;8:141–151.

28. Jain R, Frederick JP, Huang EY, et al. MicroRNAs enable mRNA therapeutics to selectively program cancer cells to self-destruct. *Nucleic Acid Therapeutics*. 2018;28(5):285–296.

29. Fu Q, Zhao X, Hu J, et al. mRNA vaccines in the context of cancer treatment: From concept to applica-tion. *Journal of Translational Medicine*. 2025;23(1):12.

30. Apavaloaei A, Hardy M-P, Thibault P, et al. The origin and immune recognition of tumor-specific anti-gens. *Cancers*. 2020;12(9):2607.

31. Deng Z, Tian Y, Song J, et al. mRNA vaccines: The dawn of a new era of cancer immunotherapy. *Frontiers in Immunology*. 2022;13:887125.

32. Gao Y, Yang L, Li Z, et al. mRNA vaccines in tumor targeted therapy: Mechanism, clinical application, and development trends. *Biomarker Research*. 2024;12(1):93.

33. Galant N, Krzyżanowska N, Krawczyk P. mRNA vaccines in the treatment of cancer. *Oncology in Clinical Practice*. 2024.

34. Vishweshwaraiah Y, Dokholyan N. mRNA vaccines for cancer immunotherapy. *Frontiers in Immunology*. 2022;13:1029069.

35. Stadler CR, Bähr-Mahmud H, Celik L, et al. Elimination of large tumors in mice by mRNA-encoded bispecific antibodies. *Nature Medicine*. 2017;23(7):815–817.

36. Sahin U, Oehm P, Derhovanessian E, et al. An RNA vaccine drives immunity in checkpoint-inhibitor-treated melanoma. *Nature*. 2020;585(7823):107–112.

37. Wang Y, Zhang L, Xu Z, et al. mRNA vaccine with antigen-specific checkpoint blockade induces an enhanced immune response against established melanoma. *Molecuar Therapeutics*. 2018;26(2):420–434.

38. Hollingsworth RE, Jansen K. Turning the corner on therapeutic cancer vaccines. *npj Vaccines*. 2019;4(1):7.

39. Bright RK, Bright JD, Byrne JA. Overexpressed oncogenic tumor-self antigens. *Human Vaccine Immunotherapy*. 2014;10(11):3297–3305.

40. Rojas LA, Sethna Z, Soares KC, et al. Personalized RNA neoantigen vaccines stimulate T cells in pan-creatic cancer. *Nature*. 2023;618(7963):144–150.

41. Sethna Z, Guasp P, Reiche C, et al. RNA neoantigen vaccines prime long-lived CD8+ T cells in pancre-atic cancer. *Nature*. 2025;639(8056):1042–1051.

42. Ni L. Advances in mRNA-based cancer vaccines. *Vaccines*. 2023;11(10):1599.

43. Smolarz B, Durczyński A, Romanowicz H, et al. miRNAs in cancer (review of literature). *International Journal of Molecular Sciences*. 2022;23(5):2805.

44. Baghi M, Yadegari E, Rostamian Delavar M, et al. MiR-193b deregulation is associated with Parkinson's disease. *Journal of Cellular and Molecular Medicine*. 2021;25(13):6348–6360.

45. Hu Z, Gao S, Lindberg D, et al. Temporal dynamics of miRNAs in human DLPFC and its association with miRNA dysregulation in schizophrenia. *Translational Psychiatry*. 2019;9(1):196–217.

46. Calore M, Lorenzon A, Vitiello L, et al. A novel murine model for arrhythmogenic cardiomyopathy points to a pathogenic role of Wnt signalling and miRNA dysregulation. *Cardiovascular Research*. 2019;115(4):739–751.

47. Cai X, Hagedorn C, Cullen B. Human microRNAs are processed from capped, polyadenylated tran-scripts that can also function as mRNAs. *RNA*. 2004;10(12):1957–1966.

48. Lee Y. MicroRNA maturation: Stepwise processing and subcellular localization. *The EMBO Journal*. 2002;21(17):4663–4670.

49. Lee Y, Ahn C, Han J, et al. The nuclear RNase III Drosha initiates microRNA processing. *Nature*. 2003;425(6956):415–419.

50. Denli A, Tops BJ, Plasterk RA, et al. Processing of primary microRNAs by the microprocessor complex. *Nature*. 2004;432(7014):231–235.

51. Okada C, Yamashita E, Lee S, et al. A High-resolution structure of the pre-microRNA nuclear export machinery. *Science*. 2009;326(5957):1275–1279.

52. Bernstein E, Caudy A, Hammond S, et al. Role for a bidentate ribonuclease in the initiation step of RNA interference. *Nature*. 2001;409(6818):363–366.

53. Lee Y-Y, Lee H, Kim H, et al. Structure of the human DICER–pre-miRNA complex in a dicing state. *Nature*. 2023;615(7951):331–338.

54. Hammond S, Bernstein E, Beach D, et al. An RNA-directed nuclease mediates post-transcriptional gene silencing in Drosophila cells. *Nature*. 2000;404(6775):293–296.

55. Liu J, Carmell M, Rivas F, et al. Argonaute2 is the catalytic engine of mammalian RNAi. *Science*. 2004;305(5689):1437–1441.

56. Chendrimada T, Gregory R, Kumaraswamy E, et al. TRBP recruits the Dicer complex to Ago2 for microRNA processing and gene silencing. *Nature*. 2005;436(7051):740–744.

57. Wang H-W, Noland C, Siridechadilok B, et al. Structural insights into RNA processing by the human RISC-loading complex. *Nature Structural & Molecular Biology*. 2009;16(11):1148–1153.

58. Iwasaki S, Kobayashi M, Yoda M, et al. Hsc70/Hsp90 chaperone machinery mediates ATP-dependent RISC loading of small RNA duplexes. *Molecular Cell*. 2010;39(2):292–299.

59. Kawamata T, Tomari Y. Making RISC. *Trends in Biochemical Sciences*. 2010;35(7):368–376.

60. Medley J, Panzade G, Zinovyeva A. microRNA strand selection: Unwinding the rules. *Wiley Interdisciplinary Reviews: RNA*. 2020;12(3):e1627.

61. Braun J, Huntzinger E, Fauser M, et al. GW182 proteins directly recruit cytoplasmic deadenylase complexes to miRNA targets. *Molecular Cell*. 2011;44(1):120–133.

62. Behm-Ansmant I, Rehwinkel J, Doerks T, et al. mRNA degradation by miRNAs and GW182 requires both CCR4:NOT deadenylase and DCP1:DCP2 decapping complexes. *Genes & Development*. 2006;20(14):1885–1898.

63. Geisler S, Coller J. Chapter Five – XRN1: A major 5′ to 3′ exoribonuclease in eukaryotic cells. In: Chanfreau GF, Tamanoi F, editors. *The Enzymes*. Vol. 31. Academic Press, 2012. pp. 97–114.

64. Fukao A, Mishima Y, Takizawa N, et al. MicroRNAs trigger dissociation of eIF4AI and eIF4AII from target mRNAs in humans. *Molecular Cell*. 2014;56(1):79–89.

65. Naeli P, Winter T, Hackett AP, et al. The intricate balance between microRNA-induced mRNA decay and translational repression. *The FEBS Journal*. 2023;290(10):2508–2524.

66. Bai J, Shi Z, Wang S, et al. MiR-21 and let-7 cooperation in the regulation of lung cancer. *Frontiers in Oncology*. 2022;12:950043.

67. Li L, Zhang X, Lin Y, et al. Let-7b-5p inhibits breast cancer cell growth and metastasis via repression of hexokinase 2-mediated aerobic glycolysis. *Cell Death Discovery*. 2023;9(1):114.

68. Jin B, Wang W, Meng X-X, et al. Let-7 inhibits self-renewal of hepatocellular cancer stem-like cells through regulating the epithelial-mesenchymal transition and the Wnt signaling pathway. *BMC Cancer*. 2016;16(1):863.

69. Wells AC, Hioki KA, Angelou CC, et al. Let-7 enhances murine anti-tumor CD8 T cell responses by promoting memory and antagonizing terminal differentiation. *Nature Communications*. 2023;14(1):5585.

70. Park EY, Chang E, Lee EJ, et al. Targeting of miR34a–NOTCH1 axis reduced breast cancer stemness and chemoresistance. *Cancer Research*. 2014;74(24):7573–7582.

71. Tang Y, Tang Y, Cheng Y-S. miR-34a inhibits pancreatic cancer progression through Snail1-mediated epithelial–mesenchymal transition and the Notch signaling pathway. *Scientific Reports*. 2017;7(1):38232.

72. Si W, Li Y, Shao H, et al. MiR-34a inhibits breast cancer proliferation and progression by targeting Wnt1 in Wnt/β-catenin signaling pathway. *The American Journal of the Medical Sciences*. 2016;352(2):191–199.

73. Duan J, Zhou K, Tang X, et al. MicroRNA-34a inhibits cell proliferation and induces cell apoptosis of glioma cells via targeting of Bcl-2. *Molecular Medicine Reports*. 2016;14(1):432–438.

74. Hong DS, Kang Y-K, Borad M, et al. Phase 1 study of MRX34, a liposomal miR-34a mimic, in patients with advanced solid tumours. *British Journal of Cancer*. 2020;122(11):1630–1637.

75. Lam JK, Chow MY, Zhang Y, et al. siRNA versus miRNA as therapeutics for gene silencing. *Molecular Therapy Nucleic Acids*. 2015;4(9):e252.

76. Charbe NB, Amnerkar ND, Ramesh B, et al. Small interfering RNA for cancer treatment: Overcoming hurdles in delivery. *Acta Pharmaceutica Sinica B*. 2020;10(11):2075–2109.

77. Hattab D, Gazzali A, Bakhtiar A. Clinical advances of siRNA-based nanotherapeutics for cancer treatment. *Pharmaceutics*. 2021;13(7):1009.

78. Fujita Y, Takeshita F, Mizutani T, et al. A novel platform to enable inhaled naked RNAi medicine for lung cancer. *Scientific Reports*. 2013;3(1):3325.

79. Haluck-Kangas A, Patel M, Paudel B, et al. DISE/6mer seed toxicity-a powerful anti-cancer mechanism with implications for other diseases. *Journal of Experimental & Clinical Cancer Research*. 2021;40(1):389–412.

80. Patel M, Bartom ET, Paudel B, et al. Identification of the toxic 6mer seed consensus for human cancer cells. *Scientific Reports*. 2022;12(1):5130.

81. Putzbach W, Gao QQ, Patel M, et al. Many si/shRNAs can kill cancer cells by targeting multiple survival genes through an off-target mechanism. *Elife*. 2017;6:e29702.

82. Taniguchi H, Suzuki Y, Imai K, et al. Antitumoral RNA-targeted oligonucleotide therapeutics: The third pillar after small molecule inhibitors and antibodies. *Cancer Science*. 2022;113(9):2952–2961.

83. Dhuri K, Bechtold C, Quijano E, et al. Antisense oligonucleotides: An emerging area in drug discovery and development. *Journal of Clinical Medicine*. 2020;9(6):2004.

84. Gagliardi M, Ashizawa AT. The challenges and strategies of antisense oligonucleotide drug delivery. *Biomedicines*. 2021;9(4):433.

85. Çakan E, Lara OD, Szymanowska A, et al. Therapeutic antisense oligonucleotides in oncology: From bench to bedside. *Cancers*. 2024;16(17):2940.

86. Singh NN, Luo D, Singh RN. Pre-mRNA Splicing Modulation by Antisense Oligonucleotides. *Methods in Molecular Biology*. 2018;1828:415–437.

87. Lauffer M, van Roon-Mom W, Aartsma-Rus A. Possibilities and limitations of antisense oligonucleotide therapies for the treatment of monogenic disorders. *Communications Medicine*. 2024;4(1):6.

88. Morelli E, Biamonte L, Federico C, et al. Therapeutic vulnerability of multiple myeloma to MIR17PTi, a first-in-class inhibitor of pri-miR-17–92. *Blood*. 2018;132(10):1050–1063.

89. Mahmoudian F, Ahmari A, Shabani S, et al. Aptamers as an approach to targeted cancer therapy. *Cancer Cell International*. 2024;24(1):108.

90. Khan S, Hussain A, Fahimi H, et al. A review on the therapeutic applications of aptamers and aptamer-conjugated nanoparticles in cancer, inflammatory and viral diseases. *Arabian Journal of Chemistry*. 2022;15(2):103626.

91. Xie X, Zhang Y, Ma W, et al. Potent anti-angiogenesis and anti-tumour activity of pegaptanib-loaded tetrahedral DNA nanostructure. *Cell Proliferation*. 2019;52(5):e12662.

92. Cox A, Lim SA, Chung EJ. Strategies to deliver RNA by nanoparticles for therapeutic potential. *Molecular Aspects of Medicine*. 2022;83:100991.

93. Houseley J, Tollervey D. The many pathways of RNA degradation. *Cell*. 2009;136(4):763–776.

94. Teodori L, Omer M, Kjems J. RNA nanostructures for targeted drug delivery and imaging. *RNA Biology*. 2024;21(1):391–409.

95. Crooke ST, Vickers TA, Liang X-H. Phosphorothioate modified oligonucleotide–protein interactions. *Nucleic Acids Research*. 2020;48(10):5235–5253.

96. Zhu Y, Zhu L, Wang X, et al. RNA-based therapeutics: An overview and prospectus. *Cell Death & Disease*. 2022;13(7):644.

97. Mainini F, Eccles MR. Lipid and polymer-based nanoparticle siRNA delivery systems for cancer therapy. *Molecules*. 2020;25(11):2692.

98. Mehta MJ, Kim HJ, Lim SB, et al. Recent progress in the endosomal escape mechanism and chemical structures of polycations for nucleic acid delivery. *Macromolecular Bioscience*. 2024;24(4):2300366.

99. Gustà MF, Edel MJ, Salazar VA, et al. Exploiting endocytosis for transfection of mRNA for cytoplasmatic delivery using cationic gold nanoparticles [Original Research]. *Frontiers in Immunology*. 2023;14:1128582.

100. Leng Q, Woodle MC, Mixson AJ. Targeted delivery of siRNA therapeutics to malignant tumors. *Journal of Drug Delivery*. 2017;2017(1):6971297.

101. Ireson CR, Kelland LR. Discovery and development of anticancer aptamers. *Molecular Cancer Therapeutics*. 2006;5(12):2957–2962.

102. Panigaj M, Reiser J. Aptamer guided delivery of nucleic acid-based nanoparticles. *DNA and RNA Nanotechnology*. 2016;2(1):42–52.

103. Zhou J, Swiderski P, Li H, et al. Selection, characterization and application of new RNA HIV gp 120 aptamers for facile delivery of Dicer substrate siRNAs into HIV infected cells. *Nucleic Acids Research*. 2009;37(9):3094–3109.

104. McNamara J, Andrechek E, Wang Y, et al. Cell type–specific delivery of siRNAs with aptamer-siRNA chimeras. *Nature Biotechnology*. 2006;24(8):1005–1015.

105. Thiel K, Hernandez L, Dassie J, et al. Delivery of chemo-sensitizing siRNAs to HER2+-breast cancer cells using RNA aptamers. *Nucleic Acids Research*. 2012;40(13):6319–6337.

106. Chu TC. Aptamer mediated siRNA delivery. *Nucleic Acids Research*. 2006;34(10):e73.

107. Xie S, Sun W, Fu T, et al. Aptamer-based targeted delivery of functional nucleic acids. *Journal of the American Chemical Society*. 2023;145(14):7677–7691.

108. Aaldering LJ, Tayeb H, Krishnan S, et al. Smart functional nucleic acid chimeras: Enabling tissue specific RNA targeting therapy. *RNA Biology*. 2015;12(4):412–425.

109. Zhou J, Rossi JJ. Aptamer-targeted cell-specific RNA interference. *Silence*. 2010;1(1):4.

110. Hao Y, Yang J, Liu D, et al. Construction of aptamer-siRNA chimera and glutamine modified carboxymethyl-β-cyclodextrin nanoparticles for the combination therapy against lung squamous cell carcinoma. *Biomedicine & Pharmacotherapy*. 2024;174:116506.

111. Jeong H, Lee SH, Hwang Y, et al. Multivalent aptamer–RNA conjugates for simple and efficient delivery of doxorubicin/siRNA into multidrug-resistant cells. *Macromolecular Bioscience*. 2017;17(4):1600343.

112. Bhardwaj H, Jangde R. Current updated review on preparation of polymeric nanoparticles for drug delivery and biomedical applications. *Next Nanotechnology*. 2023;2:100013.

113. Huang J, Ali S. PLGA – A versatile copolymer for design & development of nanoparticles for drug delivery. *Drug Development & Delivery*. 2023;12(2):72–78.

114. Elmowafy EM, Tiboni M, Soliman ME. Biocompatibility, biodegradation and biomedical applications of poly(lactic acid)/poly(lactic-co-glycolic acid) micro and nanoparticles. *Journal of Pharmaceutical Investigation*. 2019;49(4):347–380.

115. Kapadia CH, Ioele SA, Day ES. Layer-by-layer assembled PLGA nanoparticles carrying miR-34a cargo inhibit the proliferation and cell cycle progression of triple-negative breast cancer cells. *Journal of Biomedical Materials Research Part A*. 2020;108(3):601–613.

116. Risnayanti C, Jang YS, Lee J, et al. PLGA nanoparticles co-delivering MDR1 and BCL2 siRNA for overcoming resistance of paclitaxel and cisplatin in recurrent or advanced ovarian cancer. *Sci Rep*. 2018;8(1):7498.

117. Camorani S, Tortorella S, Agnello L, et al. Aptamer-functionalized nanoparticles mediate PD-L1 siRNA delivery for effective gene silencing in triple-negative breast cancer cells. *Pharmaceutics*. 2022;14(10):2225.

118. Sanchez Armengol E, Unterweger A, Laffleur F. PEGylated drug delivery systems in the pharmaceutical field: Past, present and future perspective. *Drug Development and Industrial Pharmacy*. 2022;48(4):129–139.

119. Liu X, Dong C, Ma S, et al. Nanocomplexes loaded with miR-128-3p for enhancing chemotherapy effect of colorectal cancer through dual-targeting silence the activity of PI3K/AKT and MEK/ERK pathway. *Drug Delivery*. 2020;27(1):323–333.

120. Zhang T, Xue X, He D, et al. A prostate cancer-targeted polyarginine-disulfide linked PEI nanocarrier for delivery of microRNA. *Cancer Letters*. 2015;365(2):156–165.

121. Zalba S, ten Hagen TM, Burgui C, et al. Stealth nanoparticles in oncology: Facing the PEG dilemma. *Journal of Controlled Release*. 2022;351:22–36.

122. Hong L, Wang Z, Wei X, et al. Antibodies against polyethylene glycol in human blood: A literature review. *Journal of Pharmacological and Toxicological Methods*. 2020;102:106678.

123. Liu Y, Li Y, Keskin D, et al. Poly(β-amino esters): Synthesis, formulations, and their biomedical applications. *Advanced Healthcare Materials*. 2019;8(2):1801359.

124. Karlsson J, Rhodes KR, Green JJ, et al. Poly(beta-amino ester)s as gene delivery vehicles: Challenges and opportunities. *Expert Opinion in Drug Delivery*. 2020;17(10):1395–1410.

125. Dosta P, Ramos V, Borrós S. Stable and efficient generation of poly(β-amino ester)s for RNAi delivery. *Molecular Systems Design and Engineering*; 2018;3(4):677–689.

126. Kozielski KL, Tzeng SY, Hurtado De Mendoza BA, et al. Bioreducible cationic polymer-based nanoparticles for efficient and environmentally triggered cytoplasmic siRNA delivery to primary human brain cancer cells. *ACS Nano*. 2014;8(4):3232–3241.

127. Grewal AK, Salar RK. Chitosan nanoparticle delivery systems: An effective approach to enhancing efficacy and safety of anticancer drugs. *Nano TransMed*. 2024;3:100040.

128. Zhang W, Xu W, Lan Y, et al. Antitumor effect of hyaluronic-acid-modified chitosan nanoparticles loaded with siRNA for targeted therapy for non-small cell lung cancer. *International Journal of Nanomedicine*. 2019;14:5287–5301.

129. Masjedi A, Ahmadi A, Atyabi F, et al. Silencing of IL-6 and STAT3 by siRNA loaded hyaluronate-N, N,N-trimethyl chitosan nanoparticles potently reduces cancer cell progression. *International Journal of Biological Macromolecules*. 2020;149:487–500.

130. Bastaki S, Aravindhan S, Ahmadpour Saheb N, et al. Codelivery of STAT3 and PD-L1 siRNA by hyaluronate-TAT trimethyl/thiolated chitosan nanoparticles suppresses cancer progression in tumor-bearing mice. *Life Sciences*. 2021;266:118847.

131. Wan Yusof WR, Awang NYF, Azhar Laile MA, et al. Chemically modified water-soluble chitosan derivatives: Modification strategies, biological activities, and applications. *Polymer-Plastics Technology and Materials*. 2023;62(16):2182–2220.

132. Amaldoss M, Yang J-L, Koshy P, et al. Inorganic nanoparticle-based advanced cancer therapies: Promising combination strategies. *Drug Discovery Today*. 2022;27(12):103386.

133. Gomathi AC, Xavier Rajarathinam SR, Mohammed Sadiq A, et al. Anticancer activity of silver nanoparticles synthesized using aqueous fruit shell extract of *Tamarindus indica* on MCF-7 human breast cancer cell line. *Journal of Drug Delivery Science and Technology*. 2020;55:101376.

134. Chithrani DB, Jelveh S, Jalali F, et al. Gold nanoparticles as radiation sensitizers in cancer therapy. *Radiation Research*. 2010;173(6):719–728.

135. Zhuang M, Jiang S, Gu A, et al. Radiosensitizing effect of gold nanoparticle loaded with small interfering RNA-SP1 on lung cancer. *Translational Oncology*. 2021;14(12):101210.

136. Vines J, Yoon J-H, Ryu N-E, et al. Gold nanoparticles for photothermal cancer therapy. *Frontiers in Chemistry*. 2019;7:167.

137. Cai R, Wang M, Liu M, et al. An iRGD-conjugated photothermal therapy-responsive gold nanoparticle system carrying siCDK7 induces necroptosis and immunotherapeutic responses in lung adenocarcinoma. *Bioengineering & Translational Medicine*. 2022;8(4):e10430.

138. Kanungo A, Tripathy NS, Sahoo L, et al. Theranostic siRNA loaded mesoporous silica nanoplatforms: A game changer in gene therapy for cancer treatment. *OpenNano*. 2024;15:100195.

139. Zhang B-C, Luo B-Y, Zou J-J, et al. Co-delivery of sorafenib and CRISPR/Cas9 based on targeted core–shell hollow mesoporous organosilica nanoparticles for synergistic HCC therapy. *ACS Applied Materials & Interfaces*. 2020;12(51):57362–57372.

140. Wang Y, Xie Y, Kilchrist KV, et al. Endosomolytic and tumor-penetrating mesoporous silica nanoparticles for siRNA/miRNA combination cancer therapy. *ACS Applied Materials and Interfaces*. 2020;12(4):4308–4322.

141. Jung HN, Lee SY, Lee S, et al. Lipid nanoparticles for delivery of RNA therapeutics: Current status and the role of in vivo imaging. *Theranostics*. 2022;12(17):7509–7531.

142. Xue HY, Guo P, Wen WC, et al. Lipid-based nanocarriers for RNA delivery. *Current Pharmaceutical Design*. 2015;21(22):3140–3147.

143. Zhang R, Tang L, Tian Y, et al. DP7-C-modified liposomes enhance immune responses and the antitumor effect of a neoantigen-based mRNA vaccine. *Journal of Controlled Release*. 2020;328:210–221.

144. Zhang X, Hai L, Gao Y, et al. Lipid nanomaterials-based RNA therapy and cancer treatment. *Acta Pharmaceutica Sinica B*. 2023;13(3):903–915.

145. Gao Y, Men K, Pan C, et al. Functionalized DMP-039 hybrid nanoparticle as a novel mRNA vector for efficient cancer suicide gene therapy. *International Journal of Nanomedicine*. 2021;16:5211–5232.

146. Uren AG, Kool J, Berns A, et al. Retroviral insertional mutagenesis: Past, present and future. *Oncogene*. 2005;24(52):7656–7672.

147. Luo Q, Song H, Deng X, et al. A triple-regulated oncolytic adenovirus carrying microRNA-143 exhibits potent antitumor efficacy in colorectal cancer. *Molecular Therapy Oncolytics*. 2020;16:219–229.

148. Chen X, Zhou Y, Wang J, et al. Dual silencing of Bcl-2 and Survivin by HSV-1 vector shows better antitumor efficacy in higher PKR phosphorylation tumor cells in vitro and in vivo. *Cancer Gene Therapy*. 2015;22(8):380–386.

149. Ottaviano G, Georgiadis C, Gkazi SA, et al. Phase 1 clinical trial of CRISPR-engineered CAR19 universal T cells for treatment of children with refractory B cell leukemia. *Science Translational Medicine*. 2022;14(668):eabq3010.

150. Weber JS, Carlino MS, Khattak A, et al. Individualised neoantigen therapy mRNA-4157 (V940) plus pembrolizumab versus pembrolizumab monotherapy in resected melanoma (KEYNOTE-942): A randomised, phase 2b study. *The Lancet*. 2024;403(10427):632–644.

151. Beg MS, Brenner AJ, Sachdev J, et al. Phase I study of MRX34, a liposomal miR-34a mimic, administered twice weekly in patients with advanced solid tumors. *Investigational New Drugs*. 2017;35(2):180–188.

152. Kotecki N, Opdam F, Robbrecht D, et al. Phase I/Ib study with INT-1B3, a novel LNP-formulated micro-RNA (miR-193a-3p mimic) therapeutic for patients with advanced solid cancer. *Journal of Clinical Oncology*. 2021;39(15_suppl):TPS2666.
153. van Zandwijk N, Pavlakis N, Kao SC, et al. Safety and activity of microRNA-loaded minicells in patients with recurrent malignant pleural mesothelioma: A first-in-man, phase 1, open-label, dose-escalation study. *The Lancet Oncology*. 2017;18(10):1386–1396.
154. Davis ME, Zuckerman JE, Choi CH, et al. Evidence of RNAi in humans from systemically administered siRNA via targeted nanoparticles. *Nature*. 2010;464(7291):1067–1070.
155. Giordano FA, Layer JP, Leonardelli S, et al. L-RNA aptamer-based CXCL12 inhibition combined with radiotherapy in newly-diagnosed glioblastoma: Dose escalation of the phase I/II GLORIA trial. *Nature Communications*. 2024;15(1):4210.

9 Chimeric Antigen Receptor (CAR)-T Cell Therapy

State of the Edge and the Recent Strategies Using CRISPR and mRNA

Abdellatif Bouazzaoui

9.1 DEFINITION AND BACKGROUND

The increased cancer cases and their high mortality have become a worldwide problem that requires continuous development of new efficacious therapeutic methods to treat this disease. CAR-T cell therapy (CAR-T therapy) is considered the newest and most powerful anti-cancer treatment. Cancer cells have high heterogeneity and multiple lineages; however, they possess some common antigens like CD19, CD20, and CD22. Recently, new CAR-T cells have been developed that can target specific antigens on cancer cells [1]. The reprogramed CAR-T cells have the ability to recognize and target specific antigens and kill cancer cells. This strategy is termed immunotherapy, cancer therapy, or gene therapy [2], and it was used for the first time in 1989 [3,4]. Since that time, this therapy has been developed into the fifth generation (Figure 9.1). The human immune system (HIS) is able to identify cancer cells via the presence of foreign antigens [5]. However, cancer cells also developed strategies to escape the HIS, leading to tumor survival and progression [6]. In the previous years, CAR-T cell therapy was useful against hematological malignancies cells including multiple myeloma (MM), lymphoma, and leukemia. Until now, there have been six products approved in the USA (Table 9.1) targeting either B cell mature antigen or CD19 [7]. Furthermore, a growing number of clinical trials should be registered or recruited annually (https://clinicaltrials.gov/). Recently, CD19 CAR-T cell therapy has also been applied as treatment for patients with systemic lupus erythematosus [8]. But the CAR-T cell therapy still has problems against solid tumors. However, hope always remains, hence this technology is constantly improved and developed.

9.2 THE SUCCESSIVE GENERATIONS OF CAR-T CELLS

Since the first CAR-T cell application in 1989, the CAR technology has been developed successively until it reached the fifth generation (5th gener) (Figure 9.1). The structure of CAR contains mainly the region for the binding of tumor-associated antigen (TAA), an extracellular domain, transmembrane domain (TM), and an intracellular domain with immunoreceptor tyrosine-based activation motif (ITAM). The **1st-gener CAR-T** cell has a single-chain variable fragment (scFv), a TM, and an intracellular signaling segment, which is relatively simple and consists of three pairs of ITAM region from the CD3ζ chain as a signal's transmitter without costimulatory molecules. These 1st-gener CAR-T cells release less cytokine and cause only transient T cell proliferation; for this reason, they cannot cause a sustained anti-tumor effect. Hence, the signal transmission inefficiency, different studies showed no significant effects [9,10]. In the **2nd gener**, another costimulatory molecule like CD28, TNFRSF9 (TNF receptor superfamily member9 (previously

DOI: 10.1201/9781003517870-9

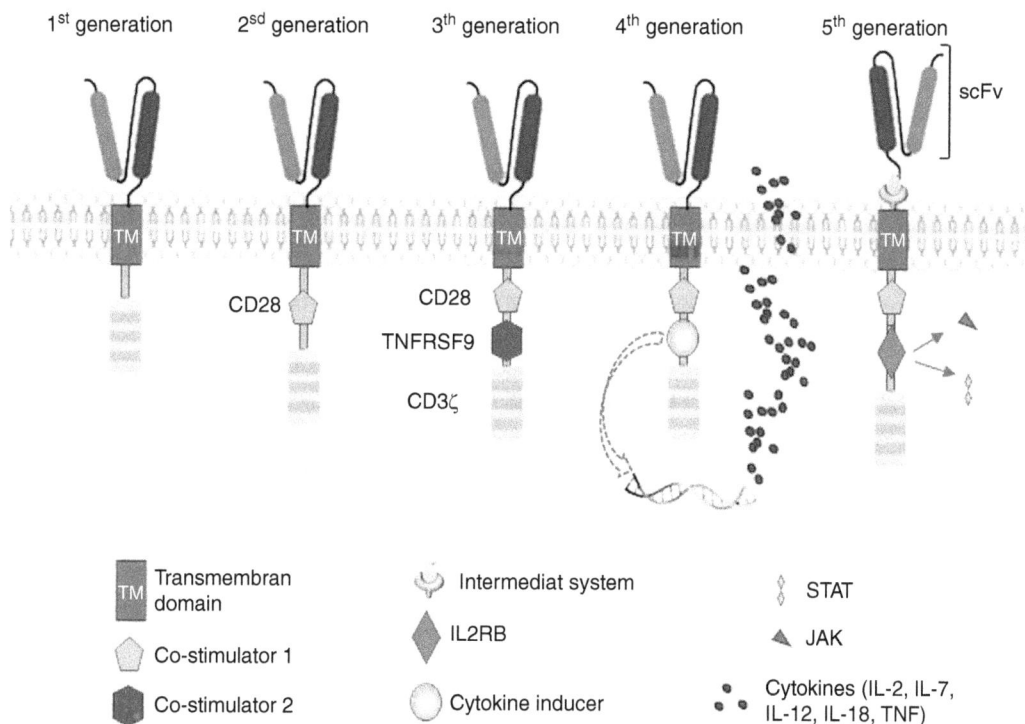

FIGURE 9.1 This figure illustrates CAR-T cell construct evolution. It shows: 1st gen: scFv, TM, CD3ζ. 2nd gen: adds CD28 or 4-1BB. 3rd gen: adds both CD28 & 4-1BB. 4th gen: includes cytokine receptor gene. 5th gen: inserts an intermediate molecule between scFv and TM, plus IL2RB between CD28 and CD3ζ for JAK/STAT activation. Each generation builds on the previous, enhancing CAR-T cell function via added signaling domains.

TABLE 9.1

The FDA-Approved CAR-T Cells Using Retro- or Lentiviral Vector for Transgene Integration

Commercial Name	Product Name	FDA Approval	Targeted Antigen	Cell Starting Population	References
Kymriah®	Tisagenlecleucel	2017	CD19	Enriched T cells	[136,137]
Yescarta®	Axicabtagene ciloleucel	2017	CD19	Ficol gradient-enriched PBMCs	[138,139]
Tecartus®	Brexucabtagene autoleucel	2020	CD19	CD19 depl. and CD4+/CD8+ enriched T cells	[140,141]
Breyanzi®	Lisocabtagene maraleucel	2021	CD19	CD8+ and CD4+ separatly	[142–144]
Abecma®	Idecabtagene vicleucel	2021	BCMA	PBMCs	[145,146]
Carvykti®	Ciltacabtagene autoleucel (cilta-cel)	2022	BCMA	Enriched T cells	[123,147]

Abbreviation: PBMCs, peripheral blood mononuclear cells; TNFRSF17, TNF receptor superfamily member 17; BCMA, B cell maturation antigen.

CD137) or TNFRSF4 (previously CD134) was attached to the CD3ζ domain of the CAR-T cell [11]. These costimulatory molecules induce NF-kappa B complex subunits (NFκB), Ephrin receptor B2 (EPHB2, previously ERK), and MAPK8 (mitogen-activated protein kinase 8, previously JNK) signaling pathways, which resulted in a more substantial killing effect [12]. In the **3th-gener of** CAR-T cells, two signaling domains such as CD28-TNFRSF4, CD28-TNFRSF9, or CD28-ICOS (inducible T cell costimulatory, previously CD278) were added beside the CD3ζ domain in order to improve the activation, extend the proliferation, increase the production of cytokines and finally extend the antitumor efficacy [13,14]. The **4th gener** of CAR-T cells was based on previous generations (2nd and 3rd), and an inducible transgenic immune modifier, such as interleukin (IL)-12, was integrated into CAR-T cells to increase the killing capacity of the CAR-T cells. After activation of the CAR-T cells, a large amount of IL-12 will be produced, which recruits different immune cells to kill the cancer cells [15,16]. Other chemokine or cytokine receptors were added to increase the killing effects in solid tumors [17–19]. In **the 5th gener of** CAR-T cells (also called Universal CAR-T), an intermediate system between the scFv and the transmembrane domain has been added in the extracellular domain. Furthermore, an IL2RB (interleukin 2 receptor subunit beta) fragment has been added between CD28 and CD3ζ domains, which induces the activation of the JAK/STAT pathway (Janus kinase/signal transducers and activators of transcription), resulting in high proliferation of the CAR-T cells [20].

9.3   STEPS FOR THE GENERATION OF CAR-T CELLS

The first step in the regeneration of CAR-T cells is the isolation of the peripheral blood from the patients (leukapheresis) [21,22]. After that, the process involves the activation, modification of genetic material, expansion, and finally the product formulation and testing.

9.3.1   LEUKAPHERESIS

This method is widely used by blood banks to collect different blood components and platelets for the treatment of renal, hematological diseases, and is considered a safe treatment [23].

9.3.2   STARTING CELL POPULATION

The selection of the starting cell population to generate CAR-T cells is very important and has a high influence on the production and the end product of CAR-T cells. After isolation of naïve T cells (Tn-cells) from peripheral blood mononuclear cells (PBMC), they could be differentiated into different cells such as exhausted (T_{EXH}), effector (T_E), effector memory (T_{EM}), central memory (T_{CM}), and stem-cell memory (T_{SCM}) cells. Furthermore, T-cells could differentiate into CD4+ helper cells (including follicular helper cells, Th1, Th2, and Th17 cells), CD8+ cytotoxic T cells, and T regulatory cells (Treg cells). The diversity in the T cell population makes the choice of the starting population to manufacture CAR-T cells a very important decision [24].

9.3.3   CELL ISOLATION

The magnetic beads or cell sorting methods are the most used techniques for the isolation of cells. The cells could be isolated by positive selection or by depletion of non-T cells. The isolated cells such as CD4+ and CD8+ could be cultured separately, which allowed the formulation of exact final ratios of the cells. Alternatively, after the isolation of the cells such as CD4+ and CD8+ cells at the beginning, these cells could be cocultured at a defined starting ratio in order to reach the desired ratio at the end of the incubation time [25].

9.3.4 T-Cell Differentiation Status

Besides the ratio of the starting cells, there are other factors that influence the antitumor efficacy such as the differentiation status of the T cells. After the stimulation with the specific antigen, the T cells begin subset differentiation into T_E cells and start a clonal expansion. The T_E cells have cytotoxic activity and release cytokines; however, they have a relatively short life. Another subset of the activated cells differentiated into memory T cells such as T_{EM}, $T_{CM,}$ and $T_{SCM,}$ which are long-living cells, have a fast and quicker response after a second antigen stimulation, and produce a lower level of cytokines, resulting in a long anti-tumor response without acute toxicities [26–28].

9.3.5 T-Cell Activation Methods

The activation method is also another factor affecting the production of CAR-T cells, which also affects the ex vivo expansion of T-cells. Commonly, the activation of T-cells could be achieved by the stimulation of CD3ζ, CD28, and other co-stimulators such as TNFRSF4, TNFRSF9, or ICOS. Moreover, a cocktail of cytokines and chemokines is also expressed to support the T-cell expansion [29,30].

9.3.6 Introducing the CAR Transgene

For the delivery of the transgene, there are different methods, which we will discuss in the following section (transgene delivery methods). The delivery method could influence the level of gene expression in CAR-T cells, the genotoxicity, as well as the safety and efficacy of the cells.

9.3.7 Ex Vivo Cell Expansion

After alteration, CAR-T cells must be amplified for 1–2 weeks until reaching the number required for administration into the patient. For the growth and expansion of CAR-T cells, other factors like IL-15, IL-7, and IL-2 must be added [29,30]. The addition of selected growth factors allowed for the enrichment of CAR-T cells, whereas non-T cells were eliminated. Furthermore, other supplements such as fetal bovine serum (FBS), tyrosine kinase inhibitor dasatinib should be added to increase the functionality and therapeutic effect of CAR-T cells [31].

9.3.8 Product Testing and Treatments

Since the pioneer treatment in 2017 using CAR-T cells, there are more than 1000 CAR-T cells in clinical trials worldwide [32], making the CAR-T cells the most important treatment against tumor. As any other proposed clinical therapy, CAR-T cells must meet defined characteristics such as **Safety** (include test for endotoxin, mycoplasma, residual viral-agent, viability, sterility, and transgene quantification); **Purity** (include viability, T-cell purity, presence of contaminating tumor cells and residual reagent and the transduction efficiency); **Potency** (include the cytotoxicity and cytokine production); **Identity** (include viability, clarity and dose); and **Stability** (include shipping, formulation and storage).

9.4 TRANSGENE DELIVERY METHODS

Transgene transfer in T-cells could be stable or transient depending on the integration or nonintegration of the genetic material into the genome of the host cell. For a successful expression of the transgene, the expression cassette must include a promoter, leader, transgene, and a polyadenylation sequence.

9.4.1 STABLE TRANSGENE TRANSFER

For a stable gene transfer, the expression cassette should integrate into the genome of the host cell. This allowed a stable and long expression of the transgene; furthermore, it will be transferred to the daughter cell during cell division. In the case of stable gene transfer, we distinguish between two methods, viral transfer and non-viral transfer.

9.4.1.1 Viral Gene Delivery

Viral gene delivery (transduction) or the use of viral vectors like the adenovirus, lentivirus or retrovirus are commonly used in gene therapy and also in CAR-T cell production. The viral gene delivery method has different advantages compared to other methods; it allows for a high integration efficiency persisting for more than 10 years after transduction [33] and a high expression of the transgene, but also has critical drawbacks such as the limited insert size and difficulties in producing high titers of vector particles. Retroviruses are generally derived from murine leukemia virus (MLV), and they integrate more frequently into promoters, in transcriptional regulatory regions, near transcription start sites (TSS), and in exon regions, whereas the lentiviral vectors integrate into active regions or exons [34–36]. Integration in exons, promotor, or regulatory regions could result in higher impact on mRNA expression [37]. However, for the lentiviral vector, the transgene integration could result in disruption of the genes, activation of oncogenes, or the gene fragments expression which leads to inactivation of tumor-suppressor genes [38].

9.4.1.2 Non-Viral Gene Delivery

Transposons are an alternative way for stable transfer of the transgene, and they include two main methods, named sleeping beauty (SB) and piggyBac (PB) [39]. The majority of transposons have element encoding for the transposase enzyme and are flanked by inverted terminal repeats (ITRs). Transposons including SB and PB have been evaluated previously [40–42], and they use a cut-and-paste mechanism to move from one place to another randomly. The insertion of the transgene by this means is not specific; therefore, it can integrate into transcriptional region sites similar to retro- or lentiviral vectors [43], which raises mutagenesis and could affect the gene expression. Therefore, the use of transposons must be studied more deeply [44]. Interestingly, the size of the transposons is very big; this important feature can address some disadvantages of the viral methods like the limited length of the transgene.

9.4.2 NON-STABLE GENE TRANSFER USING TRANSFECTION

As an alternative for the stable integration of transgene in the host cell, the introduction of naked nucleic acids such as DNA, small interfering RNA (siRNA), messenger RNA (mRNA), and single-guide RNA (sgRNA) can be used in combination with a transfection reagent like nanoparticles or nanocarriers. Especially, the transfection of mRNA for a transient expression of the transgene has high importance. Hence, the plasmid vectors and mRNA are not integrated into the cell's genome, and then successive divisions lead to the loss of the transferred genes; therefore, the expression will be lost. But since the patients receive multiple doses of CAR-T cells, we can assume that the use of mRNA has more positive than negative characteristics. Recently, mRNA technology in combination with different transfection methods has been used extensively as a vaccine against coronavirus disease 2019 (COVID-19) and reached impressive clinical results; also, the use of sgRNA in CRISPR for the exact and precise genome modification is considered an important new option for next-generation CAR-T cell production.

9.4.2.1 Utilization of mRNA as a Transgene Vector

The introduction of mRNA as an alternative for the stable transgene integration in the process of CAR-T cells production is based on recent as well as early applications. More than 30 years ago, Malone et al. showed that mRNA in combination with lipids can be transported into the cells. Without the need to move into the nucleus, the mRNA can be translated into protein [45] and therefore, mRNA could be used as a drug. Since that time, the mRNA has emerged as an important therapeutic reagent [46,47] and has been used to knock out genes, change the cell phenotype, restore the expression of genes, or encode antigens [48,49]. Also, in the production of CAR-T cells and to use a stable transgene integration, previous work used mRNA for a transient expression of the transgene [50]. The transient expression has many advantages compared to the stable integration, besides the elimination of mutagenesis, the transient transfection reduces the toxicity [51] caused by high expression of the transgene. Generally, the transgene expression declines each day after transfection and lasts for about 1 week only [50,52]. However, it is possible to stabilize the mRNA if a long expression is needed. Recently, the mRNA technology has been used extensively for the production of vaccines and reached exceptional successes [53,54]. Thanks to the severe acute respiratory syndrome coronavirus 2 (SARS-CoV-2) vaccine research, it was possible to develop different protocols for clinical translation of the mRNA technology.

9.4.2.2 Techniques Used for the Transfection

Because of the negative charge of naked DNA and RNA, it is not possible for nucleic acids to pass through the cell membrane. To reach the cytoplasm, a transfection method is needed. This could be physical, using electroporation [55], which disrupts the cell membranes and causes pores, or chemical, which causes endocytosis of mRNA in the cell. mRNAs have been used in vitro and in vivo to introduce the transgene in CAR-T cells for testing in chronic lymphocytic leukemia (CLL), hematological tumors, acute myeloid leukemia (AML), acute lymphoblastic leukemia (ALL), and solid tumors. The methods for transfection were electroporation or nanoparticles [56]. cationic polymers [57] or cationic lipids [58]. In a previous study, Smith et al. showed that polymeric nanocarriers are able to deliver a transgene to leukemia-specific CAR-T cells [59]. In a recent work, the author used polymeric nanoparticles to introduce a transgene into CAR-T cells [60]. The use of mRNA in combination with different transfection methods for the delivery of a transgene in CAR-T cells has proved to be more advantageous to replace viral vectors [61].

9.5 APPLICATION OF CRISPR/Cas9 TECHNOLOGY

Genome editing including DNA modification, deletion, or replacement could be conducted using different tools such as zinc finger nucleases (ZFNs), transcriptional activator-like effector nucleases (TALENs), and CRISPR-associated nuclease 9 (Cas9) system [62]. The CRISPR system has been found first time in prokaryotes as an immune system against phage infections [63–65]. For the TALENs and ZFNs, both tools used proteins to target the DNA modification sites; however, CRISPR/Cas9 used a single-guided RNA (sgRNA). This is about 20 nucleotides long and complementary to the target DNA, which leads to directing the Cas9 proteins (endonuclease enzyme) to the specific site of DNA [66]. The sgRNA includes a protospacer adjacent motif (PAM) sequence, which is important for the recognition of the target DNA by sgRNA and Cas9 enzyme [67,68]. After binding, the complete system including sgRNA, CRISPR, and Cas9 caused double-stranded breaks (DSBs) in the target DNA sites (Figure 9.2), allowing us to conduct genome modifications with high precision and efficiency [69]. The CRISPR/Cas9 system is considered as an efficient and simple method for precise gene modifications. For the introduction of CRISPR/Cas9, besides the introduction of plasmid DNA coding for sgRNA and Cas9 enzyme, the other safer and more efficient method is the combination of sgRNA and mRNA for Cas9 enzyme. The CRISPR/Cas9 strategy was used for the first time in the year 2012 to modify genes in prokaryotes [70]. Later, the group of Zhang Feng et al. and others used the CRISPR/Cas9 technology in mammalian cells [71–73] and also in the engineering and improvement of the antitumor activity, safety, and persistence of CAR-T cells [74,75]. The use of CAR-T cell therapy has improved significantly in the last decade. However, their effect remains unsatisfying in solid tumors and hematologic malignancy, as the CAR-T cells

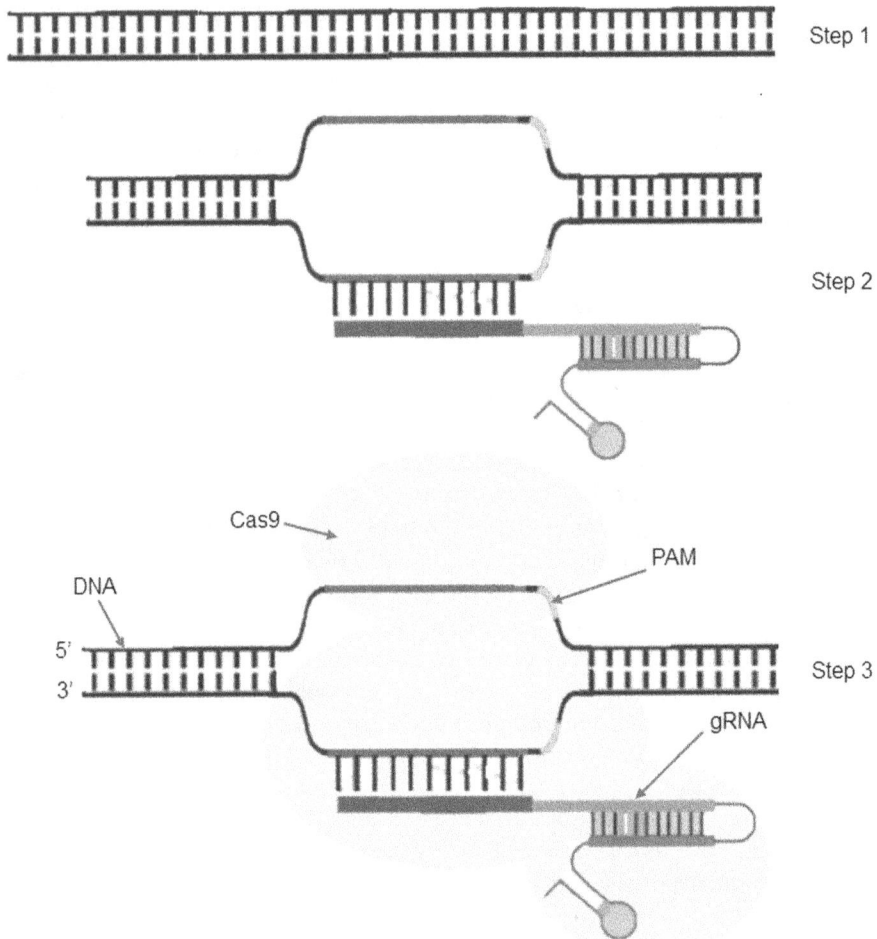

FIGURE 9.2 This figure depicts the CRISPR/Cas9 system. It shows the PAM sequence, crucial for sgRNA and Cas9 enzyme binding to target DNA. Once bound, Cas9 creates double-strand breaks (DSBs) in the DNA. This process enables precise and efficient genome modifications.

show poor expansion and persistence after infusions. To improve the technology of the CAR-T cells, the CRISPR/Cas9 techniques was used effectively.

9.5.1 Prevention of CAR-T Cell Exhaustion

The state of dysfunction and hypo-response is named T-cell exhaustion, which leads to low anti-tumor activity due to the less proliferation of the cells. The reason for this decrease in activity is the expression of the inhibitory receptors like lymphocyte activating 3 (LAG3), hepatitis A virus cellular receptor 2 (HAVCR2, previously TIM3), programmed cell death 1 (PDCD1, previously PD-1), and the cytotoxic T-lymphocyte associated protein 4 (CTLA4). Previous studies find that the binding of PDCD1 to its ligand, which inhibits the activation of T cells and reduces the cytokines (interferon-γ (IFN-γ), IL-2, and TNF-α) production [76], leads to T cell exhaustion [77,78]. However, the exact mechanisms causing T-cell exhaustion are not completely understood. One method to prevent CAR-T cell exhaustion is the disruption of the immune checkpoint using a CRISPR/Cas9 strategy [79–81]. Other studies inhibit the negative regulators like SRY-box transcription factor 4 (SOX4), inhibitor of DNA binding 3 (ID3), nuclear receptor subfamily 4 group A (NR4A) [82,83], protein tyrosine phosphatase non-receptor type 1 (PTPN1) [84], and tumor Cbl proto-oncogene B (CBLB) [85], which results in a better inhibition of the tumor cells.

9.5.2 Increasing Resistance to Cytokines

Immunosuppressive cells and tumors secrete different cytokines like transforming growth factor beta superfamily (TGF-β), IL-10, and IL-6 in the tumor microenvironment, which cause the inhibition and dysfunction of CAR-T cells [86,87]. TGF-β is very important in the inhibition of CD8+ cytotoxic T cells. The elimination of its receptor II (TGFβR2) in CAR-T cells using CRISPR/Cas9 increases the central memory and effector memory subsets and prevents exhaustion, which resulted in better killing activity and long-term efficacy against tumors [88,89]. Another protein influencing the CAR-T activity is the cytokine-inducible SH2-containing protein (CISH). The elimination of this protein reduced the expression of PDCD1, preventing the exhaustion of CAR-T cells [90]. Furthermore, it improves the cytokine release and prolongs the survival and the anti-tumor activity of CAR-T cells.

9.5.3 Epigenetic Reprogramming Strategies

Another strategy to prevent exhaustion of CAR-T cells is the deletion of Tet methylcytosine dioxygenase 2 (TET2). Deleting the TET gene resulted in hyperproliferation of CAR-T cells [91] and caused a central memory phenotype and an epigenetic profile, leading to a prolonged cancer remission [92]. In the same context, PR/SET domain 1 (PRDM1) protein plays an important role in the epigenetic regulation of CAR-T cells during terminal differentiation. The destruction of PRDM1 using CRISPR/Cas9 causes increment in genes that regulate the permanence of CAR-T cells at the early memory stage by increasing chromatin accessibility. The remaining CAR-T cells in the early memory T cell stage improve the durability of CAR-T cells and enhance their anti-tumor activity [93]. Also, the methylation of DNA by DNA methyltransferase (DNMT) influences the epigenetics of T-cells and causes T-cell exhaustion, resulting in reduced immunotherapy of CAR-T cells [94]. Previous research showed that the inhibition of DNA methyltransferase 3 alpha (DNMT3A) results in high proliferation potential and increased anti-tumor capacity of CAR-T cells [95].

9.5.4 New Target Genes in CAR-T Cells

CRISPR/Cas9 is an effective gene editing tool, used in recent research to identify new targets and regulators in the regulation of CAR-T cell memory cell formation and exhaustion. One of the newly discovered factors is RAS p21 protein activator 2 (RASA2). In a recent study, the author found that the elimination of RASA2 increased the metabolic activity and cytokine production. This prolongs the persistence and the function of CAR-T cells, which resulted in long survival of cancer patients [96]. Further new genes enhancing the activation of CAR-T cells are basic leucine zipper ATF-like transcription factor, cyclin C [97], cBAF complex, INO80 complex ATPase subunit (INO80), sorting nexin 9 (SNX9), and mediator complex subunit 12 (MED12) [98–100]. Other genes like the lymphotoxin beta receptor [101], TLE family member 4, transcriptional corepressor (TLE4), and IKAROS family zinc finger 2 (IKZF2) [102], protein disulfide isomerase family A member (3Pdia3), alpha-1,6-mannosylglycoprotein 6-beta-N-acetylglucosaminyltransferase (Mgat5), epithelial membrane protein 1 (Emp1), and proline dehydrogenase 2 (PRODH2) [103–105] are also important genes in the activation and persistence effect of CAR-T cells.

9.6 CHALLENGES AND SAFETY ISSUES IN USING CRISPR/Cas9

Despite all the positives and advantages of CRISPR/Cas9 technology, however, there are some safety concerns and challenges. Previous studies showed that the use of CRISPR/Cas9 in cells or human embryos results in the fragmentation of chromosomes [106] or the elimination of chromosomal segments [107–109]. The CRISPR/Cas9 technology increases the expression of tumor protein p53 (TP53), which subsequently reduces the CRISPR/Cas9 activity and specificity of CRISPR/Cas9 technology, generates single-stranded breaks (SSBs), and leads to DNA damage [110–112]. The CRISPR/Cas9 technology can also induce duplications, deletions, and insertions [113]. Mismatch and mutation resulted in dysregulation of protein expression and disruption of the gene function [114], which leads to malignancy [115]. All these DNA damages, off-target effects, and immunogenicity [111,116–118] raised potential issues for CAR-T cell therapy, which should be solved.

9.7 CAR-T THERAPY, RELEVANT CASES, AND ONGOING DEVELOPMENTS

In the last years, different CAR-T cell-based products have been approved for use in clinical cases. CD19 CAR-T cells were used against B-cell leukemia [119–121]. The treatment of multiple myeloma (MM) was done using B-cell maturation antigen (BCMA) CAR-T cells [122,123]. The CAR-T cell therapy showed promising results in hematological malignancies. Moreover, there are different ongoing studies of CAR-T cells as treatment against gastrointestinal cancers [124], lung cancer [42], prostate cancer [125], liver cancer [126], glioblastoma [127–129], and breast cancer [130,131].

9.7.1 VIRAL VECTOR

The successful CAR-T cell drugs available in the market use a viral vector and include KTE-C19 (Yescarta) [132] and CTL019 (Kymriah) [133] (Table 9.1). Both therapies have been developed by Kite Pharma/Gilead Sciences and Novartis [134] and were approved by the FDA in 2017 for clinical use. The global trial reported showed a high success rate, with a 6-month survival rate in 89% of the cases and a 3-month complete remission rate in 83% of the cases [135].

9.7.2 SLEEPING BEAUTY (SB)

There are also other publications that reported the feasibility of CAR-T cells using SB methods [148,149]. The authors demonstrated that SB transposase could be delivered as mRNA or DNA in addition to the transposon plasmid using electroporation to generate functional CAR-T cells. These cells proved proof of concept; furthermore, two clinical trials (NCT00968760 and NCT01497184) demonstrated the safety of the generated CAR-T cells in persons with non-Hodgkin's Lymphoma and B-cell ALL (B-ALL) after allogeneic or autologous hematopoietic stem cell transplant (HSCT) [150].

9.7.3 PB SYSTEM

Two clinical studies (one in China and one in Japan) investigated the feasibility and safety of anti-CD19 CAR-T cells produced using the PB strategy. In the study in Japan (ID: UMIN000030984), after lymphodepletion and HSCT, three patients received autologous anti-CD19 CAR-T cells. After administration of the cells, no toxicities were shown in all patients [151]. In a recent study, nine patients were treated with anti-epidermal growth factor receptor (anti-EGFR) CAR-T, which were generated using the PB system. The study showed the persistence of circulating CAR-T cells. However, only one person presented a partial response, whereas the majority had persistent disease [42].

9.7.4 mRNA SYSTEM

In two studies, the authors generated CAR-T cells using mRNA-based technology to demonstrate the cytotoxicity and anti-tumor activity of the CAR-T cells [152,153]. The studies have shown a reduced toxicity and high safety profile due to the absence of integration in the host genome. After transfection of CAR-T cells, the expression lasts for 7 days [51].

9.8 CHALLENGES FACING CAR-T CELL THERAPY

The transfer of CAR-T cells is a powerful and promising strategy against cancer; it shows excellent clinical results, represented in long-lasting and complete responses of recurrent diseases, which are revolutionary events in the fight against malignancies and hematological diseases and open new opportunities for the treatment of solid tumors. However, the CAR-T cell strategy is still facing different challenges, especially limited persistence. Previous results showed that the persistence and proliferation of CAR-T cells in the blood circulation correlate positively with long anti-tumor response [33,154–157]. In contrast, patients with undetectable CAR-T cells in the blood relapsed [158]. The factors influencing the persistence of the CAR-T cells are exhaustion status and the differentiation phenotype [159,160]. Other challenges facing CAR-T cell therapy are the restricted trafficking, poor tumor infiltration, reduced anti-tumor activity and efficacy, life-threatening toxicities, limited proliferation capacity, and antigen escape [21,161]. In addition to the challenges, there are also side effects that appear with the use of CAR-T cell therapy. The first critical side effect is cytokine release syndrome (CRS) caused by the CAR-T cell multiplication and the secretion of cytokines to kill the tumor cells, but consequently cause CRS symptoms such as capillary leakage, low blood pressure, chills, fever, headache, nausea, fatigue, and tachycardia. Other side effects are tumor lysis syndrome, anaphylaxis, B cell aplasia, and neurological toxicity. Another important drawback in the CAR-T cell therapy is the specific targeting; on the one hand, CAR-T cells target not only cancer cells, but also the normal cells, when the antigen is present on the surface of normal cells [162]. On the other hand, the absence of specific antigens for tumor results in limited infiltration of CAR-T cells [163]. To overcome all these difficulties, there are several research projects to address these problems and challenges [164,165] using different technologies like CRISPR/Cas9 and others, as we mentioned above.

9.9 CONCLUSION

The CAR-T cell therapy is useful against hematological malignancies cells, including MM, lymphoma, leukemia, and other diseases; however, this technology is still facing different challenges such as the limited persistence and proliferation of CAR-T cells in the blood circulation, which limits the efficiency of this therapy, especially against solid tumor. To improve and understand the mechanisms of CAR-T cell therapy, different strategies have been used. One of the most used methods is the CRISPR/Cas9 technology. This technique helps to improve the CAR-T cell therapy but also presents some drawbacks and safety challenges such as the fragmentation of chromosomes, elimination of chromosomal segments, induction of duplications, deletions, and insertions, and also increase in the expression of tumor protein. Taking them all together, the CAR-T cell therapy is a very interesting technique against cancers and other diseases; however, there are more challenges that await this technology until it reaches the desired success.

REFERENCES

1. Mirzaei HR, Mirzaei H, Namdar A, et al. Predictive and therapeutic biomarkers in chimeric antigen receptor T-cell therapy: A clinical perspective. *J Cell Physiol.* 2019;234(5):5827–5841. doi: 10.1002/jcp.27519.
2. Restifo NP, Dudley ME, Rosenberg SA. Adoptive immunotherapy for cancer: Harnessing the T cell response. *Nat Rev Immunol.* 2012;12(4):269–281. doi: 10.1038/nri3191.
3. Gross G, Gorochov G, Waks T, et al. Generation of effector T cells expressing chimeric T cell receptor with antibody type-specificity. *Transplant Proc.* 1989;21(1 Pt 1):127–130.

4. Gross G, Waks T, Eshhar Z. Expression of immunoglobulin-T-cell receptor chimeric molecules as functional receptors with antibody-type specificity. *Proc Natl Acad Sci USA*. 1989;86(24):10024–10028. doi: 10.1073/pnas.86.24.10024.

5. Galluzzi L, Martin P. CARs on a highway with roadblocks. *Oncoimmunology*. 2017;6(12):e1388486. doi: 10.1080/2162402x.2017.1388486.

6. Perales MA, Kebriaei P, Kean LS, et al. Building a safer and faster CAR: Seatbelts, airbags, and CRISPR. *Biol Blood Marrow Transplant*. 2018;24(1):27–31. doi: 10.1016/j.bbmt.2017.10.017.

7. Cappell KM, Kochenderfer JN. Long-term outcomes following CAR T cell therapy: What we know so far. *Nat Rev Clin Oncol*. 2023;20(6):359–371. doi: 10.1038/s41571-023-00754-1.

8. Mackensen A, Müller F, Mougiakakos D, et al. Anti-CD19 CAR T cell therapy for refractory systemic lupus erythematosus. *Nat Med*. 2022;28(10):2124–2132. doi: 10.1038/s41591-022-02017-5.

9. Siegler EL, Wang P. Preclinical models in chimeric antigen receptor-engineered T-cell therapy. *Hum Gene Ther*. 2018;29(5):534–546. doi: 10.1089/hum.2017.243.

10. Grigor EJM, Fergusson D, Kekre N, et al. Risks and benefits of chimeric antigen receptor T-cell (CAR-T) therapy in cancer: A systematic review and meta-analysis. *Transf Med Rev*. 2019;33(2):98–110. doi: 10.1016/j.tmrv.2019.01.005.

11. Brentjens R, Yeh R, Bernal Y, et al. Treatment of chronic lymphocytic leukemia with genetically targeted autologous T cells: Case report of an unforeseen adverse event in a phase I clinical trial. *Mol Ther*. 2010;18(4):666–668. doi: 10.1038/mt.2010.31.

12. Chen L, Xie T, Wei B, et al. Current progress in CAR-T cell therapy for tumor treatment. *Oncol Lett*. 2022;24(4):358. doi: 10.3892/ol.2022.13478.

13. Till BG, Jensen MC, Wang J, et al. CD20-specific adoptive immunotherapy for lymphoma using a chimeric antigen receptor with both CD28 and 4-1BB domains: pilot clinical trial results. *Blood*. 2012;119(17):3940–3950. doi: 10.1182/blood-2011-10-387969.

14. Lock D, Mockel-Tenbrinck N, Drechsel K, et al. Automated manufacturing of potent CD20-directed chimeric antigen receptor T cells for clinical use. *Hum Gene Ther*. 2017;28(10):914–925. doi: 10.1089/hum.2017.111.

15. Kueberuwa G, Kalaitsidou M, Cheadle E, et al. CD19 CAR T cells expressing IL-12 eradicate lymphoma in fully lymphoreplete mice through induction of host immunity. *Mol Ther Oncolytics*. 2018;8:41–51. doi: 10.1016/j.omto.2017.12.003.

16. Kerkar SP, Muranski P, Kaiser A, et al. Tumor-specific CD8+ T cells expressing interleukin-12 eradicate established cancers in lymphodepleted hosts. *Cancer Res*. 2010;70(17):6725–6734. doi: 10.1158/0008-5472.Can-10-0735.

17. Abramson JS. Anti-CD19 CAR T-cell therapy for B-cell non-Hodgkin lymphoma. *Transf Med Rev*. 2020;34(1):29–33. doi: 10.1016/j.tmrv.2019.08.003.

18. Chmielewski M, Abken H. TRUCKs: The fourth generation of CARs. *Expert Opin Biol Ther*. 2015;15(8):1145–1154. doi: 10.1517/14712598.2015.1046430.

19. Kagoya Y, Tanaka S, Guo T, et al. A novel chimeric antigen receptor containing a JAK-STAT signaling domain mediates superior antitumor effects. *Nat Med*. 2018;24(3):352–359. doi: 10.1038/nm.4478.

20. Lin H, Cheng J, Mu W, et al. Advances in universal CAR-T cell therapy. *Front Immunol*. 2021;12:744823. doi: 10.3389/fimmu.2021.744823.

21. Miliotou AN, Papadopoulou LC. CAR T-cell therapy: A new era in cancer immunotherapy. *Curr Pharm Biotechnol*. 2018;19(1):5–18. doi: 10.2174/1389201019666180418095526.

22. Mirzaei HR, Jamali A, Jafarzadeh L, et al. Construction and functional characterization of a fully human anti-CD19 chimeric antigen receptor (huCAR)-expressing primary human T cells. *J Cell Physiol*. 2019;234(6):9207–9215. doi: 10.1002/jcp.27599.

23. Vormittag P, Gunn R, Ghorashian S, et al. A guide to manufacturing CAR T cell therapies. *Curr Opin Biotechnol*. 2018;53:164–181. doi: 10.1016/j.copbio.2018.01.025.

24. Golubovskaya V, Wu L. Different subsets of T cells, memory, effector functions, and CAR-T immunotherapy. *Cancers*. 2016;8(3). doi: 10.3390/cancers8030036.

25. Ayala Ceja M, Khericha M, Harris CM, et al. CAR-T cell manufacturing: Major process parameters and next-generation strategies. *J Exp Med*. 2024;221(2). doi: 10.1084/jem.20230903.

26. Chang JT, Wherry EJ, Goldrath AW. Molecular regulation of effector and memory T cell differentiation. *Nat Immunol*. 2014;15(12):1104–1115. doi: 10.1038/ni.3031.

27. Farber DL, Yudanin NA, Restifo NP. Human memory T cells: Generation, compartmentalization and homeostasis. *Nat Rev Immunol*. 2014;14(1):24–35. doi: 10.1038/nri3567.

28. Kishton RJ, Sukumar M, Restifo NP. Metabolic regulation of T cell longevity and function in tumor immunotherapy. *Cell Metabol*. 2017;26(1):94–109. doi: 10.1016/j.cmet.2017.06.016.

29. Arcangeli S, Falcone L, Camisa B, et al. Next-generation manufacturing protocols enriching T(SCM) CAR T cells can overcome disease-specific T cell defects in cancer patients. *Front Immunol.* 2020;11:1217. doi: 10.3389/fimmu.2020.01217.

30. Künkele A, Brown C, Beebe A, et al. Manufacture of chimeric antigen receptor T cells from mobilized cyropreserved peripheral blood stem cell units depends on monocyte depletion. *Biol Blood Marrow Transplant.* 2019;25(2):223–232. doi: 10.1016/j.bbmt.2018.10.004.

31. Weber EW, Parker KR, Sotillo E, et al. Transient rest restores functionality in exhausted CAR-T cells through epigenetic remodeling. *Science.* 2021;372(6537). doi: 10.1126/science.aba1786.

32. Wang V, Gauthier M, Decot V, et al. Systematic review on CAR-T cell clinical trials up to 2022: Academic center input. *Cancers.* 2023;15(4). doi: 10.3390/cancers15041003.

33. Melenhorst JJ, Chen GM, Wang M, et al. Decade-long leukaemia remissions with persistence of CD4⁺ CAR T cells. *Nature.* 2022;602(7897):503–509. doi: 10.1038/s41586-021-04390-6.

34. Schröder AR, Shinn P, Chen H, et al. HIV-1 integration in the human genome favors active genes and local hotspots. *Cell.* 2002;110(4):521–529. doi: 10.1016/s0092-8674(02)00864-4.

35. Wu X, Li Y, Crise B, et al. Transcription start regions in the human genome are favored targets for MLV integration. *Science.* 2003;300(5626):1749–1751. doi: 10.1126/science.1083413.

36. Wang GP, Levine BL, Binder GK, et al. Analysis of lentiviral vector integration in HIV+ study subjects receiving autologous infusions of gene modified CD4+ T cells. *Mol Ther.* 2009;17(5):844–850. doi: 10.1038/mt.2009.16.

37. Shao L, Shi R, Zhao Y, et al. Genome-wide profiling of retroviral DNA integration and its effect on clinical pre-infusion CAR T-cell products. *J Transl Med.* 2022;20(1):514. doi: 10.1186/s12967-022-03729-5.

38. Moretti A, Ponzo M, Nicolette CA, et al. The past, present, and future of non-viral CAR T cells. *Front Immunol.* 2022;13:867013. doi: 10.3389/fimmu.2022.867013.

39. Muñoz-López M, García-Pérez JL. DNA transposons: Nature and applications in genomics. *Curr Genomics.* 2010;11(2):115–128. doi: 10.2174/138920210790886871.

40. Monjezi R, Miskey C, Gogishvili T, et al. Enhanced CAR T-cell engineering using non-viral Sleeping Beauty transposition from minicircle vectors. *Leukemia.* 2017;31(1):186–194. doi: 10.1038/leu.2016.180.

41. Prommersberger S, Reiser M, Beckmann J, et al. CARAMBA: A first-in-human clinical trial with SLAMF7 CAR-T cells prepared by virus-free Sleeping Beauty gene transfer to treat multiple myeloma. *Gene Ther.* 2021;28(9):560–571. doi: 10.1038/s41434-021-00254-w.

42. Zhang Y, Zhang Z, Ding Y, et al. Phase I clinical trial of EGFR-specific CAR-T cells generated by the piggyBac transposon system in advanced relapsed/refractory non-small cell lung cancer patients. *J Cancer Res Clin Oncol.* 2021;147(12):3725–3734. doi: 10.1007/s00432-021-03613-7.

43. Gogol-Döring A, Ammar I, Gupta S, et al. Genome-wide profiling reveals remarkable parallels between insertion site selection properties of the MLV retrovirus and the piggyBac transposon in primary human CD4(+) T cells. *Mol Ther.* 2016;24(3):592–606. doi: 10.1038/mt.2016.11.

44. Ramamoorth M, Narvekar A. Non viral vectors in gene therapy – An overview. *J Clin Diagn Res.* 2015;9(1):Ge01– Ge06. doi: 10.7860/jcdr/2015/10443.5394.

45. Malone RW, Felgner PL, Verma IM. Cationic liposome-mediated RNA transfection. *Proc Natl Acad Sci USA.* 1989;86(16):6077–6081. doi: 10.1073/pnas.86.16.6077.

46. Dolgin E. The tangled history of mRNA vaccines. *Nature.* 2021;597(7876):318–324. doi: 10.1038/d41586-021-02483-w.

47. Kuhn AN, Diken M, Kreiter S, et al. Determinants of intracellular RNA pharmacokinetics: Implications for RNA-based immunotherapeutics. *RNA Biol.* 2011;8(1):35–43. doi: 10.4161/rna.8.1.13767.

48. Simon B, Harrer DC, Schuler-Thurner B, et al. The siRNA-mediated downregulation of PD-1 alone or simultaneously with CTLA-4 shows enhanced in vitro CAR-T-cell functionality for further clinical development towards the potential use in immunotherapy of melanoma. *Exp Dermatol.* 2018;27(7):769–778. doi: 10.1111/exd.13678.

49. Siddiqi T, Soumerai JD, Wierda WG, et al. Rapid MRD-negative responses in patients with relapsed/refractory CLL treated with Liso-Cel, a CD19-directed CAR T-cell product: Preliminary results from transcend CLL 004, a phase 1/2 study including patients with high-risk disease previously treated with Ibrutinib. *Blood.* 2018;132:300. doi: 10.1182/blood-2018-99-110462.

50. Yoon SH, Lee JM, Cho HI, et al. Adoptive immunotherapy using human peripheral blood lymphocytes transferred with RNA encoding Her-2/neu-specific chimeric immune receptor in ovarian cancer xenograft model. *Cancer Gene Therapy.* 2009;16(6):489–497. doi: 10.1038/cgt.2008.98.

51. Zhao Y, Moon E, Carpenito C, et al. Multiple injections of electroporated autologous T cells expressing a chimeric antigen receptor mediate regression of human disseminated tumor. *Cancer Res.* 2010;70(22):9053–9061. doi: 10.1158/0008-5472.Can-10-2880.

52. Zhao Y, Zheng Z, Cohen CJ, et al. High-efficiency transfection of primary human and mouse T lymphocytes using RNA electroporation. *Mol Ther.* 2006;13(1):151–159. doi: 10.1016/j.ymthe.2005.07.688.

53. Baden LR, El Sahly HM, Essink B, et al. Efficacy and safety of the mRNA-1273 SARS-CoV-2 Vaccine. *N Engl J Med.* 2021;384(5):403–416. doi: 10.1056/NEJMoa2035389.

54. Bouazzaoui A, Abdellatif AAH. Vaccine delivery systems and administration routes: Advanced biotechnological techniques to improve the immunization efficacy. *Vaccine: X.* 2024;19:100500.

55. Yarmush ML, Golberg A, Serša G, et al. Electroporation-based technologies for medicine: Principles, applications, and challenges. *Ann Rev Biomed Eng.* 2014;16:295–320. doi: 10.1146/annurev-bioeng-071813-104622.

56. Kowalski PS, Rudra A, Miao L, et al. Delivering the messenger: Advances in technologies for therapeutic mRNA delivery. *Mol Ther.* 2019;27(4):710–728. doi: 10.1016/j.ymthe.2019.02.012.

57. Huth S, Hoffmann F, von Gersdorff K, et al. Interaction of polyamine gene vectors with RNA leads to the dissociation of plasmid DNA-carrier complexes. *J Gene Med.* 2006;8(12):1416–1424. doi: 10.1002/jgm.975.

58. Zohra FT, Chowdhury EH, Tada S, et al. Effective delivery with enhanced translational activity synergistically accelerates mRNA-based transfection. *Biochem Biophys Res Commun.* 2007;358(1):373–378. doi: 10.1016/j.bbrc.2007.04.059.

59. Smith TT, Stephan SB, Moffett HF, et al. In situ programming of leukaemia-specific T cells using synthetic DNA nanocarriers. *Nat Nanotechnol.* 2017;12(8):813–820. doi: 10.1038/nnano.2017.57.

60. Smith TT, Moffett HF, Stephan SB, et al. Biopolymers codelivering engineered T cells and STING agonists can eliminate heterogeneous tumors. *J Clin Invest.* 2017;127(6):2176–2191. doi: 10.1172/jci87624.

61. Dowaidar M, Abdelhamid HN, Hällbrink M, et al. Magnetic nanoparticle assisted self-assembly of cell penetrating peptides-oligonucleotides complexes for gene delivery. *Sci Rep.* 2017;7(1):9159. doi: 10.1038/s41598-017-09803-z.

62. Gaj T, Gersbach CA, Barbas CF, 3rd. ZFN, TALEN, and CRISPR/Cas-based methods for genome engineering. *Trends Biotechnol.* 2013;31(7):397–405. doi: 10.1016/j.tibtech.2013.04.004.

63. Ishino Y, Shinagawa H, Makino K, et al. Nucleotide sequence of the iap gene, responsible for alkaline phosphatase isozyme conversion in Escherichia coli, and identification of the gene product. *J Bacteriol.* 1987;169(12):5429–5433. doi: 10.1128/jb.169.12.5429-5433.1987.

64. Li T, Yang Y, Qi H, et al. CRISPR/Cas9 therapeutics: Progress and prospects. *Signal Transduct Target Ther.* 2023;8(1):36. doi: 10.1038/s41392-023-01309-7.

65. Mojica FJ, Díez-Villaseñor C, García-Martínez J, et al. Intervening sequences of regularly spaced prokaryotic repeats derived from foreign genetic elements. *J Mol Evol.* 2005;60(2):174–182. doi: 10.1007/s00239-004-0046-3.

66. Jinek M, Jiang F, Taylor DW, et al. Structures of Cas9 endonucleases reveal RNA-mediated conformational activation. *Science.* 2014;343(6176):1247997. doi: 10.1126/science.1247997.

67. Gleditzsch D, Pausch P, Müller-Esparza H, et al. PAM identification by CRISPR-Cas effector complexes: Diversified mechanisms and structures. *RNA Biol.* 2019;16(4):504–517. doi: 10.1080/15476286.2018.1504546.

68. Haurwitz RE, Jinek M, Wiedenheft B, et al. Sequence- and structure-specific RNA processing by a CRISPR endonuclease. *Science.* 2010;329(5997):1355–1358. doi: 10.1126/science.1192272.

69. Jiang F, Doudna JA. CRISPR-Cas9 structures and mechanisms. *Ann Rev Biophys.* 2017;46:505–529. doi: 10.1146/annurev-biophys-062215-010822.

70. Jinek M, Chylinski K, Fonfara I, et al. A programmable dual-RNA-guided DNA endonuclease in adaptive bacterial immunity. *Science.* 2012;337(6096):816–821. doi: 10.1126/science.1225829.

71. Cong L, Ran FA, Cox D, et al. Multiplex genome engineering using CRISPR/Cas systems. *Science.* 2013;339(6121):819–823. doi: 10.1126/science.1231143.

72. Chen M, Mao A, Xu M, et al. CRISPR-Cas9 for cancer therapy: Opportunities and challenges. *Cancer Lett.* 2019;447:48–55. doi: 10.1016/j.canlet.2019.01.017.

73. Malech HL. Treatment by CRISPR-Cas9 gene editing – A proof of principle. *N Engl J Med.* 2021;384(3):286–287. doi: 10.1056/NEJMe2034624.

74. Rafii S, Tashkandi E, Bukhari N, et al. Current status of CRISPR/Cas9 application in clinical cancer research: Opportunities and challenges. *Cancers.* 2022;14(4). doi: 10.3390/cancers14040947.

75. Razeghian E, Nasution MKM, Rahman HS, et al. A deep insight into CRISPR/Cas9 application in CAR-T cell-based tumor immunotherapies. *Stem Cell Res Ther.* 2021;12(1):428. doi: 10.1186/s13287-021-02510-7.

76. Tang Q, Chen Y, Li X, et al. The role of PD-1/PD-L1 and application of immune-checkpoint inhibitors in human cancers. *Front Immunol.* 2022;13:964442. doi: 10.3389/fimmu.2022.964442.

77. Bagchi S, Yuan R, Engleman EG. Immune checkpoint inhibitors for the treatment of cancer: Clinical impact and mechanisms of response and resistance. *Ann Rev Pathol.* 2021;16:223–249. doi: 10.1146/annurev-pathol-042020-042741.

78. He X, Xu C. Immune checkpoint signaling and cancer immunotherapy. *Cell Res.* 2020;30(8):660–669. doi: 10.1038/s41422-020-0343-4.

79. Choi BD, Yu X, Castano AP, et al. CRISPR-Cas9 disruption of PD-1 enhances activity of universal EGFRvIII CAR T cells in a preclinical model of human glioblastoma. *J Immunother Cancer.* 2019;7(1):304. doi: 10.1186/s40425-019-0806-7.

80. Nakazawa T, Natsume A, Nishimura F, et al. Effect of CRISPR/Cas9-mediated PD-1-disrupted primary human third-generation CAR-T cells targeting EGFRvIII on in vitro human glioblastoma cell growth. *Cells.* 2020;9(4). doi: 10.3390/cells9040998.

81. Hu W, Zi Z, Jin Y, et al. CRISPR/Cas9-mediated PD-1 disruption enhances human mesothelin-targeted CAR T cell effector functions. *Cancer Immunol Immunother.* 2019;68(3):365–377. doi: 10.1007/s00262-018-2281-2.

82. Good CR, Aznar MA, Kuramitsu S, et al. An NK-like CAR T cell transition in CAR T cell dysfunction. *Cell.* 2021;184(25):6081–6100.e26. doi: 10.1016/j.cell.2021.11.016.

83. Chen J, López-Moyado IF, Seo H, et al. NR4A transcription factors limit CAR T cell function in solid tumours. *Nature.* 2019;567(7749):530–534. doi: 10.1038/s41586-019-0985-x.

84. Wiede F, Lu KH, Du X, et al. PTP1B is an intracellular checkpoint that limits T-cell and CAR T-cell antitumor immunity. *Cancer Disc.* 2022;12(3):752–773. doi: 10.1158/2159-8290.Cd-21-0694.

85. Kumar J, Kumar R, Kumar Singh A, et al. Deletion of Cbl-b inhibits CD8(+) T-cell exhaustion and promotes CAR T-cell function. *J Immunother Cancer.* 2021;9(1). doi: 10.1136/jitc-2020-001688.

86. Tie Y, Tang F, Wei YQ, et al. Immunosuppressive cells in cancer: Mechanisms and potential therapeutic targets. *J Hematol Oncol.* 2022;15(1):61. doi: 10.1186/s13045-022-01282-8.

87. Jiang W, He Y, He W, et al. Exhausted CD8+T cells in the tumor immune microenvironment: New pathways to therapy. *Front Immunol.* 2020;11:622509. doi: 10.3389/fimmu.2020.622509.

88. Tang N, Cheng C, Zhang X, et al. TGF-β inhibition via CRISPR promotes the long-term efficacy of CAR T cells against solid tumors. *JCI Insight.* 2020;5(4). doi: 10.1172/jci.insight.133977.

89. Alishah K, Birtel M, Masoumi E, et al. CRISPR/Cas9-mediated TGFβRII disruption enhances anti-tumor efficacy of human chimeric antigen receptor T cells in vitro. *J Transl Med.* 2021;19(1):482. doi: 10.1186/s12967-021-03146-0.

90. Lv J, Qin L, Zhao R, et al. Disruption of CISH promotes the antitumor activity of human T cells and decreases PD-1 expression levels. *Mol Ther Oncolytics.* 2023;28:46–58. doi: 10.1016/j.omto.2022.12.003.

91. Jain N, Zhao Z, Feucht J, et al. TET2 guards against unchecked BATF3-induced CAR T cell expansion. *Nature.* 2023;615(7951):315–322.

92. Fraietta JA, Nobles CL, Sammons MA, et al. Disruption of TET2 promotes the therapeutic efficacy of CD19-targeted T cells. *Nature.* 2018;558(7709):307–312. doi: 10.1038/s41586-018-0178-z.

93. Yoshikawa T, Wu Z, Inoue S, et al. Genetic ablation of PRDM1 in antitumor T cells enhances therapeutic efficacy of adoptive immunotherapy. *Blood.* 2022;139(14):2156–2172. doi: 10.1182/blood.2021012714.

94. Ghoneim HE, Fan Y, Moustaki A, et al. De novo epigenetic programs inhibit PD-1 blockade-mediated T cell rejuvenation. *Cell.* 2017;170(1):142–157.e19. doi: 10.1016/j.cell.2017.06.007.

95. Prinzing B, Zebley CC, Petersen CT, et al. Deleting DNMT3A in CAR T cells prevents exhaustion and enhances antitumor activity. *Sci Transl Med.* 2021;13(620):eabh0272. doi: 10.1126/scitranslmed.abh0272.

96. Carnevale J, Shifrut E, Kale N, et al. RASA2 ablation in T cells boosts antigen sensitivity and long-term function. *Nature.* 2022;609(7925):174–182. doi: 10.1038/s41586-022-05126-w.

97. Freitas KA, Belk JA, Sotillo E, et al. Enhanced T cell effector activity by targeting the mediator kinase module. *Science.* 2022;378(6620):eabn5647. doi: 10.1126/science.abn5647.

98. Belk JA, Yao W, Ly N, et al. Genome-wide CRISPR screens of T cell exhaustion identify chromatin remodeling factors that limit T cell persistence. *Cancer Cell.* 2022;40(7):768–786.e7. doi: 10.1016/j.ccell.2022.06.001.

99. Trefny MP, Kirchhammer N, Auf der Maur P, et al. Deletion of SNX9 alleviates CD8 T cell exhaustion for effective cellular cancer immunotherapy. *Nat Commun.* 2023;14(1):86. doi: 10.1038/s41467-022-35583-w.

100. Zhang X, Zhang C, Qiao M, et al. Depletion of BATF in CAR-T cells enhances antitumor activity by inducing resistance against exhaustion and formation of central memory cells. *Cancer Cell.* 2022;40(11):1407–1422.e7. doi: 10.1016/j.ccell.2022.09.013.

101. Legut M, Gajic Z, Guarino M, et al. A genome-scale screen for synthetic drivers of T cell proliferation. *Nature*. 2022;603(7902):728–735. doi: 10.1038/s41586-022-04494-7.

102. Wang D, Prager BC, Gimple RC, et al. CRISPR screening of CAR T cells and cancer stem cells reveals critical dependencies for cell-based therapies. *Cancer Discov*. 2021;11(5):1192–1211. doi: 10.1158/2159-8290.Cd-20-1243.

103. Ye L, Park JJ, Dong MB, et al. In vivo CRISPR screening in CD8 T cells with AAV-Sleeping Beauty hybrid vectors identifies membrane targets for improving immunotherapy for glioblastoma. *Nature Biotechnol*. 2019;37(11):1302–1313. doi: 10.1038/s41587-019-0246-4.

104. Geiger R, Rieckmann JC, Wolf T, et al. L-Arginine modulates T cell metabolism and enhances survival and anti-tumor activity. *Cell*. 2016;167(3):829–842.e13. doi: 10.1016/j.cell.2016.09.031.

105. Ye L, Park JJ, Peng L, et al. A genome-scale gain-of-function CRISPR screen in CD8 T cells identifies proline metabolism as a means to enhance CAR-T therapy. *Cell Metabol*. 2022;34(4):595–614.e14. doi: 10.1016/j.cmet.2022.02.009.

106. Leibowitz ML, Papathanasiou S, Doerfler PA, et al. Chromothripsis as an on-target consequence of CRISPR-Cas9 genome editing. *Nat Genet*. 2021;53(6):895–905. doi: 10.1038/s41588-021-00838-7.

107. Kosicki M, Tomberg K, Bradley A. Repair of double-strand breaks induced by CRISPR-Cas9 leads to large deletions and complex rearrangements. *Nat Biotechnol*. 2018;36(8):765–771. doi: 10.1038/nbt.4192.

108. Zuccaro MV, Xu J, Mitchell C, et al. Allele-specific chromosome removal after Cas9 cleavage in human embryos. *Cell*. 2020;183(6):1650–1664.e15. doi: 10.1016/j.cell.2020.10.025.

109. Alanis-Lobato G, Zohren J, McCarthy A, et al. Frequent loss of heterozygosity in CRISPR-Cas9-edited early human embryos. *Proc Natl Acad Sci USA*. 2021;118(22). doi: 10.1073/pnas.2004832117.

110. Kang SH, Lee WJ, An JH, et al. Prediction-based highly sensitive CRISPR off-target validation using target-specific DNA enrichment. *Nat Commun*. 2020;11(1):3596. doi: 10.1038/s41467-020-17418-8.

111. Ihry RJ, Worringer KA, Salick MR, et al. p53 inhibits CRISPR-Cas9 engineering in human pluripotent stem cells. *Nat Med*. 2018;24(7):939–946. doi: 10.1038/s41591-018-0050-6.

112. Haapaniemi E, Botla S, Persson J, et al. CRISPR-Cas9 genome editing induces a p53-mediated DNA damage response. *Nat Med*. 2018;24(7):927–930. doi: 10.1038/s41591-018-0049-z.

113. Hunt JMT, Samson CA, Rand AD, et al. Unintended CRISPR-Cas9 editing outcomes: A review of the detection and prevalence of structural variants generated by gene-editing in human cells. *Hum Genet*. 2023;142(6):705–720. doi: 10.1007/s00439-023-02561-1.

114. Höijer I, Emmanouilidou A, Östlund R, et al. CRISPR-Cas9 induces large structural variants at on-target and off-target sites in vivo that segregate across generations. *Nat Commun*. 2022;13(1):627. doi: 10.1038/s41467-022-28244-5.

115. Ghaffari S, Khalili N, Rezaei N. CRISPR/Cas9 revitalizes adoptive T-cell therapy for cancer immuno-therapy. *J Exp Clin Cancer Res*. 2021;40(1):269. doi: 10.1186/s13046-021-02076-5.

116. Zhang XH, Tee LY, Wang XG, et al. Off-target Effects in CRISPR/Cas9-mediated genome engineering. *Mol Ther Nucleic Acids*. 2015;4(11):e264. doi: 10.1038/mtna.2015.37.

117. Charlesworth CT, Deshpande PS, Dever DP, et al. Identification of preexisting adaptive immunity to Cas9 proteins in humans. *Nat Med*. 2019;25(2):249–254. doi: 10.1038/s41591-018-0326-x.

118. Mehta A, Merkel OM. Immunogenicity of Cas9 protein. *J Pharm Sci*. 2020;109(1):62–67. doi: 10.1016/j.xphs.2019.10.003.

119. Pan J, Zuo S, Deng B, et al. Sequential CD19–22 CAR T therapy induces sustained remission in children with r/r B-ALL. *Blood*. 2020;135(5):387–391. doi: 10.1182/blood.2019003293.

120. Pasquini MC, Hu ZH, Curran K, et al. Real-world evidence of tisagenlecleucel for pediatric acute lym-phoblastic leukemia and non-Hodgkin lymphoma. *Blood Adv*. 2020;4(21):5414–5424. doi: 10.1182/bloodadvances.2020003092.

121. Shargian L, Raanani P, Yeshurun M, et al. Chimeric antigen receptor T-cell therapy is superior to stan-dard of care as second-line therapy for large B-cell lymphoma: A systematic review and meta-analysis. *Brit J Haematol*. 2022;198(5):838–846. doi: 10.1111/bjh.18335.

122. Martino M, Canale FA, Alati C, et al. CART-cell therapy: Recent advances and new evidence in mul-tiple myeloma. *Cancers*. 2021;13(11). doi: 10.3390/cancers13112639.

123. Berdeja JG, Madduri D, Usmani SZ, et al. Ciltacabtagene autoleucel, a B-cell maturation antigen-directed chimeric antigen receptor T-cell therapy in patients with relapsed or refractory multiple myeloma (CARTITUDE-1): A phase 1b/2 open-label study. *Lancet*. 2021;398(10297):314–324. doi: 10.1016/s0140-6736(21)00933-8.

124. Qi C, Gong J, Li J, et al. Claudin18.2-specific CAR T cells in gastrointestinal cancers: Phase 1 trial interim results. *Nat Med*. 2022;28(6):1189–1198. doi: 10.1038/s41591-022-01800-8.

125. Narayan V, Barber-Rotenberg JS, Jung IY, et al. PSMA-targeting TGFβ-insensitive armored CAR T cells in metastatic castration-resistant prostate cancer: A phase 1 trial. *Nat Med.* 2022;28(4):724–734. doi: 10.1038/s41591-022-01726-1.

126. Dai H, Tong C, Shi D, et al. Efficacy and biomarker analysis of CD133-directed CAR T cells in advanced hepatocellular carcinoma: A single-arm, open-label, phase II trial. *Oncoimmunology.* 2020;9(1):1846926. doi: 10.1080/2162402x.2020.1846926.

127. Liu Z, Zhou J, Yang X, et al. Safety and antitumor activity of GD2-Specific 4SCAR-T cells in patients with glioblastoma. *Mol Cancer.* 2023;22(1):3. doi: 10.1186/s12943-022-01711-9.

128. O'Rourke DM, Nasrallah MP, Desai A, et al. A single dose of peripherally infused EGFRvIII-directed CAR T cells mediates antigen loss and induces adaptive resistance in patients with recurrent glioblastoma. *Sci Transl Med.* 2017;9(399). doi: 10.1126/scitranslmed.aaa0984.

129. Brown CE, Rodriguez A, Palmer J, et al. Off-the-shelf, steroid-resistant, IL13Rα2-specific CAR T cells for treatment of glioblastoma. *Neuro-oncology.* 2022;24(8):1318–1330. doi: 10.1093/neuonc/noac024.

130. Yang YH, Liu JW, Lu C, et al. CAR-T cell therapy for breast cancer: From basic research to clinical application. *Int J Biol Sci.* 2022;18(6):2609–2626. doi: 10.7150/ijbs.70120.

131. Tchou J, Zhao Y, Levine BL, et al. Safety and efficacy of intratumoral injections of chimeric antigen receptor (CAR) T cells in metastatic breast cancer. *Cancer Immunol Res.* 2017;5(12):1152–1161. doi: 10.1158/2326-6066.Cir-17-0189.

132. Fala L. Yescarta (Axicabtagene Ciloleucel) second CAR T-cell therapy approved for patients with certain types of large B-cell lymphoma. 2018. https://jons-online.com/browse-by-topic/fda-approvals-news-updates/1829-yescarta-axicabtagene-ciloleucel-second-car-t-cell-therapy-approved-for-patients-with-certain-types-of-large-b-cell-lymphoma

133. Bach PB, Giralt SA, Saltz LB. FDA approval of Tisagenlecleucel: Promise and complexities of a $475 000 cancer drug. *JAMA.* 2017;318(19):1861–1862. doi: 10.1001/jama.2017.15218.

134. Rotolo A, Caputo V, Karadimitris A. The prospects and promise of chimeric antigen receptor immunotherapy in multiple myeloma. *Brit J Haematol.* 2016;173(3):350–64. doi: 10.1111/bjh.13976.

135. Almond LM, Charalampakis M, Ford SJ, et al. Myeloid sarcoma: Presentation, diagnosis, and treatment. *Clin Lymph Myeloma Leuk.* 2017;17(5):263–267. doi: 10.1016/j.clml.2017.02.027.

136. Fowler NH, Dickinson M, Dreyling M, et al. Tisagenlecleucel in adult relapsed or refractory follicular lymphoma: The phase 2 ELARA trial. *Nat Med.* 2022;28(2):325–332. doi: 10.1038/s41591-021-01622-0.

137. Schuster SJ, Bishop MR, Tam CS, et al. Tisagenlecleucel in adult relapsed or refractory diffuse large B-cell lymphoma. *N Engl J Med.* 2019;380(1):45–56. doi: 10.1056/NEJMoa1804980.

138. Jacobson CA, Chavez JC, Sehgal AR, et al. Axicabtagene ciloleucel in relapsed or refractory indolent non-Hodgkin lymphoma (ZUMA-5): A single-arm, multicentre, phase 2 trial. *Lancet Oncol.* 2022;23(1):91–103. doi: 10.1016/s1470-2045(21)00591-x.

139. Locke FL, Miklos DB, Jacobson CA, et al. Axicabtagene Ciloleucel as second-line therapy for large B-cell lymphoma. *N Engl J Med.* 2022;386(7):640–654. doi: 10.1056/NEJMoa2116133.

140. Mian A, Hill BT. Brexucabtagene autoleucel for the treatment of relapsed/refractory mantle cell lymphoma. *Expert Opin Biol Ther.* 2021;21(4):435–441. doi: 10.1080/14712598.2021.1889510.

141. Wang M, Munoz J, Goy A, et al. KTE-X19 CAR T-cell therapy in relapsed or refractory mantle-cell lymphoma. *N Engl J Med.* 2020;382(14):1331–1342. doi: 10.1056/NEJMoa1914347.

142. Kamdar M, Solomon SR, Arnason J, et al. Lisocabtagene maraleucel versus standard of care with salvage chemotherapy followed by autologous stem cell transplantation as second-line treatment in patients with relapsed or refractory large B-cell lymphoma (TRANSFORM): results from an interim analysis of an open-label, randomised, phase 3 trial. *Lancet.* 2022;399(10343):2294–2308. doi: 10.1016/s0140-6736(22)00662-6.

143. Sehgal A, Hoda D, Riedell PA, et al. Lisocabtagene maraleucel as second-line therapy in adults with relapsed or refractory large B-cell lymphoma who were not intended for haematopoietic stem cell transplantation (PILOT): An open-label, phase 2 study. *Lancet Oncol.* 2022;23(8):1066–1077. doi: 10.1016/s1470-2045(22)00339-4.

144. Teoh J, Brown LF. Developing lisocabtagene maraleucel chimeric antigen receptor T-cell manufacturing for improved process, product quality and consistency across CD19(+) hematologic indications. *Cytotherapy.* 2022;24(9):962–973. doi: 10.1016/j.jcyt.2022.03.013.

145. Al Hadidi S, Szabo A, Esselmann J, et al. Clinical outcome of patients with relapsed refractory multiple myeloma listed for BCMA directed commercial CAR-T therapy. *Bone Marrow Transplant.* 2023;58(4):443–445. doi: 10.1038/s41409-022-01905-1.

146. Hansen DK, Sidana S, Peres LC, et al. Idecabtagene vicleucel for relapsed/refractory multiple myeloma: Real-world experience from the myeloma CAR T consortium. *J Clin Oncol.* 2023;41(11):2087–2097. doi: 10.1200/jco.22.01365.

147. San-Miguel J, Dhakal B, Yong K, et al. Cilta-cel or standard care in lenalidomide-refractory multiple myeloma. *N Engl J Med.* 2023;389(4):335–347. doi: 10.1056/NEJMoa2303379.

148. Huang X, Wilber AC, Bao L, et al. Stable gene transfer and expression in human primary T cells by the Sleeping Beauty transposon system. *Blood.* 2006;107(2):483–491. doi: 10.1182/blood-2005-05-2133.

149. Singh H, Manuri PR, Olivares S, et al. Redirecting specificity of T-cell populations for CD19 using the Sleeping Beauty system. *Cancer Res.* 2008;68(8):2961–2971. doi: 10.1158/0008-5472.Can-07-5600.

150. Kebriaei P, Singh H, Huls MH, et al. Phase I trials using Sleeping Beauty to generate CD19-specific CAR T cells. *J Clin Invest.* 2016;126(9):3363–3376. doi: 10.1172/jci86721.

151. Nishio N, Hanajiri R, Ishikawa Y, et al. A phase I study of CD19 chimeric antigen receptor-T cells generated by the PiggyBac Transposon vector for acute lymphoblastic leukemia. *J Blood.* 2021;138:3831.

152. Soundara Rajan T, Gugliandolo A, Bramanti P, et al. In vitro-transcribed mRNA chimeric antigen receptor T cell (IVT mRNA CAR T) therapy in hematologic and solid tumor management: A preclinical update. *Int J Mol Sci.* 2020;21(18). doi: 10.3390/ijms21186514.

153. Beatty GL, Haas AR, Maus MV, et al. Mesothelin-specific chimeric antigen receptor mRNA-engineered T cells induce anti-tumor activity in solid malignancies. *Cancer Immunol Res.* 2014;2(2):112–120. doi: 10.1158/2326-6066.Cir-13-0170.

154. Porter DL, Hwang WT, Frey NV, et al. Chimeric antigen receptor T cells persist and induce sustained remissions in relapsed refractory chronic lymphocytic leukemia. *Sci Transl Med.* 2015;7(303):303ra139. doi: 10.1126/scitranslmed.aac5415.

155. Jafarzadeh L, Masoumi E, Fallah-Mehrjardi K, et al. Prolonged persistence of chimeric antigen receptor (CAR) T cell in adoptive cancer immunotherapy: Challenges and ways forward. *Front Immunol.* 2020;11:702. doi: 10.3389/fimmu.2020.00702.

156. Maude SL, Frey N, Shaw PA, et al. Chimeric antigen receptor T cells for sustained remissions in leukemia. *N Engl J Med.* 2014;371(16):1507–1517. doi: 10.1056/NEJMoa1407222.

157. Hsieh EM, Scherer LD, Rouce RH. Replacing CAR-T cell resistance with persistence by changing a single residue. *J Clin Invest.* 2020;130(6):2806–2808. doi: 10.1172/jci136872.

158. Hay KA, Gauthier J, Hirayama AV, et al. Factors associated with durable EFS in adult B-cell ALL patients achieving MRD-negative CR after CD19 CAR T-cell therapy. *Blood.* 2019;133(15):1652–1663. doi: 10.1182/blood-2018-11-883710.

159. López-Cantillo G, Urueña C, Camacho BA, et al. CAR-T cell performance: How to improve their persistence? *Front Immunol.* 2022;13:878209. doi: 10.3389/fimmu.2022.878209.

160. Liu Y, An L, Huang R, et al. Strategies to enhance CAR-T persistence. *Biomarker Res.* 2022;10(1):86. doi: 10.1186/s40364-022-00434-9.

161. Maalej KM, Merhi M, Inchakalody VP, et al. CAR-cell therapy in the era of solid tumor treatment: Current challenges and emerging therapeutic advances. *Mol Cancer.* 2023;22(1):20. doi: 10.1186/s12943-023-01723-z.

162. Brudno JN, Kochenderfer JN. Toxicities of chimeric antigen receptor T cells: Recognition and management. *Blood.* 2016;127(26):3321–3330. doi: 10.1182/blood-2016-04-703751.

163. Wei W, Chen ZN, Wang K. CRISPR/Cas9: A powerful strategy to improve CAR-T cell persistence. *Int J Mol Sci.* 2023;24(15). doi: 10.3390/ijms241512317.

164. Kebriaei PJCLM. CAR T-cell therapies: Overcoming the challenges and new strategies. *J Clin Lymph Myeloma Leuk.* 2017;17:S74–S78.

165. Bonifant CL, Jackson HJ, Brentjens RJ, et al. Toxicity and management in CAR T-cell therapy. *Mol Ther Oncolytics.* 2016;3:16011. doi: 10.1038/mto.2016.11.

10 Clinical Translation of Nanomedicines; Keys of Cancer Targeting

Mahmoud A. Younis

10.1 INTRODUCTION

Cancer has been a leading cause of death globally for decades. Nevertheless, the COVID-19 pandemic has stolen attention during the past 5 years, which might have negatively impacted the screening and healthcare programs related to cancer. Subsequently, the American Cancer Society forecasts an increase in cancer incidence as well as cancer-related mortality in the near future [1]. According to a survey conducted in 115 countries, the World Health Organization (WHO) has estimated 20 million new cases and 9.7 million deaths from cancer in 2022. Furthermore, the WHO forecasts a 20% cancer incidence rate in humans during their lifetime, with approximately 11.11% and 8.33% mortality rates in males and females, respectively [2]. These horrible figures necessitate an urgent development of effective therapeutic approaches that can cope with the rapid increase in cancer patients. However, till now, classic approaches including surgical resection, chemotherapy, and radiotherapy are dominating the clinical guidelines [3]. While surgical removal of tumor lesions is the first choice in the clinical management of cancer, it has a limited efficacy in the case of advanced-stage or metastatic tumors. Meanwhile, chemotherapy has been associated with inevitable off-target effects as well as the emergence of multidrug resistance (MDR) [4,5].

The third millennium has witnessed massive progress in the application of nanomedicines in the diagnosis or treatment of cancer, where nanomedicines have demonstrated a great promise in overcoming the challenges that are commonly encountered in the classic therapeutic modalities and improving the efficacy of anticancer chemotherapy [6–8]. Moreover, nanomedicines have enabled the introduction of innovative therapeutic strategies for cancer such as gene therapy, cell therapy, and anticancer vaccines [9–12].

Despite their promise, the translation of nanomedicine technology from the bench into the clinics is far below expectations, where only a few candidates among thousands of publications could eventually reach the market. A recent article published in 2022 has shed light on the general problem of poor clinical translatability of nanomedicines, comparing publications, clinical trials, and FDA approvals and analyzing potential reasons for the clinical failure of a substantial proportion of nanomedicines [13]. Although some anticancer nanomedicines have already been approved by the FDA, they have failed to achieve a breakthrough in the survival rates or exert a substantial influence on clinical practices [14]. Subsequently, the area of clinical translation has attracted increasing attention from researchers as an essential requirement for research to receive funding or adoption by the pharmaceutical industry.

The present chapter deeply focuses on the challenges hampering the clinical application of cancer-targeted nanomedicines. First, the clinical status of cancer-targeted nanomedicines is highlighted in light of the current FDA approvals and registered clinical trials. Subsequently, the technical, industrial, economic, and regulatory issues that hinder the clinical applicability of such modalities are discussed, with a particular focus on the biosafety issue as a critical limiting factor of the clinical fate of any new therapeutic modality. Then, some recent technologies that can tackle

DOI: 10.1201/9781003517870-10

these challenges are reported. Lastly, the key considerations in the design of cancer-targeted nano-medicines to avoid clinical failure and the future perspectives in this area of endeavor are discussed. We hope that this chapter will increase the researchers' awareness of this critical topic and contribute to the promotion of the clinical fate of anticancer nanomedicines for a successful transfer from academia to industry and clinics.

10.2 CLINICAL STATUS OF NANOMEDICINES IN CANCER

The year 1995 marked an evolutionary step in the development of cancer-targeted nanomedicines, when Doxil® received its FDA approval as the world's first nanomedicine to reach the market. Doxil is a polyethylene glycol (PEG)-modified liposome-encapsulating doxorubicin, which relies on a PEGylation approach to prolong its lifetime in the blood post intravenous administration. Subsequently, it can passively accumulate in the tumor tissue via the "enhanced permeability and retention (EPR) effect" [15]. Nevertheless, the subsequent three decades have witnessed the FDA approval of only 12 cancer-related nanomedicines, which dropped far below expectations [16]. Table 10.1 lists the FDA-approved anticancer nanomedicines as of 2024. Meanwhile, the National Institute of Health (NIH) database lists 321 clinical trials recruiting various nanomedicine modalities for the treatment of cancer as of 2024 [17]. On the other hand, the PubMed database revealed 23,373 anticancer nanomedicine-related publications as of November 2024 [18]. The marked differences between the above numbers indicate the formidable obstacles that slow down the rate of

TABLE 10.1

List of FDA-Approved Anticancer Nanomedicines as of 2024

Product Name	Description	Indication	Approval Year	Manufacturer
Doxil®	PEGylated liposomal doxorubicin	AIDS-related Kaposi's sarcoma, Breast cancer, Ovarian cancer	1995	Janssen
DaunoXome®	Liposomal daunorubicin	AIDS-related Kaposi's sarcoma	1996	Galen
DepoCyt®	Liposomal cytarabine	Lymphomatous meningitis	1999	Sigma-Tau
Ontak®	Engineered protein combining IL-2 and Diphtheria toxin	Cutaneous T-cell lymphoma	1999	Eisai
Eligard®	Leuprolide acetate-loaded polymeric NPs	Advanced prostate cancer	2002	Tolmar
Abraxane®	nab-PTX	Advanced NSCLC, Metastatic breast cancer, Metastatic pancreatic cancer	2005	Celgene
Oncaspar®	PEGylated aspargase	Acute lymphoblastic leukemia	2006	Baxalta U.S.
Myocet®	Non-PEGylated liposomal doxorubicin	HER2-positive metastatic breast cancer	2010	Sopherion Therapeutics
Marqibo®	Liposomal vincristine	Philadelphia chromosome-negative acute lymphoblastic leukemia	2012	Acrotech
Onivyde®	Liposomal irinotecan	Metastatic pancreatic cancer	2015	IPSEN
Vyxeos®	Liposomal cytarabine-daunorubicin combination	Acute myeloid leukemia	2017	Jazz Pharmaceuticals
Nano-therm®	Iron oxide NPs	Thermal therapy of recurrent glioblastoma, prostate cancer	2018	MagForce

Abbreviations: AIDS, acquired immunodeficiency syndrome; IL-2, interleukin-2; nab-PTX, nanoparticle albumin–bound paclitaxel; NSCLC, non-small cell lung cancer; HER2, human epidermal growth factor receptor 2.

transfer of the promising laboratory findings in this area of interest into clinical applications, and thus, the issue of clinical translation of cancer-targeted nanomedicines comes to the forefront.

10.3   CHALLENGES LIMITING THE CLINICAL APPLICATION OF CANCER-TARGETED NANOMEDICINES

As mentioned above, there is a complex network of challenges that hamper the clinical translation of cancer-targeted nanomedicines, which can be roughly classified into *technical* (related to the nature of nanomedicines, their mode of action, and the relevant biological barriers), *toxicological* (related to the interaction of nanomedicines with the human body), *industrial* (related to the large-scale production of nanomedicines), *economic*, and *regulatory* challenges. Herein, these challenges and how they impact the clinical potential of nanomedicines as anticancer therapeutics are reported.

10.3.1   TECHNICAL CHALLENGES

10.3.1.1   Ex Vivo Stability

Unlike other classic dosage forms, nanomedicines possess a high surface-to-volume ratio that acts as a two-edged weapon. While such a huge surface area contributes to the improvement of the solubility, dissolution, and bioavailability of drugs, it also exerts a negative impact on the stability of these modalities, with a high susceptibility to aggregation, contamination, moisture adsorption, and degradation [19]. Thus, the vast majority of nanomedicines cannot withstand the ambient storage conditions and need intensive conditions such as cooling, freezing, or drying. Additionally, the huge surface area of nanomedicines can accelerate the rate of physico-chemical changes in the drug cargo, which may have a detrimental effect on its therapeutic efficacy. For instance, nanoparticles encapsulating drugs in the amorphous state demonstrate an amorphous-to-crystalline transitions of the drug payload during storage, which subsequently reduce the solubility, dissolution, and bioavailability of the drugs in question [20]. Since amorphous formulations have been reported for a wide diversity of anticancer drugs, including doxorubicin [21], cisplatin [22], and paclitaxel [23], such an undesired conversion should be carefully taken into consideration.

10.3.1.2   In Vivo Barriers

To reach their target, anticancer nanomedicines need to overcome successive biological barriers while maintaining their integrity. Figure 10.1 summarizes the *in vivo* barriers that anticancer nanomedicines encounter during their journey to the target cancer cells.

Following their administration, the nanomedicines inevitably come into contact with the biological fluids. Since most anticancer therapeutics are administered intravenously, interaction with blood is a pivotal factor that determines the *in vivo* fate of cancer-targeted nanomedicines. Owing to their high surface area, nanoparticles tend to adsorb a large amount of serum proteins, which leads to their aggregation or induces a premature release of their cargo into the blood before they reach their intended site of action [24]. Moreover, the adsorption of certain serum proteins (e.g. opsonins and complement components) promotes the recognition of the nanoparticles by the "reticulo-endothelial system (RES)" (e.g. monocytes, macrophages, and dendritic cells) as an innate immune response that subsequently reduces the fraction of the administered dose that can reach the tumor and increases the accumulation of the nanoparticles in the off-target tissues, particularly the liver and spleen, which is often associated with side effects and toxicities [25,26]. Considering the fact that most anticancer therapeutics are administered parenterally, the extravasation of nanomedicines into the tumor region is an essential requirement for their delivery. In 1986, Matsumura and Maeda reported on the increased vascular permeability and the reduced lymphatic drainage in the tumors, which can be exploited to deliver and retain anticancer therapeutics in the tumor region, a concept that was referred to as "EPR effect" [27]. Subsequently, the concept of

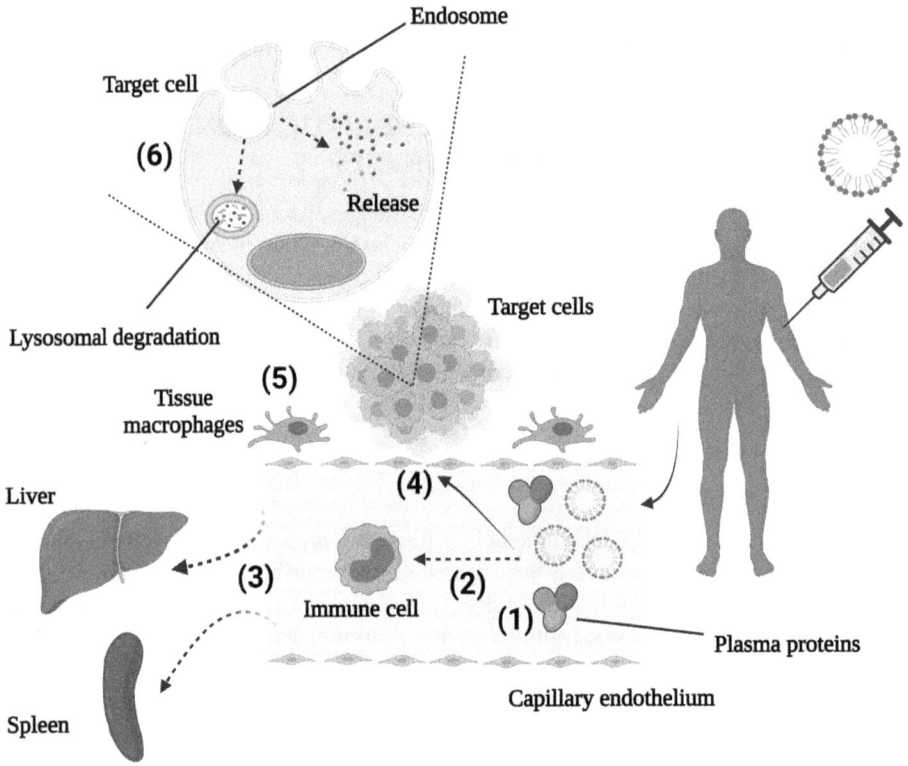

FIGURE 10.1 A summary of the in vivo barriers that cancer-targeted nanomedicines encounter. (1) The adsorption of plasma proteins to the surface of the nanoparticles may adversely affect their stability or trigger a premature release of their drug cargo. (2) The recognition of the nanoparticles by the immune cells of the reticulo-endothelial system (RES) leads to an increased phagocytosis of the administered formulations before reaching their target cells. (3) A substantial proportion of the administered nanoparticles is usually taken up by the off-target organs such as the liver and the spleen, which reduces their efficiency and may result in undesired side effects. (4) The nanoparticles need to extravasate into the tumor region and penetrate through the tumor microenvironment that may be rich in stroma, which hinders the access of nanoparticles to the tumor cells. (5) The nanoparticles may be engulfed by the tumor-resident macrophages before they reach the target cancer cells. (6) The nanoparticles need to be efficiently taken up by the cancer cells and escape from the lysosomal degradation for an efficient intracellular delivery of their cargo. (Adapted from Younis et al. [13], with permission from Elsevier. © 2021 Elsevier B.V.)

"passive targeting" has been coined, which implements the formulation of long-circulating nanomedicines through their coating with a hydrophilic polymer (e.g., PEG) to mask their recognition by RES and enable their delivery to the tumor region via the EPR effect [28]. Although such a concept has been adopted both clinically (e.g., Doxil) and in a huge volume of preclinical studies, recent reports have criticized its broad applicability in human tumors [29]. For instance, a meta-analysis by Petersen and co-workers revealed that unlike preclinical data, 14 randomized controlled clinical trials (RCT) recruiting 2589 patients showed no significant differences in objective response, overall survival, or progression-free survival rates upon directly comparing liposomal chemotherapies and their equivalent conventional modalities. A possible explanation pointed to the differences in EPR effect between humans and mice models, where the murine tumor models are directly implanted to grow rapidly within a short period of time leading to an extraordinary rate of neoangiogenesis that exaggerates the EPR effect, in contrast to human tumors that are initiated and grow over a prolonged time period. A second explanation attributed such discrepancies to the fact that the EPR effect does not take into consideration the impact of the tumor microenvironment on the efficacy of the nanomedicines [30]. In this context, it is noteworthy to highlight that nanomedicines are often demonstrating a lower therapeutic efficiency in stroma-rich tumors (e.g. hepatocellular carcinoma, HCC, and pancreatic cancer) owing to the impact of stroma in hindering the access of nanomedicines to the cancer cells [26]. Eventually, the nanomedicines should bypass the healthy cells and be efficiently taken up by the target cancer cells. Unfortunately, the stealth coating of nanomedicines adopted in passive targeting usually reduces their cellular uptake efficiency, a situation that is often referred to as the "PEG dilemma" [8]. Therefore, "active targeting" has appeared to tackle the above challenge through decorating the nanoparticles with targeting moieties that can be recognized by differentially expressed receptors or unique stimuli in the tumor cells [31]. Despite promise, ligand-targeted nanomedicines suffer from several drawbacks, including reduced stability, altered pharmacokinetics, and poor scalability [32]. Following their uptake, the nanoparticles are subjected to lysosomal degradation and need to be equipped by endosomal escape functionalities to enable the efficient intracellular delivery of their drug cargo [33]. Collectively, the ability of cancer-targeted nanomedicines to overcome the successive biological barriers alluded to above is the key limiting factor of their *in vivo* performance and therapeutic efficiency.

10.3.2 Toxicological Challenges

While the small size of nanomedicines can be considered "a blessing", it can also act as a "curse". The ultra-fine nature of nanoparticles imparts an enormous surface area that exaggerates their interactions with the biological components and membranes compared to the classic formulations, which has led to the evolution of the concept of "nanotoxicity" [34]. In addition, the distribution of the administered nanoparticles to off-target tissues is usually associated with harm to the healthy tissues in spite of the researchers' efforts to minimize it through various targeting strategies. For example, the hepatic accumulation of several types of nanomedicines can result in liver injury that is manifested by elevated serum levels of the hepatic enzymes, such as aspartate transaminase (AST), alanine aminotransferase (ALT), and alkaline phosphatase (ALP) [35]. Moreover, some nanomedicines have been reported to associate with nephrotoxicity through a variety of mechanisms, including degeneration of the renal tubules (e.g. epithelial cells, cellular fragments, and proteinaceous liquid in tubule lumen), renal interstitial fibrosis, swollen glomeruli, and alterations in Bowman's space and proliferation of mesangial cells [36].

Furthermore, a substantial proportion of anticancer nanomedicines are composed of non-biodegradable materials, which increases the risks associated with their restricted excretion and chronic accumulation. For instance, metallic nanoparticles, including gold nanoparticles (AuNPs) and silver nanoparticles (AgNPs) have been widely investigated as either an anticancer diagnostic/therapeutic modality or a delivery system for other modalities, including cytotoxic drugs [37,38]. Yet, owing to their nature, such nanoparticles are non-biodegradable, which raises several biosafety concerns. The toxicological profile of metallic nanoparticles is highly dependent on their physico-chemical properties. Cho and co-workers reported on a higher acute toxicity by 10-nm AgNPs compared to their 60-nm or 100-nm counterparts post intraperitoneal administration to BALB/c mice [39]. A similar finding was reported by Lopez-Chaves *et al.* using AuNPs, where traces of AuNPs were detected in the major tissues, such as liver, spleen, kidney, and intestine. Small-size AuNPs showed increased DNA damage and generation of reactive oxygen species (ROS), leading to genotoxicity and cytotoxicity [40].

Since exogenous nanoparticles are recognized by living bodies as xenobiotics, they can induce immune responses that are primarily mediated through the innate immunity. The immunogenicity of nanoparticles is a topic of concern from both therapeutic and toxicological points of view. Therapeutically, the recognition of anticancer nanomedicines by RES leads to a phagocytic fate for a substantial proportion of the administered dose, which subsequently reduces their therapeutic efficacy [41]. Toxicologically, the excessive accumulation of anticancer nanomedicines in off-target tissues results in increased tissue damage and toxicities. In addition, the interaction between nanoparticles and the immune cells can result in the induction of immune signaling that may end up with serious inflammations or allergies [42]. For instance, lipid nanoparticles (LNPs), a widely-used nanomedicine platform, have been reported to interact with the immune system through three major pathways. First, PEG, a universal excipient in LNPs, can induce the production of anti-PEG IgE antibodies. Subsequently, these antibodies bind to FcεRI receptors on the surface of mast cells and basophils, leading to the release of hypersensitivity mediators, including histamine, prostaglandins, and leukotrienes [43]. Second, PEG-induced anti-PEG IgM activates complement pathway, leading to a series of inflammatory reactions that are referred to as complement-associated pseudoallergy (CARPA) [44]. Third, binding of LNPs to pattern recognition receptors (PRRs) (e.g. Toll-like receptors, TLRs) stimulates type I interferon, leading to an undesired autoimmune response [45]. Figure 10.2 summarizes the three aforementioned mechanisms [42].

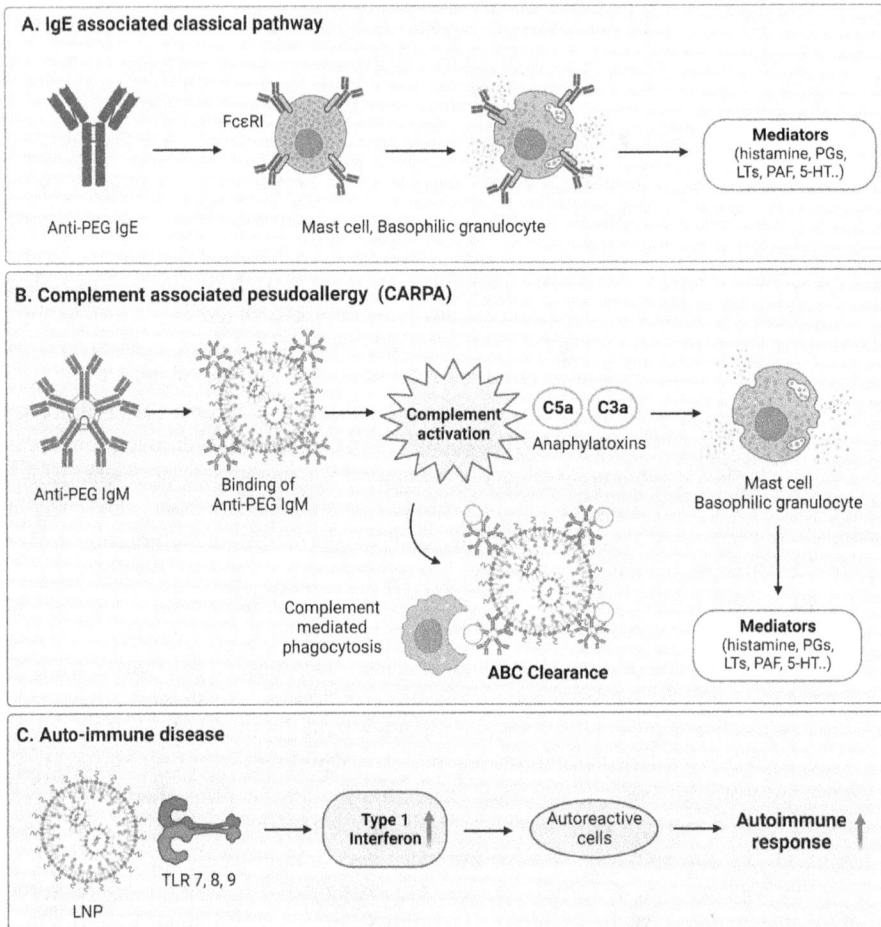

FIGURE 10.2 A summary of the major pathways through which LNPs, as a model widely used nanomedicine platform, interact with the immune system. (a) The induction of anti-PEG IgE by PEGylated LNPs leads to the stimulation of hypersensitivity via binding to FcεRI receptors on mast cells and basophils, which subsequently release mediators including histamine, prostaglandins, and leukotrienes. (b) The binding of anti-PEG IgM to LNPs leads to complement activation through the classic pathway with a subsequent anaphylatoxins-mediated stimulation of mast cells and basophils, the situation that ends up with CARPA. (c) The interaction of LNPs with toll-like receptors, TLR 7,8,9, stimulates a type I interferon-mediated autoimmune response (reproduced from Lee et al. [42], under a Creative Commons Attribution 4.0 International License (http://creativecommons.org/licenses/by/4.0/).)

10.3.3 INDUSTRIAL CHALLENGES

The flexible nature of nanoparticles has inspired researchers to create interesting modes of functionalization to increase the nanoparticle efficiency and selectivity. Examples include decoration with targeting ligands, antibodies, stimuli-responsive moieties, creation of multilayered nanosystems, and so forth. Despite the smart nature of these sophisticated systems and their success during proof-of-concept studies, scaling up these systems to the mass production scale is really a challenging process [13]. Increasing the number of preparation steps results in a subsequent increase in the production cost as well as the risks associated with contamination, loss, and the generation of undesired byproducts. In addition, the presence of minor contaminants in the raw materials or minor variations in the processing operations is aggravated upon scale-up, leading to a dramatic effect on the final product. For example, Dormont *et al.* reported that the scale-up of squalene-adenosine nanoparticles resulted in the generation of a series of squalene analogs that resulted in variations in the physico-chemical characteristics of the resultant nanoparticles, including the crystal structure, protein adsorption, and cytotoxicity [46]. Colombo *et al.* showed that minor variations in the impeller speed or agitation time during scale-up of nanocapsules prepared via the emulsification–diffusion technique affected the particle size of the resultant nanocapsules [47]. Furthermore, the increase in the components of the nanoparticle increases the cost and challenges of the quality control and good manufacturing practice [13].

10.3.4 ECONOMIC CHALLENGES

It is wise to take into consideration that not only is the clinical translation of anticancer nanomedicines affected by the aforementioned challenges associated with their nature and fabrication, but also economic interests and challenges exert a massive impact on the fate of this technology. For instance, the development of novel anticancer nanomedicines requires expensive research and development (R&D) costs compared to other classic therapeutic modalities. Moreover, the expensive production cost of anticancer nanomedicines alluded to above is passed to the consumers in the form of unaffordable prices, which restricts the marketability of these products [48]. Furthermore, the poor clinical translatability of nanomedicines discourages big pharma from investing in the development of cancer-targeted nanomedicines and shifts the paradigm toward more economically favorable choices such as the mass production modalities (e.g. vaccines) or personalized nanomedicines (e.g. for the treatment of orphan diseases), which subsequently slows down the rate of development of anticancer nanomedicines and limits the opportunities to achieve future breakthroughs in this area of endeavor.[49]

10.3.5 REGULATORY CHALLENGES

According to several literature reviews, one of the biggest hassles that hamper the translation of nanomedicines into clinics is the lack of regulatory protocols that can cope with the unique features of nanomedicines as a distinct class of therapeutics [50]. For example, despite the promising properties of AgNPs as an anticancer modality, the regulatory protocols adopted in the USA so far do not distinguish AgNPs from other silver-based products and do not acknowledge specific characterization methods for them [37]. In contrast, the European Union (EU) has established a system for the "Registration, Evaluation, Authorization, and Restriction of Chemicals (REACH)," which attempts to deal with nanomaterials as distinct chemical substances independent of their bulk counterparts. While this system represents a promising progressive step toward a deeper understanding as well as better regulatory procedures for nanomedicines, it still has a shortcoming of exempting materials that are produced in a scale below 1 ton from the reporting requirements, which subsequently excludes a substantial proportion of anticancer nanomedicines from the system [51]. Moreover, owing to the intricate composition of anticancer nanomedicines, they are subjected to overlapping regulatory laws that deal with the individual components of these nanosystems rather than considering them as a whole distinct unit [13].

10.4 APPROACHES TO PROMOTE THE CLINICAL POTENTIAL OF CANCER-TARGETED NANOMEDICINES

Herein, we shed light on some innovative strategies developed to cope with the challenges that restrict the clinical applicability of anticancer nanomedicines, with a focus on the successive steps of the development of new anticancer nanomedicines: design, production, and evaluation.

10.4.1 LIGAND-FREE TARGETING

As alluded to in the previous section, ligand modification of nanoparticles is associated with multiple challenges when shifting from laboratory to clinical settings. Therefore, there has been a growing interest recently in the development of ligand-free targeting methods that can achieve selective delivery of the therapeutic cargo, while overcoming the drawbacks of the classic ligand-decorated nanosystems. For instance, it has been reported that tweaking the composition and properties of nanocarriers affects the nature of the "protein corona" adsorbed on them *in vivo*, which subsequently determines their biodistribution and *in vivo* performance. Hence, such a strategy can be exploited to enable the harnessing of endogenous ligands to deliver the cargo in question to the desired target cells rather than attaching exogenous ligands to the nanocarrier [32,52]. In a recent study, the optimization of pH-responsive LNPs containing an ionizable cationic lipid, 1,2-dioleoyl-3-dimethylammonium propane (DODAP), successfully delivered a small-interfering RNA (siRNA) against monocarboxylate transporter 4 (MCT4) to the breast cancer cells with a knockdown efficiency of ~90% and a subsequent ~80% eradication of the cancer cells through remodeling their cytokine profile. In contrast, the same formulation had a fivefold lower uptake in the normal cells [10]. In another study, harnessing the composition of LNPs, composed of ionizable lipids and helper phospholipids, enabled a significant control on their *in vivo* tropism to the various tissues following their intravenous administration. An optimum formulation, referred to as CL15H6 LNPs, selectively delivered antigen-encoding messenger RNA (mRNA) to the splenic dendritic cells with a subsequent recruitment of the cytotoxic T lymphocytes (CTL) to protect mice from a tumor challenge, a concept that is referred to as "anticancer vaccination" [53]. Furthermore, capping AgNPs with chitosan of a medium molecular weight (190–310 kDa) created a therapeutic modality with an intrinsic anticancer activity against the breast cancer cells. It was hypothesized that chitosan ionizes in the low pH of the tumor microenvironment to increase the interaction of AgNPs with the polyanionic heparan sulfate proteoglycans (HSPGs) that are overexpressed on the surface of the cancer cells, which in turn enhances the cellular uptake of AgNPs. Subsequently, chitosan controls the release of silver ions intracellularly as well as their accumulation in either the nucleus, causing genotoxicity, or in the mitochondria, causing an impairment of energy production, which end up with the apoptosis of the cancer cells. Figure 10.3 illustrates the proposed mechanism of such a novel therapeutic modality [6].

10.4.2 SCALABLE PRODUCTION OF NANOPARTICLES

As alluded to above, the scale-up of anticancer nanomedicines has been one of the formidable obstacles that hamper their large-scale production and subsequently limit their clinical translatability. In this area of endeavor, the development of scalable production methods of nanocarriers can tackle such a challenge and improve the industrial acceptance of nanomedicines. In the area of LNPs/polymeric nanoparticles, the recruitment of microfluidics technology enabled not only scalable preparation of cancer-targeted nanoparticles, but also a high reproducibility of nanoparticle characteristics

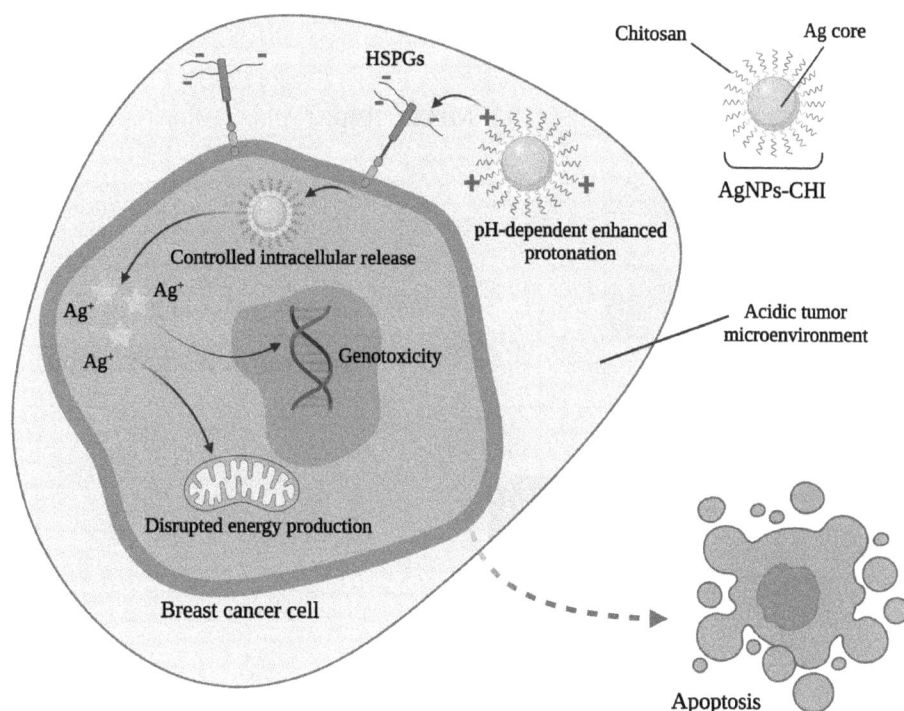

FIGURE 10.3 Chitosan-capped AgNPs exert a potent intrinsic activity against the breast cancer cells. Chitosan protonates in the acidic tumor microenvironment and subsequently interacts with the polyanionic HSPGs on the surface of the cancer cell to promote the cellular uptake of AgNPs. Intracellularly, chitosan modulates the release of silver ions from AgNPs and their subsequent accumulation in the nucleus, causing genotoxicity, or in the mitochondria, causing a disruption in the energy production, which collectively end up with inducing cellular apoptosis (reproduced from Abdellatif et al. [6], under a Creative Commons Attribution 4.0 International License (http://creativecommons.org/licenses/by/4.0/).)

and a precise control of their physico-chemical properties. The concept of microfluidic mixers relies on the rapid mixing of aqueous and organic phases containing the components in question through micrometer-sized channels plated on a chip. Meanwhile, the flow rates of the involved fluids are precisely controlled using syringe pumps [54]. Figure 10.4a illustrates the principle of using microfluidic mixers in the preparation of nanoparticles [13]. In a previous study, a novel microfluidic mixer referred to as "invasive lipid nanoparticle production device" was recruited for the preparation of "ultra-small LNPs (usLNPs)" co-loaded with a cytotoxic drug, sorafenib (SOR), and siRNA targeting Midkine, which is known to be overexpressed in HCC. The "usLNPs" had a particle diameter of 60 nm, which enabled their penetration through the HCC microenvironment, rich in stroma, following systemic administration. Subsequently, the "usLNPs" achieved a high knockdown efficiency of the target gene in mice with an aggressive HCC model. Moreover, the usLNPs demonstrated a high loading capacity of the hydrophobic drug, SOR, up to 10 mol% of the total lipid amount with a higher encapsulation efficiency (EE) compared to the thin lipid film hydration method, a classic method that is commonly used in the preparation of LNPs. Furthermore, the combination of chemotherapy and gene therapy delivered by these usLNPs re-sensitized HCC cells to SOR and eradicated a SOR-resistant HCC model *in vivo* [33]. In another study, Ghasemi Toudeshkchouei and co-workers applied a fork-shaped microfluidic mixer to fabricate poly (lactic-co-glycolic) acid nanoparticles (PLGA NPs) loaded with 5-fluorouracil (5-FU) with an EE as high as 95%. The resultant nanoparticles demonstrated enhanced anticancer activities against colorectal carcinoma (CRC) cell lines, SW-480 [55]. Furthermore, several studies have reported the extension of microfluidics

A

B

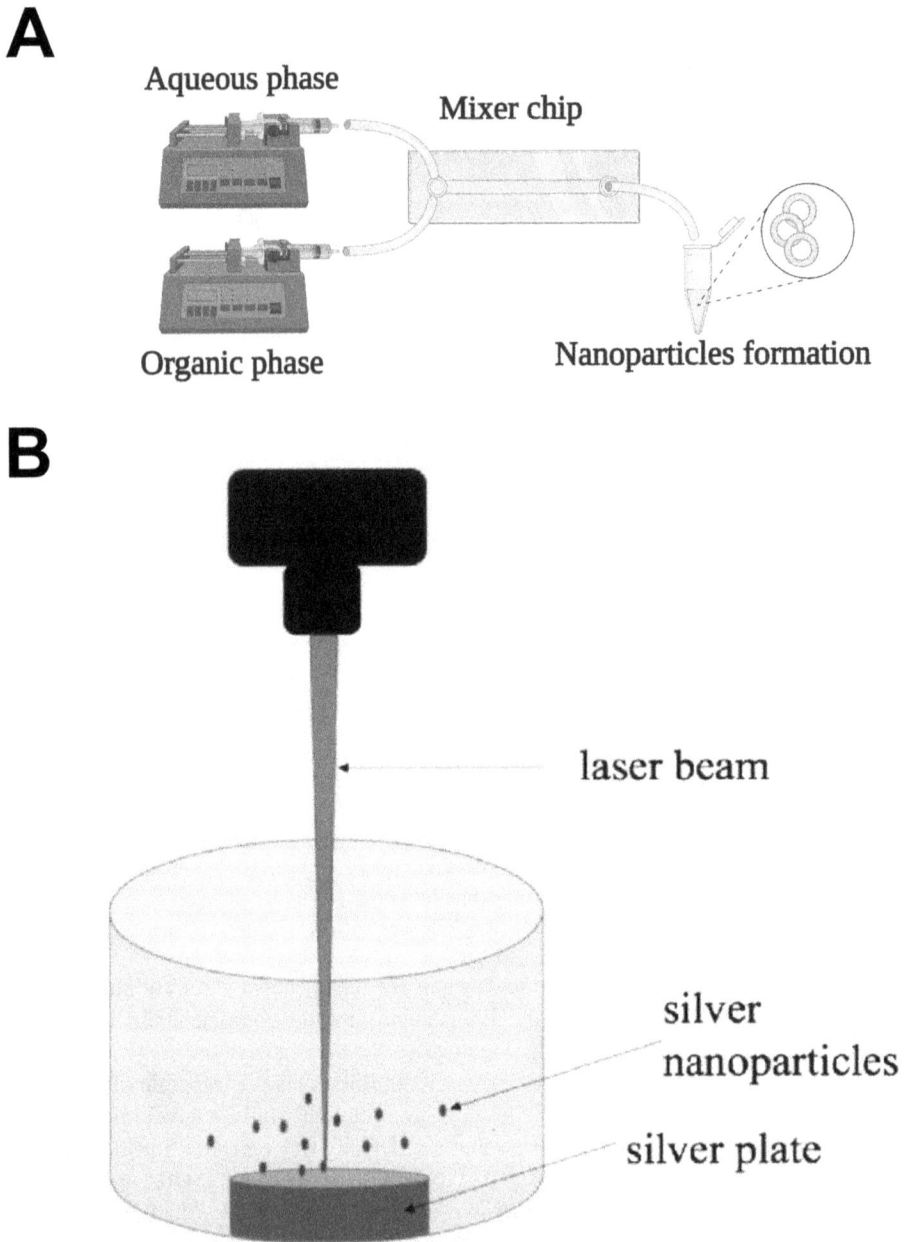

FIGURE 10.4 Examples of novel methods for the scalable preparation of anticancer nanomedicines. (a) A microfluidic mixer for the preparation of LNPs/polymeric nanoparticles. The aqueous and organic phases are rapidly mixed through a microchip mixer, while their flow rates are precisely controlled via syringe pumps. (Adapted from Younis et al. [13], with permission from Elsevier. © 2021 Elsevier B.V.) (b) Scalable preparation of metallic nanoparticles via laser ablation. A diagram for the preparation of AgNPs is illustrated as an example. (Reproduced from Zhang et al. [63], under a Creative Commons Attribution 4.0 International License (http://creativecommons.org/licenses/by/4.0/).)

technology to other types of anticancer nanomedicines, including mesoporous silica nanoparticles [56], quantum dots [57], and metal organic frameworks [58].

In the area of metallic nanoparticles, the application of physical synthetic methods has been shown to be effective from both scalability and economic points of view. For instance, a laser ablation technique has been used in the preparation of diverse metallic nanoparticles such as AgNPs [59], AuNPs [60], silicon nanoparticles [61], or hybrid platforms of them [62]. The principle of such a technique involves the use of a laser beam to stimulate the precipitation of colloidal metallic nanoparticles from their metallic precursor in a solution containing stabilizers [37]. Figure 10.4b illustrates the principle of laser ablation in the synthesis of AgNPs, an example of metal-based anti-cancer nanomedicines [63].

10.4.3 GREEN SYNTHESIS

Sustainability issues have received an increasing focus from the pharmaceutical industry in the past few years, so as to secure a continuous supply of raw materials, reduce energy consumption, as well as to minimize the generation of environment-polluting waste. In this context, the concept of "green synthesis" has evolved. Green synthesis depends on the design of eco-friendly synthetic method-ologies that leverage natural resources (e.g., plant extracts and microorganisms) for the preparation of nanoparticles rather than the classic methods that rely on the use of organic solvents or energy-consuming processes (e.g., heating) [7]. For instance, Susanto *et al.* harnessed the extract of *Moringa oleifera* leaf powder as a natural reducing agent to prepare AgNPs from a $AgNO_3$ precursor. Unlike the classic methods, such a green synthesis enabled the synthesis of AgNPs at room temperature, eliminating the heating steps. In addition, the prepared AgNPs induced cell-cycle arrest in the CRC cell line, HT-29 [64]. Solanki and co-workers fabricated nanoparticles based on the naturally occur-ring protein, bovine serum albumin, and loaded them with the plant-derived isoquinoline alkaloid, berberine. These green nanoparticles demonstrated an enhanced cytotoxicity to the breast cancer cells [65].

10.4.4 CLINICALLY RELEVANT EXPERIMENTAL MODELS

One of the greatest disadvantages of anticancer nanomedicines is their limited ability to reproduce the promising results obtained in the laboratory when proceeding to the clinical trial stage. This complicated problem can be divided into two main issues: the low *in vitro–in vivo* correlation and the poor animal–human correlation. Regarding the first issue, the vast majority of *in vitro* research on anticancer nanomedicine is performed in two-dimensional cell culture models, which lack the dynamic *in vivo* conditions such as three-dimensional geometry, blood supply, immune responses, and cross-talk between the various types of cells. Therefore, the results obtained from such mod-els are less-likely to be reproduced *in vivo* [13]. Although testing the nanomedicines *in vivo* using mice models is widely accepted in academia as a more reliable tool compared to cell cultures, it also raises a second issue of poor clinical relevance. Since a substantial proportion of murine tumor models are produced using laboratory-grown cancer cells, such models may lack some features of real human tumors. In addition, these tumors are usually induced via the injection of a large num-ber of cancer cells into mice, leading to fast tumor growth within a short time period (~2 weeks), which completely differs from human tumors that are initiated and grown over time periods that may extend for several years. Moreover, most murine tumor models are created in immunocompro-mised mice (to avoid graft rejection), which lack the important interactions between the tumor and the immune system. Furthermore, tumors inoculated ectopically (e.g., in the flanks) do not reflect the real interaction of the tumor with the concerned organ from which it has been derived [26,30].

To tackle the above challenges, the application of "three-dimensional *in vitro* models" such as spheroids and organoids has recently received growing interest in comparison with the classic cell culture models. These elegant models can mimic the three-dimensional morphology of the tissues in question and simulate the interaction between various cellular populations [11]. In addition, some models have further delved into mimicking the blood supply of tissues via supplementing organ-oids with microfluidic networks, resulting in the generation of an "organ-on-chip" for a deeper simulation of the dynamic *in vivo* conditions [66]. In the area of animal models, researchers have developed advanced tumor models for better clinical relevance. For example, orthotopic models that induce a primary tumor in its tissue of origin (e.g., HCC in the liver) have been reported to provide more clinically relevant data compared to their ectopic counterparts inoculated subcutaneously [9,26]. Furthermore, creating tumor models based on cancer cells retrieved from a patient specimen (i.e., patient-derived xenografts) has also demonstrated a higher clinical relevance compared to those generated using cultured cell lines, which opens the way to their application in designing personalized anticancer nanomedicines [67].

10.5   KEY CONSIDERATIONS IN THE DESIGN OF CANCER-TARGETED NANOMEDICINES TO AVOID CLINICAL FAILURE

After discussing the challenges that hamper the clinical translation of anticancer nanomedicines and highlighting some emerging technologies to cope with them, it is a rational strategy to pick up the key tips that should be considered during the design of anticancer nanomedicines to avoid their potential clinical failure. First, the composition of these nanosystems needs to be simplified by reducing the number of components, avoiding intricate designs, and replacing multi-step preparation methods with single-step methodologies [13,19,37]. Second, adopting rational experimental designs is crucial to save the effort and expenses of research and development, which will subsequently encourage pharmaceutical industries to invest in such platforms. For example, the application of the "Design of Experiments approach" for the optimization of nanocarriers can substantially help in investigating multiple formulation parameters simultaneously and assess the interaction between them, while reducing the number of required experiments, in contrast to the classic single factor-based designs [68]. In addition, the recent breakthrough in artificial intelligence and its associated models can facilitate benefiting from the cumulative published data in the formulation and optimization steps of anticancer nanomedicines to avoid re-inventing the wheel [69,70]. Third, the introduction of innovative technologies that enable high throughput screening of formulations can speed up the discovery of novel anticancer nanotherapeutics at a reasonable cost. For instance, Dahlman et al. developed an elegant method for the high throughput screening of LNPs via a "barcoding" tool. Through labeling each LNP formulation with a distinct short DNA sequence (i.e., the barcode), it was possible to track the in vivo fate of each formulation by detecting its specific barcode using next-generation sequencing. Thus, dozens of formulations can be injected into a single mouse and screened simultaneously. As a proof of concept, the authors applied such a technology to simultaneously track the biodistribution of 30 LNP formulations to 8 tissues with high accuracy. Therefore, it can be extrapolated to enable the screening of a massive number of formulations in a short time and using a limited number of experimental animals [71]. Fourth, it is necessary to recruit clinically relevant testing models for the evaluation of the proposed anticancer nanomedicines to ensure that they really reflect the characteristic features of the target tumor. In addition to the advanced models alluded to above, the application of in silico models can also contribute to reducing the time and cost and saving the lives of experimental animals [72]. Furthermore, such in silico models can also be used to predict the nanotoxicity of the proposed formulations and their biological properties such as permeation, aggregation, and interaction with the biological membranes [73]. Fifth, a comprehensive biosafety assessment of the proposed formulations during the preclinical research stage will encourage more individuals to participate in the subsequent clinical trials, which will increase the reliability of the data, taking into consideration that most clinical trials are currently suffering from an insufficient number of subjects [74,75]. In this area, an in-depth assessment of the long-term effects of nanomedicines on the immune responses, cytokine profiles, organ functions, and genome is essential to be considered rather than the currently adopted simple evaluations [76]. Lastly, the current methods adopted in the clinical trials involving anticancer nanomedicines need to be revised for a better consideration of the specific nature of nanomedicines. In addition to evaluating the tumor volume and the overall survival rate of subjects, implementing biomarkers in the evaluation criteria will allow for a better understanding of the biological impacts of these nanoformulations and how they can be invested properly. Moreover, it can be applied to drive the design of the clinical trials for precise and personalized anticancer medicines [77,78].

10.6 CONCLUSION

The recruitment of nanomedicines in the battle against cancer holds a great promise either as theranostic modalities or as versatile drug delivery vehicles. Nevertheless, the translation of such promising findings from the bench into the clinics is still encountering formidable obstacles, including technical, toxicological, industrial, economic, and regulatory challenges. The smart design of anticancer nanomedicines through the introduction of novel targeting approaches apart from the classic ligand-based modifications, the application of eco-friendly scalable production together with high throughput screening methodologies, and the patient-oriented assessment of the proposed nanotherapies using clinically relevant experimental models would increase the clinical translatability of such modalities, which would subsequently encourage small pharmaceutical firms and venture companies to invest in them, and finally, attract the attention of big pharma and regulatory authorities to adopt them. Instead of waiting for support from governments and industry leaders, researchers in this area of endeavor should take the initiative to understand the challenges discussed in this chapter and reformulate their research to cope with these challenges so as to impose their research output on the market and help the mankind to get rid of cancer as a terrible health threat that lasted too long.

REFERENCES

1. R.L. Siegel, K.D. Miller, N.S. Wagle, A. Jemal, Cancer statistics, 2023, *CA: A Cancer Journal for Clinicians*, 73 (2023) 17–48.
2. World Health Organization (WHO), Global cancer burden growing, amidst mounting need for services. https://www.who.int/news/item/01-02-2024-global-cancer-burden-growing--amidst-mounting-need-for-services. Accessed on 2024/11/07.
3. American Society of Clinical Oncology (ASCO), Guidelines by clinical area. https://society.asco.org/practice-patients/guidelines. Accessed on 2024/11/07.
4. B. Katta, C. Vijayakumar, S. Dutta, B. Dubashi, V.P. Nelamangala Ramakrishnaiah, The incidence and severity of patient-reported side effects of chemotherapy in routine clinical care: A prospective observational study, *Cureus*, 15 (2023) e38301.
5. K. Bukowski, M. Kciuk, R. Kontek, Mechanisms of multidrug resistance in cancer chemotherapy, *International Journal of Molecular Sciences*, 21 (2020) 3233.
6. A.A.H. Abdellatif, A. Abdelfattah, M.A. Younis, S.M. Aldalaan, H.M. Tawfeek, Chitosan-capped silver nanoparticles with potent and selective intrinsic activity against the breast cancer cells, *Nanotechnology Reviews*, 12 (2023) 20220546.
7. A.A.H. Abdellatif, M.A.H. Mostafa, H. Konno, M.A. Younis, Exploring the green synthesis of silver nanoparticles using natural extracts and their potential for cancer treatment, *3 Biotech*, 14 (2024) 274.
8. A.A.H. Abdellatif, A.S. Alshubrumi, M.A. Younis, Targeted nanoparticles: The smart way for the treatment of colorectal cancer, *AAPS PharmSciTech*, 25 (2024) 23.
9. M.A. Younis, H. Harashima, Understanding gene involvement in hepatocellular carcinoma: Implications for gene therapy and personalized medicine, *Pharmacogenomics and Personalized Medicine*, 17 (2024) 193–213.
10. A.A.H. Abdellatif, A. Bouazzaoui, H.M. Tawfeek, M.A. Younis, MCT4 knockdown by tumor microenvironment-responsive nanoparticles remodels the cytokine profile and eradicates aggressive breast cancer cells, *Colloids and Surfaces B: Biointerfaces*, 238 (2024) 113930.
11. A.A.H. Abdellatif, G. Scagnetti, M.A. Younis, A. Bouazzaoui, H.M. Tawfeek, B.N. Aldosari, A.S. Almurshedi, M. Alsharidah, O.A. Rugaie, M.P.A. Davies, T. Liloglou, K. Ross, I. Saleem, Non-coding RNA-directed therapeutics in lung cancer: Delivery technologies and clinical applications, *Colloids and Surfaces B: Biointerfaces*, 229 (2023) 113466.

12. A.A.H. Abdellatif, M.A. Younis, A.F. Alsowinea, E.M. Abdallah, M.S. Abdel-Bakky, A. Al-Subaiyel, Y.A.H. Hassan, H.M. Tawfeek, Lipid nanoparticles technology in vaccines: Shaping the future of prophylactic medicine, *Colloids and Surfaces B: Biointerfaces*, 222 (2023) 113111.

13. M.A. Younis, H.M. Tawfeek, A.A.H. Abdellatif, J.A. Abdel-Aleem, H. Harashima, Clinical translation of nanomedicines: Challenges, opportunities, and keys, *Advanced Drug Delivery Reviews*, 181 (2022) 114083.

14. L. Salvioni, M.A. Rizzuto, J.A. Bertolini, L. Pandolfi, M. Colombo, D. Prosperi, Thirty years of cancer nanomedicine: Success, frustration, and hope, *Cancers*, 11 (2019) 1855.

15. Y. Barenholz, Doxil®--the first FDA-approved nano-drug: Lessons learned, *Journal of Controlled Release*, 160 (2012) 117–134.

16. Food and Drug Administration (FDA), Drug approvals and databases. https://www.fda.gov/drugs/devel opment-approval-process-drugs/drug-approvals-and-databases. Accessed on 2024/11/12.

17. National Instiute of Health (NIH), Cancer | nanoparticles. https://clinicaltrials.gov/search?cond=Cancer &intr=nanoparticles. Accessed on 2024/11/12.

18. PubMed, Cancer | nanoparticles. https://pubmed.ncbi.nlm.nih.gov/?term=nanomedicine%3B+cancer. Accessed on 2024/11/12.

19. A.A.H. Abdellatif, M.A. Younis, M. Alsharidah, O. Al Rugaie, H.M. Tawfeek, Biomedical applications of quantum dots: Overview, challenges, and clinical potential, *International Journal of Nanomedicine*, 17 (2022) 1951–1970.

20. L. Minqian, H. Weili, D. Juan, W. Yuanfeng, G. Yuan, Z. Jianjun, Crystallization of amorphous drugs and inhibiting strategies, *Progress in Chemistry*, 33 (2021) 2116–2127.

21. M.U. Akram, N. Abbas, M. Farman, S. Manzoor, M.I. Khan, S.M. Osman, R. Luque, A. Shanableh, Tumor micro-environment sensitive release of doxorubicin through chitosan based polymeric nanoparticles: An in-vitro study, *Chemosphere*, 313 (2023) 137332.

22. M.P. Marques, R. Valero, S.F. Parker, J. Tomkinson, L.A. Batista de Carvalho, Polymorphism in cisplatin anticancer drug, *The Journal of Physical Chemistry B*, 117 (2013) 6421–6429.

23. C. Qin, X. Xin, X. Pei, L. Yin, W. He, Amorphous nanosuspensions aggregated from paclitaxel-hemoglobulin complexes with enhanced cytotoxicity, *Pharmaceutics*, 10 (2018) 92.

24. C. von Baeckmann, H. Kählig, M. Lindén, F. Kleitz, Irreversible adsorption of serum proteins onto nanoparticles, *Particle & Particle Systems Characterization*, 38 (2021) 2000273.

25. M.A. Younis, I.A. Khalil, M.M. Abd Elwakil, H. Harashima, A multifunctional lipid-based nanodevice for the highly specific codelivery of sorafenib and midkine siRNA to hepatic cancer cells, *Molecular Pharmaceutics*, 16 (2019) 4031–4044.

26. M.A. Younis, I.A. Khalil, H. Harashima, Gene therapy for hepatocellular carcinoma: Highlighting the journey from theory to clinical applications, *Advanced Therapeutics*, 3 (2020) 2000087.

27. Y. Matsumura, H. Maeda, A new concept for macromolecular therapeutics in cancer chemotherapy: Mechanism of tumoritropic accumulation of proteins and the antitumor agent smancs, *Cancer Research*, 46 (1986) 6387–6392.

28. J.S. Suk, Q. Xu, N. Kim, J. Hanes, L.M. Ensign, PEGylation as a strategy for improving nanoparticle-based drug and gene delivery, *Advanced Drug Delivery Reviews*, 99 (2016) 28–51.

29. K. Park, The beginning of the end of the nanomedicine hype, *Journal of Controlled Release*, 305 (2019) 221–222.

30. G.H. Petersen, S.K. Alzghari, W. Chee, S.S. Sankari, N.M. La-Beck, Meta-analysis of clinical and preclinical studies comparing the anticancer efficacy of liposomal versus conventional non-liposomal doxorubicin, *Journal of Controlled Release*, 232 (2016) 255–264.

31. W.C. Chen, A.X. Zhang, S.-D. Li, Limitations and niches of the active targeting approach for nanoparticle drug delivery, *European Journal of Nanomedicine*, 4 (2012) 89–93.

32. M.A. Younis, Y. Sato, Y.H.A. Elewa, H. Harashima, Reprogramming activated hepatic stellate cells by siRNA-loaded nanocarriers reverses liver fibrosis in mice, *Journal of Controlled Release*, 361 (2023) 592–603.

33. M.A. Younis, I.A. Khalil, Y.H.A. Elewa, Y. Kon, H. Harashima, Ultra-small lipid nanoparticles encapsulating sorafenib and midkine-siRNA selectively-eradicate sorafenib-resistant hepatocellular carcinoma in vivo, *Journal of Controlled Release*, 331 (2021) 335–349.

34. Y. Wang, A. Santos, A. Evdokiou, D. Losic, An overview of nanotoxicity and nanomedicine research: Principles, progress and implications for cancer therapy, *Journal of Materials Chemistry B*, 3 (2015) 7153–7172.

35. S.K. Das, K. Sen, B. Ghosh, N. Ghosh, K. Sinha, P.C. Sil, Molecular mechanism of nanomaterials induced liver injury: A review, *World Journal of Hepatology*, 16 (2024) 566–600.

36. I. Iavicoli, L. Fontana, G. Nordberg, The effects of nanoparticles on the renal system, *Critical Reviews in Toxicology*, 46 (2016) 490–560.

37. M.A. Younis, Clinical translation of silver nanoparticles into the market, in: P. Kesharwani (Ed.), *Silver Nanoparticles for Drug Delivery*, Academic Press, 2024, pp. 395–432.

38. A. Abdelfattah, A.E. Aboutaleb, A.B.M. Abdel-Aal, A.A.H. Abdellatif, H.M. Tawfeek, S.I. Abdel-Rahman, Design and optimization of PEGylated silver nanoparticles for efficient delivery of doxorubicin to cancer cells, *Journal of Drug Delivery Science and Technology*, 71 (2022) 103347.

39. Y.-M. Cho, Y. Mizuta, J.-I. Akagi, T. Toyoda, M. Sone, K. Ogawa, Size-dependent acute toxicity of silver nanoparticles in mice, *Journal of Toxicologic Pathology*, 31 (2018) 73–80.

40. C. Lopez-Chaves, J. Soto-Alvaredo, M. Montes-Bayon, J. Bettmer, J. Llopis, C. Sanchez-Gonzalez, Gold nanoparticles: Distribution, bioaccumulation and toxicity. *In Vitro and In Vivo Studies, Nanomedicine: Nanotechnology, Biology, and Medicine*, 14 (2018) 1–12.

41. T. Nakamura, Y. Sato, Y. Yamada, M.M. Abd Elwakil, S. Kimura, M.A. Younis, H. Harashima, Extrahepatic targeting of lipid nanoparticles in vivo with intracellular targeting for future nanomedicines, *Advanced Drug Delivery Reviews*, 188 (2022) 114417.

42. Y. Lee, M. Jeong, J. Park, H. Jung, H. Lee, Immunogenicity of lipid nanoparticles and its impact on the efficacy of mRNA vaccines and therapeutics, *Experimental & Molecular Medicine*, 55 (2023) 2085–2096.

43. M. Mohamed, A.S. Abu Lila, T. Shimizu, E. Alaaeldin, A. Hussein, H.A. Sarhan, J. Szebeni, T. Ishida, PEGylated liposomes: Immunological responses, *Science and Technology of Advanced Materials*, 20 (2019) 710–724.

44. J. Szebeni, Complement activation-related pseudoallergy: A stress reaction in blood triggered by nanomedicines and biologicals, *Molecular Immunology*, 61 (2014) 163–173.

45. L.E. Guimarães, B. Baker, C. Perricone, Y. Shoenfeld, Vaccines, adjuvants and autoimmunity, *Pharmacological Research*, 100 (2015) 190–209.

46. F. Dormont, M. Rouquette, C. Mahatsekake, F. Gobeaux, A. Peramo, R. Brusini, S. Calet, F. Testard, S. Lepetre-Mouelhi, D. Desmaële, M. Varna, P. Couvreur, Translation of nanomedicines from lab to industrial scale synthesis: The case of squalene-adenosine nanoparticles, *Journal of Controlled Release*, 307 (2019) 302–314.

47. A.P. Colombo, S. Briançon, J. Lieto, H. Fessi, Project, design, and use of a pilot plant for nanocapsule production, *Drug Development and Industrial Pharmacy*, 27 (2001) 1063–1072.

48. R. Bosetti, S.L. Jones, Cost–effectiveness of nanomedicine: Estimating the real size of nano-costs, *Nanomedicine: Nanotechnology, Biology, and Medicine*, 14 (2019) 1367–1370.

49. S. Hua, M.B.C. de Matos, J.M. Metselaar, G. Storm, Current trends and challenges in the clinical translation of nanoparticulate nanomedicines: Pathways for translational development and commercialization, *Frontiers in Pharmacology*, 9 (2018) 790.

50. S. Mühlebach, Regulatory challenges of nanomedicines and their follow-on versions: A generic or similar approach? *Advanced Drug Delivery Reviews*, 131 (2018) 122–131.

51. H. Rauscher, K. Rasmussen, B. Sokull-Klüttgen, Regulatory aspects of nanomaterials in the EU, *Chemie Ingenieur Technik*, 89 (2017) 224–231.

52. M.A. Younis, Y. Sato, Y.H.A. Elewa, Y. Kon, H. Harashima, Self-homing nanocarriers for mRNA delivery to the activated hepatic stellate cells in liver fibrosis, *Journal of Controlled Release*, 353 (2023) 685–698.

53. M.A. Younis, Y. Sato, Y.H.A. Elewa, H. Harashima, Harnessing the composition of lipid nanoparticles to selectively deliver mRNA to splenic immune cells for anticancer vaccination, *Drug Delivery and Translational Research* 15 (2025) 3626–3641, https://doi.org/10.1007/s13346-025-01824-w.

54. A. Fabozzi, F. Della Sala, M. di Gennaro, M. Barretta, G. Longobardo, N. Solimando, M. Pagliuca, A. Borzacchiello, Design of functional nanoparticles by microfluidic platforms as advanced drug delivery systems for cancer therapy, *Lab on a Chip*, 23 (2023) 1389–1409.

55. M. Ghasemi Toudeshkchouei, P. Zahedi, A. Shavandi, Microfluidic-assisted preparation of 5-fluorouracil-loaded PLGA nanoparticles as a potential system for colorectal cancer therapy, *Materials*, 13 (2020) 1483.

56. J. Yan, X. Xu, J. Zhou, C. Liu, L. Zhang, D. Wang, F. Yang, H. Zhang, Fabrication of a pH/redox-triggered mesoporous silica-based nanoparticle with microfluidics for anticancer drugs doxorubicin and paclitaxel codelivery, *ACS Applied Bio Materials*, 3 (2020) 1216–1225.

57. X. Li, M. Zha, Y. Li, J.S. Ni, T. Min, T. Kang, G. Yang, H. Tang, K. Li, X. Jiang, Sub-10 nm aggregation-induced emission quantum dots assembled by microfluidics for enhanced tumor targeting and reduced retention in the liver, *Angewandte Chemie (International Edition in English)*, 59 (2020) 21899–21903.

58. Y.L. Balachandran, X. Li, X. Jiang, Integrated microfluidic synthesis of aptamer functionalized biozeolitic imidazolate framework (BioZIF-8) targeting lymph node and tumor, *Nano Letters*, 21 (2021) 1335–1344.
59. T. Tsuji, K. Iryo, N. Watanabe, M. Tsuji, Preparation of silver nanoparticles by laser ablation in solution: Influence of laser wavelength on particle size, *Applied Surface Science*, 202 (2002) 80–85.
60. N.Z.A. Naharuddin, A.R. Sadrolhosseini, M.H. Abu Bakar, N. Tamchek, M.A. Mahdi, Laser ablation synthesis of gold nanoparticles in tetrahydrofuran, *Optical Materials Express*, 10 (2020) 323–331.
61. J.A. Serrano-Ruz, J.G. Quiñones-Galván, J. Santos Cruz, F. de moure-Flores, E. Campos González, A. Chávez-Chávez, G. Gómez-Rosas, Synthesis of silicon nanoparticles by laser ablation at low fluences in water and ethanol, *Materials Research Express*, 7 (2020) 025008.
62. J.R. González-Castillo, E. Rodriguez, E. Jimenez-Villar, D. Rodríguez, I. Salomon-García, G.F. de Sá, T. García-Fernández, D.B. Almeida, C.L. Cesar, R. Johnes, J.C. Ibarra, Synthesis of Ag@Silica nanoparticles by assisted laser ablation, *Nanoscale Research Letters*, 10 (2015) 399.
63. S. Zhang, Y. Tang, B. Vlahovic, A review on preparation and applications of silver-containing nanofibers, *Nanoscale Research Letters*, 11 (2016) 80.
64. H. Susanto, S.D.R.A. Firdaus, M. Sholeh, A.T. Endharti, A. Taufiq, N.A.N.N. Malek, H.K. Permatasari, *Moringa oleifera* leaf powder – silver nanoparticles (MOLP-AgNPs) efficiently inhibit metastasis and proliferative signaling in HT-29 human colorectal cancer cells, *Journal of Agriculture and Food Research*, 16 (2024) 101149.
65. R. Solanki, K. Patel, S. Patel, Bovine serum albumin nanoparticles for the efficient delivery of berberine: Preparation, characterization and in vitro biological studies, *Colloids and Surfaces A: Physicochemical and Engineering Aspects*, 608 (2021) 125501.
66. S. Zhang, Z. Wan, R.D. Kamm, Vascularized organoids on a chip: strategies for engineering organoids with functional vasculature, *Lab on a Chip*, 21 (2021) 473–488.
67. Y. Liu, W. Wu, C. Cai, H. Zhang, H. Shen, Y. Han, Patient-derived xenograft models in cancer therapy: Technologies and applications, *Signal Transduction and Targeted Therapy*, 8 (2023) 160.
68. M.A. Younis, M.R. El-Zahry, M.A. Tallat, H.M. Tawfeek, Sulpiride gastro-retentive floating microsponges; analytical study, in vitro optimization and in vivo characterization, *Journal of Drug Targeting*, 28 (2020) 386–397.
69. P.J. Harrison, H. Wieslander, A. Sabirsh, J. Karlsson, V. Malmsjö, A. Hellander, C. Wählby, O. Spjuth, Deep-learning models for lipid nanoparticle-based drug delivery, *Nanomedicine*, 16 (2021) 1097–1110.
70. K.P. Das, J. Chandra, Nanoparticles and convergence of artificial intelligence for targeted drug delivery for cancer therapy: Current progress and challenges, *Frontiers in Medical Technology*, 4 (2022) 1067144.
71. J.E. Dahlman, K.J. Kauffman, Y. Xing, T.E. Shaw, F.F. Mir, C.C. Dlott, R. Langer, D.G. Anderson, E.T. Wang, Barcoded nanoparticles for high throughput in vivo discovery of targeted therapeutics, *Proceedings of the National Academy of Sciences*, 114 (2017) 2060–2065.
72. N.R. Stillman, M. Kovacevic, I. Balaz, S. Hauert, In silico modelling of cancer nanomedicine, across scales and transport barriers, *npj Computational Materials*, 6 (2020) 92.
73. S. Shityakov, N. Roewer, J.-A. Broscheit, C. Förster, In silico models for nanotoxicity evaluation and prediction at the blood-brain barrier level: A mini-review, *Computational Toxicology*, 2 (2017) 20–27.
74. H. Huang, W. Feng, Y. Chen, J. Shi, Inorganic nanoparticles in clinical trials and translations, *Nano Today*, 35 (2020) 100972.
75. D.B. Fogel, Factors associated with clinical trials that fail and opportunities for improving the likelihood of success: A review, *Contemporary Clinical Trials Communications*, 11 (2018) 156–164.
76. P.L. Michael, Y.T. Lam, J. Hung, R.P. Tan, M. Santos, S.G. Wise, Comprehensive evaluation of the toxicity and biosafety of plasma polymerized nanoparticles, *Nanomaterials*, 11 (2021) 1176.
77. C. Hu, J.J. Dignam, Biomarker-driven oncology clinical trials: Key design elements, types, features, and practical considerations, *JCO Precision Oncology*, 3 (2019) PO.19.00086.
78. M. Antoniou, R. Kolamunnage-Dona, J. Wason, R. Bathia, C. Billingham, J.M. Bliss, L.C. Brown, A. Gillman, J. Paul, A.L. Jorgensen, Biomarker-guided trials: Challenges in practice, *Contemporary Clinical Trials Communications*, 16 (2019) 100493.

Index

For Product Safety Concerns and Information please contact our EU
representative GPSR@taylorandfrancis.com
Taylor & Francis Verlag GmbH, Kaufingerstraße 24, 80331 München, Germany

www.ingramcontent.com/pod-product-compliance
Lightning Source LLC
Chambersburg PA
CBHW061413210326
41598CB00035B/6193

9 781032 853758